The *Origin* Then and Now

"Jaws not so strong as [to] produce pain to finger"

This specimen, referred to by Darwin as a male stag beetle
in the genus *Lucanus*, is now classified as *Chiasognathus granti*.
Darwin encountered these beetles on Chiloe Island, off the coast
of Chile, and near Valdivia, Chile. The specimen pictured here
was photographed in native forest near Valdivia in February 2009.
The Chilean name for this beetle is *ciervo volante*, or flying deer.

The *Origin* Then and Now

An Interpretive Guide to the *Origin of Species*

David N. Reznick

With an Introduction by Michael Ruse

PRINCETON UNIVERSITY PRESS

Princeton and Oxford

4/12/10
WW
$29.95

Published by Princeton University Press, 41 William Street, Princeton, New Jersey 08540

In the United Kingdom: Princeton University Press, 6 Oxford Street, Woodstock, Oxfordshire OX20 1TW

Library of Congress Cataloging-in-Publication Data

Reznick, David N., 1952–
    The origin then and now : an interpretive guide to The origin of species / David N. Reznick ; with an introduction by Michael Ruse.
        p. cm.
    Includes bibliographical references and index.
    ISBN 978-0-691-12978-5 (hardcover : alk. paper)   1. Darwin, Charles, 1809–1882. On the origin of species.   2. Evolution (Biology)   3. Natural selection.   I. Title.
    QH365.O8R49 2010
    576.8—dc22   2009013200
British Library Cataloging-in-Publication Data is available

This book has been composed in Minion
Printed on acid-free paper. ∞
press.princeton.edu
Printed in the United States of America
10 9 8 7 6 5 4 3 2 1

I DEDICATE THIS BOOK to my wife, Melody, and daughters, Tess and Kate, for their support and tolerance during the very long time that I spent writing this book. My daughters are young enough and this project took long enough that I am sure they grew up with some properties in common with Darwin's children. Some of Darwin's children thought that all fathers had a special room where they hid away to study barnacles. Mine surely think that all fathers hide away on evenings and weekends to write books.

Contents

Preface

D arwin's *Origin of Species* has been described as one of the books that is most widely referred to, but least likely to be read. My goal is to make the *Origin* accessible to a larger audience and to do so by placing it in a continuum of science. The *Origin* was the inception of a new, unifying theory of the life sciences, but it was also strongly influenced by the science that preceded it; so understanding the *Origin* requires looking back in time to define the context in which it was written. Sometimes this also means understanding that some details that were critical to Darwin's theory were not yet known. The *Origin* was an inspiration to science because it highlighted these gaps in our knowledge and why it was important to fill them. It was much more of an inspiration because it defined so many new areas of inquiry. To appreciate the *Origin*, then, it is also important to look forward in time to see how it changed science.

I am an evolutionary biologist and have been studying evolution for over thirty years. I have specialized in the experimental evaluation of facets of the theory of evolution in nature. No one should be surprised to hear that someone like me is a fan of Charles Darwin. But being a fan of Darwin and his ideas is different from being an admirer of the *Origin of Species*. I first read the *Origin* during the summer break between completing my bachelor's degree at Washington University and beginning PhD studies at the University of Pennsylvania. I cannot recall much about what I learned from that first reading. I do recall finding it very hard going. I finished it out of a sense of obligation. I did not read the *Origin* again until I was an assistant professor at the University of California, when I decided to use it as part of a graduate class. It was not until I had read the first and sixth editions a few times, plus learned more about the development of evolution as a discipline, that I appreciated the scope and lasting importance of the book. More than a decade later I found myself, as a guest professor at the University of Konstanz in Germany, leading a seminar on the *Origin of Species* for a polyglot group of graduate students and postdocs. That is when I got a full sense of what the difficulties of the book were, but also of the potential for turning it into a document that would be accessible to a much wider audience. That is my goal with this book. I intend it to be a reader's guide to the *Origin* that addresses the issues that make Darwin's masterpiece a challenge to read and yet emphasizes the importance of reading it today, not as a historical benchmark, but as a living, relevant document.

Why is this worth doing? I think that a working knowledge of evolution can enhance anyone's understanding of any aspect of biology. For the majority of professional scientists, this knowledge represents a void that will rarely be filled; the growth of biological sciences and the extent to which we have become specialized means that most biologists get little or no exposure to evolution and often do not see its relevance. The general public hears about evolution in the news, mostly as sound bites from the opposing sides of the evolution-creation controversy. People in this wider audience can benefit as much as any student or scientist from understanding the theory of evolution because it has become a central issue in the design and delivery of science education. Understanding will lead to more informed decisions about how education is delivered.

Evolution is also an underappreciated feature of our day-to-day lives. Understanding evolution can help us understand the ever-changing risk of disease (e.g., why antibiotics and insecticides can do more harm than good if overused) or appreciate some of the risks associated with using genetic engineering to modify crops. The properties of plants and animals that make evolution inevitable are central to the development of domesticated breeds, without which life as we know it could not exist. Evolution plays a central role in understanding the impact of humans on the natural environment too, since organisms can evolve on short timescales. It plays a role in determining which species will survive and which will go extinct as climates change. It can play a key role when species are accidentally transported to new environments and become pests that invade and change local communities. Humans cause evolution by being predators—when hunting or catching fish, for example. The cod of today are genetically different from the cod that were fished fifty years ago and are less valuable as a source of food because of genetic changes promoted by human predation. All the above represents evolution that happens quickly, within a single person's lifetime, and that profoundly affects our welfare.

All these observations support an argument for understanding the basics of evolution, but these basics can be gained from any general biology textbook. They are not an argument for reading the *Origin*. One argument for reading the *Origin* is that it is the first comprehensive statement of the mechanism of natural selection and of the extent to which evolution by means of natural selection can serve as a unifying concept in biology. The *Origin* brings into its fold the entire scope of the natural sciences that become explainable by evolution. But this is still an incomplete argument; there are other sources for the same information. Reading the *Origin*, as opposed to turning to some other source, also provides a basis for understanding how science evolves and how we define the important unanswered questions to be addressed with future research. This is why we continue to include the *Origin* in our graduate classes.

But I am aiming for a bigger audience than students of the life sciences. Reading the *Origin* can teach anyone at any level important lessons about the structure of science and the meaning of the word *theory*. Another goal of this book is to define the word *theory* as it is used in the sciences and to detail how and why Darwin developed evolution as a scientific theory. Reading the *Origin* can also highlight the role that evolutionary theory played in shaping the future development of science. For example, the growth of genetics in the early twentieth century was strongly influenced by an interest in evolution. The connection between genetics and the *Origin* is that inheritance—the passing of characteristics from one generation of organisms to the next—was a critical component of the mechanism of natural selection. Darwin did his best to come to grips with how inheritance worked. What he saw and reported is a time capsule of what any acute observer would think of inheritance without the benefit of knowledge of Mendel. A good observer would rarely see anything like Mendelian inheritance governing the similarities between parents and offspring, which is why it took so long for Mendelism to be universally accepted. Darwin highlighted the importance of understanding how inheritance works. It was those who were influenced by Darwin who rediscovered and appreciated Mendel's laws.

The *Origin* can also still inspire new research because of the broad implications of some of Darwin's ideas. One example is the development of molecular evolution. Darwin knew nothing about DNA, but he was very clear about how the theory of evolution predicts that organisms will contain clues to their history. He applied this idea to the fossil record and the comparative anatomy and embryology of living organisms. It was for those who followed to realize that this same logic could be applied to the structure of DNA and proteins, and then to integrate evolutionary and molecular biology.

Reading the *Origin* also highlights basic issues that have yet to be fully resolved. Some of these include the causes of gaps in the fossil record, the relationship between micro- and macroevolution, the evolution of complex organs (what Darwin referred to as "organs of extreme complexity"), or the mechanisms of population regulation in living organisms. All these phenomena and more appear in the *Origin* as part of Darwin's development of the theory of evolution, and all represent active areas of research.

For all these reasons, the *Origin* represents far more than a book of historical interest. It is as relevant to science today as it was in 1859.

This is all an abbreviated argument for why we should read the *Origin*, yet very few people try to. Of those who try, few finish it, and even fewer understand it. I sympathize with anyone who has tried, given my own experience. I have tried to make the *Origin* more accessible by rewriting it, and by identifying and addressing the difficulties inherent in the original document. People often talk about the difficulty of Darwin's Victorian prose, but this is a minor issue. I have learned to like and admire his writing. One of the more substantive reasons that the *Origin* is difficult to read is that Darwin draws on a breadth of scientific expertise that exceeds that of virtually any layman but also most scientists. As the life sciences have expanded since 1859, we have tended to become far more specialized than were the scientists of Darwin's day. A second source of difficulty is that Darwin's presentation is rooted in the science of 1859, which is quite different from the science of today. A third source is that Darwin had multiple goals that are interwoven throughout the book, rather than being neatly separated. One main goal was to define natural selection. A second was to show how natural selection causes speciation. A third was to develop evolution as a theory, which means to show how the single mechanism of natural selection can help to explain a diversity of phenomena that were previously thought to be unrelated to one another. His fourth main goal was to anticipate the debate that his new ideas would generate, since he was attempting to overturn established dogma. These themes are often presented together as a complex fabric, rooted in the science and arguments of the past.

I have tried to simplify the presentation of the *Origin* by deconstructing the original, then presenting its themes separately as the three main sections of this book. Part 1 presents "natural selection," which Darwin proposed as the universal mechanism that causes organisms to evolve. Part 2 presents "speciation," which is one possible consequence of evolution by natural selection. Part 3 presents the "theory of evolution," which develops the explanatory power of evolution and defines its potential as the central, organizing theme of the life sciences. The fourth theme, Darwin's response to anticipated criticisms of his theory, is presented throughout this book in a fashion that is loyal to Darwin's description of the *Origin* as one long argument. In the last chapter I apply Darwin's principles to a topic he avoided in the *Origin*—the evolution of humans.

In each section, I present Darwin's arguments and use his examples, but also offer the background needed to understand his arguments. In so doing, I highlight some areas of growth in science inspired by the *Origin*. I also present observations about the expanded influence of the *Origin*; these extra topics appear as "evolution today" segments at the end of some chapters and as several independent chapters dispersed through the book. Unless I say otherwise, I present what I believe to be Darwin's understanding of evolution, which is not necessarily the same as our understanding of evolution today. I maintain the connection between this book and the *Origin* by often quoting Darwin from the facsimile of the first edition that is now available from Harvard University Press. All page references are to the first edition, unless otherwise noted. Many people may not want to read this book and the *Origin* together, so I have written this book to serve as either a stand-alone explanation of the *Origin* or a reader's guide to the *Origin*. I hope that some readers will be inspired to read the original.

Acknowledgments

I am very grateful to Melody Clark (my wife) and Meredith Clark (my sister-in-law) for reading and rereading the entire book, and to Alison Prior for reading the first section of the book. Michael Ruse reviewed and commented on the entire book, plus he was a constant source of advice and information about the history and philosophy of science that was the background to the *Origin of Species*. Nigel Hughes and David Oglesby generously provided much needed advice on the geology portions of the book. Joel Sachs reviewed and offered helpful comments and discussion on multiple chapters. Many others were very generous with their time in providing vital help and advice on portions of the book. They include Douglas Altshuler, Nathan Bailey, Francis Borella, Jerry Coyne, David Culver, Frietson Galis, John Gatesy, Cheryl Hayashi, Andrew Hendry, Durrell Kapan, Richard Highton, Sang He Lee, Tom Morton, Len Nunney, Todd Oakley, Susanne Renner, Brian Reznick, Robert Ricklefs, Raymond Rie, John Rotenberry, Clay Sassaman, Tom Smith, Mark Springer, Blaire van Valkenburgh, Rudi Verovnik, Steve Wagstaff, Robert Wayne, and Marlene Zuk. I also thank the members of my lab, including Jeff Arendt, Sonya Auer, Mandy Banet, Ron Bassar, Swanne Gordon, Andres Lopez-Sepulcre, Marcelo Pires, Bart Pollux, Mauricio Torres, Martin Turcotte, and Matt Walsh for our discussions about some chapters and for their tolerance of my frequent absences, or only semi-presence when present. All of these individuals contributed significantly to improving the quality of the final product.

I am also especially grateful to my many colleagues who generously provided photographs and helped with the captions, including Paul Bentzen, Brian Brock, Mark Chappell, Andrew Hendry, Nigel Hughes, Arnaud Jamin, Soren Jensen, Elizabeth and Rudy Raff, Boris Sket, John Surridge, Rudy Verovnik and L. T. Waserthal.

Finally, I thank all those associated with Princeton University Press, including Sam Elworthy, who accepted the book proposal and helped initiate the project, Robert Kirk, who provided much encouragement and comments on the book, Anna Pierrehumbert, who helped with the figures, Mark Bellis, who helped with the final production, and Will Hively, who did an outstanding job with the copyediting.

I am sure that there are many others who helped but whose names I have either forgotten or do not know. I thank them all for their time and efforts on my behalf.

The *Origin* Then and Now

Introduction

Charles Darwin and the *Origin of Species*

MICHAEL RUSE

Charles Robert Darwin was born on February 12, 1809, the same day as Abraham Lincoln across the Atlantic. He died on April 19, 1882. Unlike the future president, there was no log-cabin birth for the man who is known as the "father of evolution." The Darwins were an upper-middle-class family living in the town of Shrewsbury, in the British Midlands. Charles's father Robert was a physician like his father, Erasmus. In those days, physicians were university-educated men with significant social status. Robert Darwin was also a very canny money man, acting as a link between aristocrats, with money needs and land to mortgage, and the new crop of businessmen being produced by the industrial revolution, looking for safe places to park their cash. But the real source of Darwin's wealth came from his mother's family. Charles's maternal grandfather Josiah Wedgwood was the founder of the pottery firm that bore his name, and had become one of the richest magnates of his day. Charles further secured his financial independence when, in 1839, he married his first cousin, Emma Wedgwood, the daughter of Josiah's oldest son, also called Josiah Wedgwood.

Early Years

The background and the money tell us much about Charles Darwin and his place in British society. He was not an aristocrat, but he was a gentleman, with a very secure background and expectations. As a child and grandchild of the world of business and technology, he would be properly educated, starting with one of England's leading private schools; he was going to be committed to a world of change but not revolution (manufacturers appreciated societal stability); he would be liberal in a nineteenth-century sense, which meant being strongly against slavery but prepared to let the working classes labor for minimal wages as the political economy of the day demanded; and he probably would be religious but not obsessively so. The Darwins were Anglicans, something important where the elite English universities, Oxford and Cambridge, were concerned because membership in that state church was a necessary condition for graduation. The Wedgwoods were "dissenters" or "nonconformists," being Unitarians, which meant that they believed in God but denied the Trinity, thinking Jesus just a good man and not divine.

Charles was the fourth of five children. Although his mother died when he was young, his childhood was happy, with two older sisters taking charge of the care of their younger siblings. Father Robert was a large and rather forbidding person, but Charles respected him greatly. As a child, Charles did not seem particularly gifted academically, perhaps excusable in a world where education consisted of huge amounts of translation from and into Greek and Latin, interspersed with solving geometry problems from Euclid. But he was fascinated by the world of nature, and living in rural Britain was ideal training for a future naturalist. This was helped by frequent visits to Maier Hall, the home of the Wedgwoods. Uncle Josiah was a man of learning and understanding, a great favorite of the whole Darwin family. As Darwin got older, his visits with the Wedgwoods included learning to ride a horse and to become a good shot with a hunting rifle, which were skills that served him well in unexpected ways later in life. No one could have then imagined that he would someday be a provider of fresh meat to a ship's crew camping out on the unpopulated shores of South America or ride across the wild pampas of Argentina with gauchos as companions.

After school days, Charles was packed off to Edinburgh University, then one of the great European centers of learning, to follow in the family tradition of medicine. It was not to be. Young Darwin (he was only seventeen when he went north) hated the operations he was forced to watch. Even more he hated having to rise early to listen to dry old Scotsmen lecture on dry old subjects. After two years he had had enough. But the time in Edinburgh was far from wasted, for Darwin now was starting to get serious about natural history, mixing with people more learned than he, including Robert Grant, who had unorthodox ideas about the origins of species. His activities at Edinburgh included attending a natural history course that gave him an introduction to geology and paleontology, hanging out at the natural history museum to talk to the director and look at the collections, and being tutored by John Edmonston, a freed slave from the West Indies, on how to prepare bird skins. Grant mentored Darwin on his first research project, which was to describe the free-swimming larvae of a marine invertebrate. Darwin also joined the local student science club, the Plinian Society, and attended some meetings of the professional Wernerian Natural History Society. He got his first introduction to the ideas of some of the great natural scientists of his day, including Georges Cuvier, Etienne Geoffroy Saint-Hilaire, and Jean-Baptiste de Lamarck. He also studied his grandfather Erasmus's eighteenth-century writings, in which he joined Lamarck as one of the earliest proponents of evolution.

There was no question of Darwin now switching into a program of biological studies, as do so many present-day premedical students who were initially fulfilling their parents' desires rather than their own. Although there was enough family money that he would never need to work for a living, it was not part of the Darwin-Wedgwood creed that young men should grow up idle, although Darwin's slightly older brother Erasmus did precisely that, living the life of a London gentleman of leisure.

After serious family discussion, Charles Darwin was redirected toward the church—that is, toward the prospect of becoming a parson in the Church of England. There was a certain amount of cynicism in this from Robert Darwin, for it is clear that religious belief lay lightly upon him, a polite way of saying that he was an atheist or near. Charles at this time however believed in conventional Christianity, being even a bit more of a literalist than most of his set. However, the main attraction of parsonhood was not strictly theological. It was a respectable profession for a gentleman, especially one of private means who could thereby supplement his income and afford a curate to do much of the daily work. A parsonhood could leave one with abundant free time and provide an excellent forum for pursuing an interest in natural history.

Entry into the church as a man of the cloth demanded a degree from Oxford or Cambridge, and so in 1828 Charles Darwin was packed off to Christ's College, Cambridge, where he was to spend three very happy years. A university education in those days was like Darwin's earlier training—devoted to classics with a bit of mathematics mixed in, unless one was brave enough to pursue an honors degree, which Darwin was not. However, when Darwin arrived at Cambridge, things were beginning to change. There was a growing interest in science. Although there were few courses in science, there were a number of professorships. To this point, most holders of the posts had simply enjoyed the perks but felt little need to do anything. Now these professors were taking their responsibilities seriously. Adam Sedgwick, professor of geology, John Henslow, professor of botany, and William Whewell, professor of mineralogy, were uniting and creating a space for empirical studies, although it was not until the middle of the century that they managed to get degrees in scientific subjects incorporated into the university curriculum.

It is clear that these reformers were talent spotting, and when the young Darwin appeared on the scene—sharp, ambitious, well connected—he was welcomed into the group. He was invited to evening socials with professors and was able to interact with them one-on-one and gain tutoring beyond what would come in the lecture hall. For three years he attended the lectures of John Henslow, the professor of botany, becoming his pet pupil. He pursued natural history during his abundant free time, and became a fanatic for collecting beetles. During the summer after graduation, he took a walking tour through Scotland with Adam Sedgwick and received a two-week crash course in the practice and theory of geology. Above all, he was taken under the wing of one of the most remarkable men of his age, William Whewell (pronounced "Hule"), who was later to make the peculiar switch from being the professor of mineralogy to become professor of moral philosophy and later the most powerful man in Cambridge, as the crown-appointed master of Trinity College. Whewell wrote mathematics textbooks but already he was planning his magni opera, the three-volume *History of the Inductive Sciences* (1837) and the subsequent two-volume *Philosophy of the Inductive Sciences* (1840). Knowledgeable about everything, Whewell lectured everyone on science, its history and its methodology. Also, it was he who urged Darwin to read what was to become a very popular handbook on the nature of science, *A Preliminary Discourse on Natural Philosophy* (1831) by the astronomer-philosopher John F. W. Herschel. In those days, "natural philosophy" was what we call science. "Moral philosophy" was what we call philosophy. Some British universities still use the old language.

Darwin, who was now a tall and strong young man, was obviously popular with his fellows. Much of the university time was spent most enjoyably—riding, dining, playing at those things that occupied young men of his class.

But it is clear that he was starting to mature intellectually and to show the drive so typical of his class and background.

The Beagle Voyage

By 1831, when Darwin graduated from Cambridge, he was already half-trained as a scientist and recognized as an up-and-coming young man, from the right sort of background and with both talent and enthusiasm that recommended him to those with authority and influence. It was therefore fortuitous but perhaps not too surprising when Darwin was offered the chance to join a British warship, HMS *Beagle*, that was about to set out to South America with the charge to bring home much-improved maps and charts. Being captain of such a ship was an awesome task and a lonely prospect—one was of a social class separate from all others and with the responsibilities for the ship and its crew, not to mention the assigned tasks. Robert FitzRoy, the captain (and incidentally an aristocrat, a nephew of the former foreign secretary Viscount Castlereagh), was but twenty-three. He was looking for someone, a gentleman rich enough to pay for his own board, who would come along as a kind of captain's friend. Darwin's father initially opposed the voyage because he was reluctant to keep spending money on a son who seemed always to be putting off getting a real job. Uncle Josiah persuaded Darwin's father that the voyage was an opportunity that should not be missed. Charles joined the ship, but not as the ship's naturalist; the surgeon doubled as the official naturalist, as was often the case on such voyages. Darwin's talents and enthusiasm soon gained the attention of the captain and, as the captain's companion, gained him special favors and attention from the crew. The official naturalist resigned early in the voyage in protest of Darwin's favored treatment.

Much of what naturalists did in those days was collect, preserve, and catalog to create a record of expeditionary travels. Darwin spent the whole of the voyage, which stretched for some five years, from 1831 to 1836, making massive collections of rocks, plants, and animals, and at the same time studying and theorizing on the magnificent world unfurling before him. Because Darwin came officially as the captain's companion and was paying his own way, all his collections were his own, rather than the property of the Royal Navy. These valuable collections were his calling card to enter the world of science after he returned to England.

Darwin became as popular with the officers of the *Beagle* as he had been with his contemporaries in Oxford. Many of the officers shared his interests in natural history and were making collections of their own. Darwin's diverse activities and broad knowledge of natural history and geology earned him the nickname "Philos," or the philosopher, and he was sometimes referred to as the man who knew everything. His marksmanship and riding ability also made him one of the providers of fresh meat when the *Beagle* moored and the crew spent time on land in thinly populated parts of South America. When at sea, Darwin continued to pursue natural history, but also helped provide food with the trawls and fishing lines that he used to sample sea life. Hunting and natural history were two sides of the same coin.

Although Darwin was ecumenical in his tastes and interests, his central scientific motivation was geology. Geology was a very trendy science at the time, with significant industrial applications: no investor wanted to sink a shaft in a part of the world that would not yield coal or other valuable minerals or natural products. By one of those strokes of extreme good luck— Darwin tended to be lucky; his genius was to take advantage of it—just as the *Beagle* was setting off, the Scottish lawyer-turned-geologist Charles Lyell published the first volume of what was to prove a lasting and respected vision of the geological history of the earth. FitzRoy presented Darwin with this first edition as a gift just before the *Beagle*'s departure; it was a measure of the interests they would share on the voyage.

In his *Principles of Geology*, Lyell argued for what Whewell was to call "uniformitarianism." Lyell proposed that forces of a kind and intensity that we see about us today caused all the geological features of the earth. Wind, rain, deposition, erosion, earthquakes, and the like can do everything, including build mountain ranges, excavate canyons, or carve seaside cliffs from a sloping coastline, given sufficient time.

Lyell also argued for a steady-state view of the earth's history. Thus he opposed more traditional geologists like Sedgwick who embraced what Whewell was to label "catastrophism," or the belief that the earth's history was punctuated by times of extreme upheaval, and that the planet was changing progressively as it cooled from a molten state to its present form. Lyell realized that the evidence did not speak directly to a steady-state view—there are fossil palms around Paris suggesting that the climate was once much warmer—and so to defend his thinking Lyell introduced his "grand theory of climate." Apparently the earth is a bit like a water bed; some parts of the surface that are now dry land were once under the sea, while others that are now inundated were once dry land. Hence, distributions of land and water are always changing, which affects ocean currents and climate. England today is warmer than we might expect of its latitude, which is the same as that of Hudson Bay, Canada, where polar bears roam. The British can thank the Gulf Stream for their mild weather. In a similar fashion, historical changes in landmass distributions and ocean currents account for historical changes in climate.

Darwin was seduced by Lyell's easy prose and subtle arguments and was soon converted to Lyell's principles. At the very first landfall of the *Beagle*, at St. Jago in the Cape Verde Islands, he saw rocks bearing fossil shells similar to animals that could still be found living along the shoreline, but the fossils were well above sea level. He inferred that land that had once been under the sea had been elevated above the waterline.

Darwin spent the whole of the *Beagle* voyage looking for further evidence of Lyellian processes. Indeed, it was through this dedication to Lyell's ideas that Darwin made his first great scientific move, which ironically was to propose a different explanation from Lyell's for the origin of coral atolls. Lyell had argued that these curious rings of coral reefs in the middle of the ocean were simply the tops of extinct volcanoes peeking out above the waves. Darwin thought it highly improbable that so many volcanoes had all reached exactly the height of the sea surface. It seemed much more likely that the tips of the volcanoes once emerged from the ocean, then sank back into the sea as the land beneath them subsided. The reefs formed as the coral grew ever upward from the subsiding volcano. Coral can survive and grow only in shallow water. Darwin may have disagreed with Lyell's proposed mechanism for the formation of coral atolls, but he used Lyell's principles to develop his alternative. Lyell very much appreciated Darwin's initiative. Darwin has proven to be correct.

Darwin's conversion to the Lyellian perspective fit in with changes that were happening in his religious convictions. There was no road-to-Damascus experience in reverse, but the Lyellian emphasis on causes operating today as the clues to causes operating in the past was clearly one of the factors leading Darwin to question the authenticity of miracles. Miracles are causes that lie outside our day-to-day experience. As his belief in these started to go, so did Darwin's belief in Christianity. Of course, today no liberal Christians takes seriously stories about turning water into wine, but Darwin's faith was especially vulnerable here. A standard text read at that time by all students at Cambridge was *The Evidences of Christianity* by Archdeacon William Paley of Carlyle. Paley's proof of revealed religion was simply that the disciples who suffered unto death for their faith would not have done so had the miracles been inauthentic. For Darwin, with the miracles so went the faith.

This does not mean that Darwin became an atheist. This never happened. But his beliefs did start to move in the direction of a God as unmoved Mover, one who set the world in motion and who then achieved everything through unbroken law. This was a popular position among intellectuals at the end of the eighteenth century, especially those denying the divinity of Christ, and hence the miracles attributed to him.

An important by-product of Darwin's switch to uniformitarianism is that suddenly a question like the origins of organisms is transformed. In moving toward evolution, you are moving toward denial of the miraculous origins of life. But no longer are you cutting against your philosophy and your religious convictions. Your God now is one who works through unbroken law, rather than individual acts of creation. Showing that all plants and animals have a natural origin, meaning an origin that is a function of unbroken law and hence not supernatural, is a proof of your God and a testament to His great power. Tie this in to the fact that Darwin came from a family of industrialists, who saw their great achievements as doing with machines (which function in accord with the natural laws defined by Newton) what previously had been done by hand, and the recipe is virtually complete. Charles Babbage, the inventor of a kind of proto-computer, was then showing how he could program his machine to do the same thing a million times, then on the million and first do something else, and then back to the original for all time. Babbage was explicit in his claim that this was the right way to consider miracles—unexpected yes, outside of law, no.

As with religion, however, there was no immediate road-to-Damascus experience leading to Darwin's theory of evolution. It is only the cumulative effect of all that he saw during the voyage and what he learned after his return to England that can be credited with his conversion.

Much of the time that the *Beagle* spent around South America (four years in all) found Darwin engaged in regular activities of sampling wildlife in the ocean or collecting and traveling on land. He was often terribly seasick, and there was limited attraction in staying on board as the ship moved slowly up and down the coast taking readings. The geology and land life of South America beckoned, so he organized expeditions to the mainland while the *Beagle* did its surveying. He made some long, and at times thrilling, trips into the interior.

In an early outing, he spent May and June 1832 in a cottage on Botofogo Bay, near Rio de Janeiro. The equipment that he brought along on his daily trips into the rain forest reflected his diverse interests. His tool kit included a geology hammer, nets, shotgun, and pistol. Guns may seem an odd accessory for natural history studies, but a priority then was to collect and preserve specimens for storage in museum cabinets and eventual study. Darwin described his natural history studies as being very much like a hunting holiday. His experience as a beetle hunter was a key to his success, since he was well schooled in the virtues of turning over stones, peeling bark, and pawing through leaf litter. One early passion was colorful flatworms. Others were bioluminescent insects, butterflies, moths, beetles, and spiders.

During his time in Brazil and throughout his travels, Darwin remained faithful to the rigors of field science. He identified whatever plants, animals, and fossils he could from the *Beagle*'s library and got his brother Erasmus to send more reference books during the voyage. He kept detailed journals and properly preserved and cataloged all specimens. He remained in constant contact with the outside world through correspondence.

In September 1832, the *Beagle* rounded the headland of Punta Alta, near Bahia Blanca, Argentina. Men on board spotted bones and shells eroding out of the banks on the shoreline. The ship anchored for a week while Darwin and the crew excavated fossils. The fauna, represented in what Darwin described as 200 square yards of beach, included gigantic mammals in the size range of a rhinoceros or elephant. They included giant ground sloths and a glyptodont, which was like a giant armadillo. Darwin identified some fossils with use of the *Beagle* reference library and found that one, a *Megalonyx* or *Megatherium*, had been identified by Georges Cuvier thirty years earlier. A remarkable feature of these fossils, other than their size, was that they were all clearly related to the much smaller sloths and armadillos still found in South America. Darwin augmented this collection of fossils with others obtained throughout the voyage. They were Darwin's "in" to meet and collaborate with Richard Owen after he returned to England. Owen's monograph on the fossils of the *Beagle* voyage was awarded the Wollaston Medal from the Geological Society in 1838.

Darwin made five trips across Argentina and Uruguay, often in the company of soldiers or gauchos. These travels were during Indian and civil wars, through a countryside that was at times populated by soldiers, of whom Darwin said, "I should think such a villainous, banditti-like army was never before collected together" (Darwin [1860] 1962, p. 70). He traveled up the

Santa Cruz River, through the barren plains of southern Argentina, with Captain FitzRoy and a company of men from the *Beagle* (where they were shadowed by the local Indians), with the unfulfilled hope of reaching its headwaters in the Andes. He rode through the dense, temperate rain forest of Chiloe Island, off the coast of Chile, and later through the Portillo Pass, from the Pacific to Atlantic slopes of the Andes, then back again to the Pacific slope through the Upsallata Pass. He saw how the Andes were a divide between two distinct faunas, one that occupied the steep Pacific slopes of the Andes, the other the Atlantic slopes and diverse habitats that lay between the mountains and the Atlantic Ocean. All the while, he added to his collections of rocks, plants, and animals and made observations of geology.

Darwin also turned his critical eye to the diversity of peoples he encountered along the way. He saw African slaves in Brazil, Indian tribes throughout South America, the natives of Tahiti, and the Maori of New Zealand. He observed and took notes on the people he encountered during his travels in the same way that he kept notes on zoology, botany, and geology.

One formative experience for Darwin came as the *Beagle* rounded the Horn, passing from the Atlantic to the Pacific. FitzRoy had returned to England from a previous voyage with three natives from Tierra del Fuego, one of the islands at the tip of South America. These islands are persistently cold and wet, and the natives lived in a very primitive state. The three natives that FitzRoy had taken away were now aboard the *Beagle*. His idea was to return his "guests" to their homeland, together with a Christian missionary, to start an outpost of European civilization at the bottom of the world. The plan proved disastrous, for the natives robbed the missionary (who had to be rescued), and the three returnees turned, in a matter of weeks, from dandified Englishmen to the "savages," to use Darwin's Victorian descriptor, that they had once been. Darwin learned a lesson never to be forgotten: civilization is just a thin veneer that is easily stripped away. We humans are close to the apes. This is perhaps one of the most striking things about the thought of Charles Darwin. Virtually every one of his contemporaries was picking at the scab of humankind. Charles Lyell could never be really true to his principles and always wanted miracles for our appearance. He sadly sighed in later life that he could not "go the whole orang"—meaning attribute to humans a simian origin. Darwin's great supporter Thomas Henry Huxley (the grandfather of Aldous Huxley, the novelist) came from the other side, never letting anyone forget the common origins of humans and their close relatives the great apes. Darwin was calm and placid in the middle: We are part of the animal world. It is worth talking about but not worth worrying about. It is neither frightening nor amazing.

As well as collecting and cataloging, Darwin kept a detailed diary, parts of which were sent home together with lengthy letters. These were read avidly by the Cambridge circle, who arranged for the publication of some of them, as well as by the family. It was becoming clear to all that Dr. Darwin's younger son was more talented than anyone had realized. Darwin returned from the voyage to find that he was already a minor celebrity, based on the publication of some of his letters. After the voyage, his diary was used as the basis of an official account of the *Beagle* voyage and then republished in popular form as the *Voyage of the* Beagle. In the middle of the nineteenth century, people loved travel books. Long before Darwin achieved fame as an evolutionist, Darwin of the *Beagle* was a household name.

Evolution

As most everybody knows, key experiences came when the *Beagle* left South America and cut across the Pacific to a group of volcanic islands on the equator, the Galapagos archipelago. Darwin was an expert on the flora and fauna of South America when he arrived on the islands. Teeming with life, these islands proved to have inhabitants that bore similarities to those that Darwin had seen in South America, yet many were distinct species. To Darwin's great amazement, the giant tortoises that are found there differ from island to island. So also do the small birds, the finches (now known as Darwin's finches), and the mockingbirds. Darwin had just spent years in South America, where he had observed that you might find the same or closely related species in the steamy jungles of the Amazon and the snowy deserts of Patagonia. Now, apparently, God so loved the Galapagos that he put different species on little islands but a few miles from each other. There had to be something up. Darwin puzzled about this all the way home.

Upon his return home, Darwin at once plunged into detailed examinations of his collections, which included parceling prize specimens out to selected experts for identification and description. The Galapagos birds were examined by a leading ornithologist, John Gould, who at once declared them to be different species. All the other details of his journey, such as the giant fossils of extinct animals that were related to those still alive in the same place or the Andean divide between two distinct faunas, began to sink in. At some point early in the spring of 1837, all that he had seen and all that he was learning from experts like Gould and Owen caused him to slip over to being an evolutionist or, as people called it then, a transmutationist. This was no light or casual thing to do. On the one hand, Darwin knew that his scientific set would be appalled, since they strongly opposed such thinking; so he would do well to keep his ideas to himself. On the other hand, Darwin knew that discovering how species originated was the prize above all worth getting. Herschel, the same Cambridge mentor who so strongly influenced Darwin, named the origins of organisms as the "mystery of mysteries." Darwin glowed at the thought of being the one to make the big hit. Note that we do not say this in any sense of disapproval. Great science demands great talent and great effort. It does not come to the unprepared or to the indifferent. Our point is that Darwin was very human and hence, in this respect, understandable scientist.

Darwin knew that becoming an evolutionist was only the first stage. Now he had to find a mechanism, a cause. Here again Darwin was not working or thinking blind. He was a graduate of the University of Cambridge, the university of Newton. Herschel, Sedgwick, Henslow, and above all Whewell (who was now president of the London Geological Society and who had pressed Darwin onto the managing board of that organization) stressed again and again that Newton's great achievement was to find a force, gravity, that explained the universal laws of motion, including the orbits of the planets established through the Copernican Revolution and its aftermath. Darwin wanted to be the Newton of biology, so he had to find the biological equivalent of the Newtonian force of attraction.

What would such a bright, ambitious young scientist consider the right model or archetype of a scientific theory? His Cambridge mentors, Herschel and Whewell, were trying to articulate the definition of good science. They were the ones lifting Newton to a pedestal far above all others, and it was their interpretation of a Newtonian science that influenced young Darwin. He had read Herschel's work on the philosophy of science and spoke highly of its influence in his autobiography. He also knew Whewell both at college and after the *Beagle* voyage and, again in his autobiography, spoke highly of him and his books.

So what was the Newtonian ideal of a theory? In respects it was to be what today's philosophers of science call a "hypothetico-deductive system." By this is meant that it is supposed to be axiomatic, like Euclidean geometry—one starts with a number of premises or axioms and then deduces everything from them. So in Newtonian mechanics you start with the laws of motion and of gravitational attraction and then deduce Kepler's laws, such as planets going in ellipses, and Galileo's laws, including cannonballs flying in parabolic trajectories. The difference between mathematics and an empirical science is that the former makes claims that are in some sense true by virtue of their form while the latter makes claims that are true because they correctly describe the physical universe.

But there was more to the picture than this. Newton's real claim to fame was that he had come up with the cause of the precisely observed but as yet unexplained orbits of the Copernican Revolution—gravity. It was this that made him the icon of science to Darwin's mentors. However, what exactly does one mean when one talks of "cause"? It is clearly something that makes other things work or move, but how does one know that one has a cause? How does one know that one has what Newton somewhat mysteriously called a *vera causa* or "true cause"? Here the authorities divided somewhat. Herschel was more of an empiricist and wanted observational evidence, direct or indirect. Whewell was more of a rationalist and wanted a sufficient explanation of all the facts, even if we do not see the cause at all.

Thus, for Herschel, the example was of the attraction between the earth and the moon. How do we know that gravity is a vera causa keeping the two together? Because we have all seen objects drop to the ground or have swung a stone around at the end of a piece of string and felt the pull required to keep the stone from flying off. We have directly sensed the cause—or something analogous. Empiricist though he may have been, Herschel thought that all natural force was a product of God's will, as the tug on the circling stone is a product of our will. Whewell would have none of that. He wanted a kind of global explanation, with the vera causa to bring together and unite diverse phenomena under a single explanation. He referred to this unifying feature of a theory as a "consilience of inductions."

> The Consilience of Inductions takes place when an Induction obtained from one class of facts, coincides with an Induction from another different class. This Consilience is a test of the truth of the Theory in which it occurs." (William Whewell, Philosophy of the Inductive Sciences, p. 1840)

"Consilience" comes from the same root as "reconcile," and inductions are logical arguments based on observations. Thus a consilience of inductions is the bringing together of scientific observations derived from different disciplines and showing that they have a common cause. Newtonian physics met this ideal because a single entity, gravity, could be shown to explain phenomena as diverse as the movement of planets around the sun, of the moon around the earth, ocean tides, or the fall of an apple from a tree.

To grasp the difference between Herschel and Whewell's philosophy, think of a murder. We find a body lying in a pool of blood with a murder weapon, a knife, lying beside it. After an investigation, someone is charged with the murder. Why? Did the investigation reveal a vera causa? A Herschelian would be satisfied only if the accused were seen killing the victim. A Whewellian would instead be satisfied with the establishment of a means, motive, and evidence that directly linked the accused with the crime, such as his fingerprints on the knife. Why should we need eyewitness testimony? Evidence alone can establish guilt.

We can see Darwin's dedication to Herschel's emphasis on vera causa in the opening chapter of the *Origin*, where he begins with a discussion of artificial selection. Darwin, who grew up in a rural district during the British industrial revolution, had seen artificial selection in action. Think for a moment about what industrialization entails. People leave the land and move to cities, and there is a general rise in population numbers. Hence, more people need to be fed with a smaller proportion of them being farmworkers. In other words, an industrial revolution demands also an agricultural revolution. Here, the key proved to be not machines but selective breeding. Bigger and fatter pigs, cows that produce more milk, shaggier sheep. Perhaps most important of all was the development of quality root crops like turnips. Darwin, coming from rural Britain—his uncle Josiah was a gentleman farmer trying to upgrade sheep production—knew all about breeding. So, early in his quest for a mechanism, Darwin sensed that selection, which meant breeding from the good forms and rejecting the inadequate, was the key to organic change.

The practicality of artificial selection and its ability to cause change was only part of the battle. Darwin then had to address a more challenging theoretical issue. At Cambridge, Darwin was introduced to Archdeacon Paley's major work *Natural Theology*, dealing with the side of religious belief that focuses on reason and evidence for the deity. In particular, Darwin was saturated with the favorite argument of the British for the existence of God, namely the argument from design. Inviting his readers to consider the eye, Paley argued that it resembles a telescope in both structure and function. But telescopes have telescope designers and makers. Hence, eyes must have a designer and maker as well, better known as the god of Christianity. At this point in his life, the late 1830s, Darwin was not inclined to reject a Designer, although already he wanted Him to be someone who worked at a distance through law rather than individual acts. What Darwin did get from Paley was the premise that he accepted to the end of his life, namely that the organic world is as if designed. The hand, the eye, the nose, and all the other "contrivances" (a favorite word of Darwin's) or, as we call them today, the adaptations that organisms have for living and reproducing, have a design-like nature. Organisms are not just thrown together in a random way. They work, they are organized, they function, they exist as if an intelligence sat down and planned them. Random, blind forces lead to messes—remember Murphy's Law, that if things can go wrong they will. Organisms are not messes. They are the very antithesis of a mess. Hence, for Darwin, a mechanism of change had to be capable of creating design-like features. Darwin realized that selection can do all these things. The whole point of the breeders' art was to produce the features that were desired—more fat on pigs, more milk from cows, more fleece on sheep. By introducing artificial selection and then arguing to natural selection, he was providing a Herschel-type vera causa. He was arguing from a force we know, a force we cause ourselves, to a force of nature. The first chapter of the *Origin* is deceptively friendly, all about things people know about. But, at the same time, it is establishing proof for the mechanism that underlies his theory. Organisms can be changed in permanent ways through the process of selection.

The big problem now facing Darwin was how to get a natural equivalent of the breeders' artificial selection. What could make a force so strong and persistent that selection could operate efficiently in the wild?

At the end of September 1838, Charles Darwin read a late edition of Thomas Robert Malthus's *Essay on the Principle of Population*, where Malthus observed that human populations can increase in size faster than does their ability to produce food, causing a struggle for existence. At once Darwin seized on this struggle, generalizing it to animals and plants. Darwin put exponential population growth together with artificial selection, and he had his mechanism. His notebooks of 1838 show that he had arrived by then at natural selection as the mechanism that causes evolution.

Darwin was as sensitive to Whewell's concept of a consilience of inductions as he was to Herschel's concept of vera causa. After the mechanism of selection is introduced, and after some other matters are attended to—for instance, the introduction of the secondary mechanism of sexual selection, the principle of divergence, some rather unsatisfactory speculations about the causes of heredity—the *Origin* turns precisely to such a survey of biology, showing the big problems and showing how selection speaks to them. In fact, the bulk of the *Origin* is devoted to this consilience of inductions.

Darwin worked systematically through instinct and behavior, paleontology and the fossil record, biogeography and the distributions of organisms, morphology, systematics, and embryology. Again and again, Darwin argued that selection speaks to the issues and conversely the issues justify a belief in natural selection.

The Long Wait

Darwin first articulated his theory in a 35-page "Sketch" in 1842 and then a full 250-page "Essay" in 1844, neither of which was published during his lifetime. We wonder why Darwin waited fifteen more years to publish the *Origin*.

One interpretation is that the delay was just part of the fabric of his life. On a personal level, at the beginning of 1839 he got married, and he and his wife were soon plunged into the joys and cares of a family. They moved from London to a village in Kent, where eventually they had ten children, seven of whom grew to adulthood. Darwin was a devoted father, but also a persistent tinkerer, constantly doing experiments and making natural history observations. It was also around this time that Darwin started to fall sick—from being a vibrant young man who had lived rough on the Argentinean pampas with the gauchos, he now became an invalid. He had headaches, stomach cramps, insomnia, boils, bad breath, and a host of other ailments.

Despite medical attention and frequent visits to spas and other rest resorts, Darwin never again enjoyed good health. The nature of his illness is one of those great Victorian mysteries. Some think Darwin's troubles were physical; perhaps he got Chagas' disease from insect bites he suffered while traveling in South America. Others think they were psychological, caused by the anxiety he felt about his theory, its likely stormy reception, or its effects on Victorian society. Either way, Darwin really did suffer, and this slowed him down; however, he learned to use his illness to his advantage. He interacted with society only when it suited him, and fixed his life so that his work came first and last, with due exceptions made for his family and visitors in between. Health problems or not, Darwin continued to write and publish. He published his diary of the voyage in 1839 as the third of a three-volume series about the voyage, then revised it for publication by itself under the title *The Voyage of the* Beagle (1839, second edition in 1845). He edited multiple volumes of the zoology of the voyage (1838–1843) and published his three monographs on geology (1842, 1844, 1846). All this activity belies bad health alone as the cause of his waiting so long after articulating his theory in the "Essay" before publishing it in the *Origin*.

An alternative explanation is that Darwin spent the intervening fifteen years improving his status and knowledge as a biologist. After finishing his geological work in the mid-1840s, he turned to what grew into a massive study of barnacle classification that took more than seven years to complete. There is no doubt that this work solidified Darwin's standing as a scientist. In 1853, he was awarded the Royal Medal for Natural Science for his barnacle monographs. His barnacle years also gave him hands-on experience in comparative anatomy, embryology, paleontology, and systematics. All these fields were central to the argument presented in his essay of 1844. It was during this interval that he experimented with pigeon breeding to better understand selection and inheritance, performed experiments to characterize how organisms disperse across oceans, and tinkered endlessly with other problems suggested by his theory. We can see clear imprints of all of these activities in the *Origin*. So, even though he had worked out his theory by 1844, by 1859 he was able to combine it with a wealth of personal experience. In 1844 he wrote as one who had gained much of his knowledge from reading. In 1859 he wrote with the insights of a practicing scientist, and as a scientist who was widely recognized for his achievements.

Another reason for Darwin's delay was likely the publication in 1844 of the anonymously authored *Vestiges of the Natural History of Creation*, a book written for a general audience that promoted the idea of species transmutation, or evolution. Its proposal of evolution, and many other features of its popular science, caused a huge row. Although there were circles where its ideas were welcomed, all Darwin's mentors took up arms against the book. *Vestiges*, judged as a work of science, is not a great masterpiece, and Darwin certainly felt no threat from the contents. But he did not want to engage in battle, certainly not at a time when he was feeling really unwell; and so we suspect he shelved any publication plans that he had at that time.

Darwin instead settled into a comfortable regime of research. At the same time, however, realizing that a successful theory needs friends, he introduced his scientific companions to his theory. Apart from Charles Lyell, his confidants were men of a new generation, including Joseph Hooker, who would become the director of the Royal Botanical Gardens, and Thomas Huxley, who rivaled and overmastered Richard Owen as an anatomist, plus others, not all of whom were converted to his ideas. He cultivated a team of future advocates but also benefited from their arguments, which became part of the *Origin* and helped Darwin anticipate the arguments that would arise after the *Origin* was published. Darwin must have felt a need to present his ideas to a critical world in more complete form, so exhaustively documented that even the skeptics who lashed the *Vestiges* might be convinced. By 1858, he was well along in writing his "big book" about evolution, which would have run to well over one thousand pages if it had ever been published.

Meanwhile, word of Darwin's reputation as a naturalist got out, so it was not pure coincidence that in 1858 a young naturalist-collector out on the Malay Archipelago sent him an essay in which he presented his own theory for the origin of species. Receiving it in the early summer, Darwin read this essay by Alfred Russel Wallace and realized that the younger man had made exactly the same discovery he himself had made some twenty years before and was just now readying for public display in his "big book." Joseph Hooker and Charles Lyell arranged for the immediate publication of Wallace's essay alongside some extracts of Darwin's earlier writings that documented Darwin's priority in discovering natural selection. Then, for fifteen months, Darwin wrote furiously. *On the Origin of Species* was published in the late fall of 1859, the same year that Oregon was admitted as the thirty-third state of the Union, John Brown raided Harper's Ferry, and Charles Dickens wrote *A Tale of Two Cities*. Darwin always thought of the *Origin*, which ran to nearly five hundred pages, as just an abstract of the "big book," which was never published. The *Origin* had none of the citations or data that characterized his other writings.

On the Origin of Species by Means of Natural Selection, or the Preservation of Favoured Races in the Struggle for Life, was first published in London by John Murray on November 24, 1859. Twelve hundred and fifty copies were printed and bound. Five went to official organizations, the author got 12 free copies, and Darwin purchased another 90 to be distributed to friends, family, and prominent scientists throughout the world. He also sent copies to all his Cambridge mentors. Of the remaining 1,100 copies, 500 were purchased by Mudie's Subscription library, which guaranteed

Darwin a wider readership, and the remainder were purchased by book dealers, all on the day of publication. Murray at once asked Darwin to prepare a second edition.

Aftermath

In all, there were six editions, the final one appearing in 1872. Darwin listened to all the objections that were raised to his theory, so by the sixth edition the work was very much changed, with additions, subtractions, and rewritings on every page. A few more slight corrections were made in 1876, although the reprinting done in that year was not considered a new edition. It was this version that Murray reprinted again and again until the copyright expired in 1901. By then, approximately 40,000 copies had been sold in Britain and the empire. Foreign translations appeared immediately after the first edition. The *Origin* has now appeared in over forty languages, most recently in Tibetan. It is also available in braille. Traditionally, in both English and foreign languages it was the sixth edition that was printed, since this seemed to be the definitive edition and final statement by Darwin on the subject. However, in the past thirty or forty years, with the rise of the history of science as a freestanding academic discipline, there has been an increasing tendency to reprint the first edition. It is felt that this is the work that shook the world and moreover the work in its cleanest form, before Darwin started to tinker with it, often for reasons that seem less compelling today than in the 1860s.

The *Origin* caused an immediate uproar in Victorian society. In 1860, at the annual meeting of the British Association for the Advancement of Science in Oxford, Samuel Wilberforce, the bishop of Oxford, debated the *Origin* with Thomas Huxley. Although it may be apocryphal, legend has it that Wilberforce asked Huxley if he was descended from monkeys on his grandfather's or his grandmother's side. Huxley replied that he would rather be descended from a monkey than from a bishop of the Church of England. True or false, the Wilberforce-Huxley story captures one dimension of the reaction to the *Origin*. What is not apocryphal is that, at that meeting, Darwin's old captain, FitzRoy of the *Beagle*, who earlier had probably had the same liberal religious ideas as Darwin but who had by now become a bit of a religious fanatic, went around the room brandishing the Bible and condemning his old shipmate!

What is less well known is just how widespread acceptance of evolutionary ideas became in the years immediately after the *Origin*. Few were prepared to go along with natural selection, at least as a comprehensive and adequate explanation of the evolutionary process, but the idea of "descent with modification," as Darwin called it, was almost universally accepted. This was true even of practicing Christians, as we can tell from religious publications, letters, sermons, and other sources of information. John Henry Newman, the great theologian who started life as an evangelical Christian, then became the leader of the High Church faction of the Church of England, and finally converted to Rome and ended up as a cardinal, can stand for many. He had long rejected the argument from design as a major stumbling block to science. In 1870, about his seminal philosophical work, *A Grammar of Assent*, Newman wrote: "I have not insisted on the argument from *design*, because I am writing for the 19th century, by which, as represented by its philosophers, design is not admitted as proved. And to tell the truth, though I should not wish to preach on the subject, for 40 years I have been unable to see the logical force of the argument myself. I believe in design because I believe in God; not in a God because I see design" (Newman 1973, p. 97). He continued: "Design teaches me power, skill and goodness—not sanctity, not mercy, not a future judgment, which three are of the essence of religion."

When he was asked about whether Darwin should be awarded an honorary degree from the University of Oxford, Newman replied: "Is this [Darwin's theory] against the distinct teaching of the inspired text? If it is, then he advocates an Antichristian theory. For myself, speaking under correction, I don't see that it does—contradict it" (letter of June 5, 1870, in Newman 1973, p. 137). Certainly Newman, like many others, would have liked more direction to the evolutionary process than Darwin would have allowed, and when it came to human souls he would have been adamant that here there was need of miraculous intervention, but overall he like many other fellow believers wanted to let science get on with its business. When he was setting up a Catholic university in Dublin in the 1850s, Newman insisted on the need to keep scientific inquiry unfettered by religious dogma (Newman 1873, pp. 428–32). In his view, science and religion deal with different spheres and thus, properly understood, do not interact and cannot conflict. Of the natural and supernatural worlds, he said that "it will be found that, on the whole, the two worlds and the two kinds of knowledge respectively are separated off from each other; and that, therefore, as being separate, they cannot on the whole contradict each other" (p. 389).

With the *Origin* published and with the world turning toward evolution, what of Charles Darwin? He was to live another twenty-three years, and somewhat expectedly all else was a bit of an anticlimax. He kept up a punishing schedule of work. His original intention had been to publish his "big book" on evolution, but his publisher persuaded him not to. The *Origin* was making its mark and selling well. The "big book" would just cut into those sales. Darwin instead planned to write separate books that detailed various parts of the *Origin*, which he continued to think of as an abstract. For the first decade or so, this ambition was dampened by the need to keep up with the publication of new editions of the *Origin*, some of which involved extensive revisions. Partly for this reason, the grand plan got sidetracked, and in the end Darwin finished only *The Variation of Animals and Plants Under Domestication*—published in two volumes in 1868—from this intended series. Before *Variation*, and in the midst of the controversies and battles spawned by the *Origin*, Darwin published *On the Various Contrivances By Which British and Foreign Orchids Are Fertilized By Insects, and On the Good Effects of Intercrossing* (1862), which was the first of six books on plants. Darwin saw orchids as objects of beauty, but also as presenting great problems to be solved because of their complex structure and their intricate mechanisms of pollination. Orchids were a wonderful distraction from the *Origin* controversies. In the end, he also saw the orchid book as "a flank movement on the enemy" (Browne 2002, p. 174) because orchids provided such a good counterpoint to natural theology. While Bishop Paley argued that such intricate beauty was a product of design, Darwin showed that it was instead a product of natural selection that provided a mechanism to attract insect pollinators, efficiently transmit pollen to other flowers, and thus reproduce.

Darwin also produced two major works on our species and its evolution—*The Descent of Man and Selection in Relation to Sex* (2 vols., 1871) and *The Expression of Emotions in Man and Animals* (1872). These books are where Darwin argued that humans are part of the animal kingdom, which was a fight that he avoided in the *Origin*. Darwin also wrote a brief autobiography, intended for private reading by his family. It, like virtually every letter and notebook attributable to Darwin, has since been published.

For the rest, Darwin rather dabbled as his fancy took him, but these projects also often provided forums for him to expand on ideas presented in the *Origin*. His five other books on plants include one on the effects of cross- and self-fertilization, which expands on his preoccupation with the dark side of inbreeding—a personal concern of his since he had married his first cousin and worried about the health problems of some of his children. His finale, published the year before his death, is on earthworms and their role in the formation of vegetable mold. It seems particularly whimsical, but it represents a different form of his interest in how seemingly small, everyday processes can cause great change if they persist over long intervals of time.

Throughout his life, Darwin kept up a huge correspondence, and later he watched as his supporters—Huxley particularly—spread the gospel of evolution far and wide. In his last decade, Darwin's health improved somewhat, and he was able to enjoy both his large family and his fame. Finally, at the age of seventy-three, his heart gave out and he died—no real surprise, for like everyone else of his generation he was a heavy smoker. By now, what is also not surprising is the outpouring of national recognition and respect. Darwin was not just famous; he was a man to make the country proud. He was a brilliant scientist who labored on despite years of ill health; he was a beloved travel writer; he was rich but did not flaunt his money, nor did he squander it; he was a model family man (wife, children, faithful servants, not a hint of scandal); he may not have had formal religious beliefs, but he was modest in his doubts, and they never stopped him from bonding with believers including his close friend the local vicar. He was in short the archetypical Victorian gentleman in addition to being the Newton of the life sciences. There could be only one burial place for him, and there, for all eternity, he lies in Westminster Abbey, next to that other great English scientist, Isaac Newton.

References

Browne, J. 1995. *Charles Darwin: Voyaging*. Princeton, NJ: Princeton University Press. (Pp. 65–98, source of details about Darwin's time in Edinburgh; pp. 96–104, his beetle collecting in Cambridge; pp. 202–5, Darwin's ascendancy to status as unofficial naturalist of the *Beagle*; pp. 214–17, time in Botofogo Bay; p. 350, fossil collecting at Bahia Blanca.)

Browne, J. 2002. *Charles Darwin: The power of place*. Princeton, NJ: Princeton University Press.

Darwin, C. [1860] 1962. *The voyage of the* Beagle. Annotated and with an introduction by Leonard Engel. Garden City, NY: Doubleday and Company.

Keynes, R., ed. 2000. *Charles Darwin's zoology notes & specimen lists from H.M.S. Beagle*. Cambridge: Cambridge University Press.

Newman, J. H. [1873] 1999. *The idea of a university*. New York: Regnery.

Newman, J. H. 1973. *The letters and diaries of John Henry Newman, XXV*. Ed. C. S. Dessain and T. Gornall. Oxford: Clarendon Press.

Part One

Natural Selection

Chapter 1

Preamble to Natural Selection

In the opening chapters of the *Origin*, Darwin intertwined the processes of natural selection and speciation. You will see why he did so later. His not separating the two has been a source of confusion. Some readers concluded that the *Origin of Species* is all about natural selection and not really about the origin of species. Others concluded that speciation and evolution are equivalents, so that we only see evolution when we see the origin of new species. Neither of these inferences is true. As a first step in making the *Origin* accessible, I have separated Darwin's proposal of natural selection as the mechanism of evolutionary change from speciation, which is a possible consequence of natural selection. One virtue of separating natural selection from speciation is that natural selection, and the resulting evolution that it causes, does not necessarily cause speciation. Natural selection is a process that is constantly in action within all species and has important consequences in its own right. Today, many scientists study natural selection without any reference to speciation. In part 1, I describe Darwin's presentation of natural selection. I consider speciation separately in part 2. This presentation is consistent with Darwin's message, since he was explicit in addressing the importance of natural selection to evolution within a species and in stating that speciation is only an occasional outcome of evolution by natural selection.

Darwin was not the first to propose a theory of evolution. His distinction, shared by A. R. Wallace, is that he proposed natural selection as the cause of evolution. The full title for Darwin's theory is thus the "theory of evolution by natural selection." I will often abbreviate this title to just "Darwin's theory." It is important to realize that evolution and evolution by natural selection are separable. Darwin was very successful in convincing the world that evolution is a fact, but his success in promoting natural selection as the cause of evolution was delayed until the modern synthesis era (1920s–1950s). As biologists' appreciation for natural selection was rekindled, they also realized that there are other causes of evolution, to be described later. Since this book is about Darwin and the *Origin*, I will focus on natural selection.

Darwin presented the concept of natural selection over the first five chapters of the *Origin*. Because of the length and complexity of his argument, it is good to know the premises and some of the properties of evolution by natural selection at the outset. To present these, I begin in the Galapagos Islands, where Darwin gained some of his key insights about natural selection and speciation.

The Galapagos archipelago straddles the equator and lies 500–600 miles off the west coast of South America. These are "oceanic" islands, meaning that they have never had contact with other landmasses. They were formed by volcanoes that erupted under the ocean, then grew to the surface and emerged as lifeless, dry land. Prevailing winds and ocean currents control their climate. Trade winds blow persistently from east to west and drive the surface water of the ocean in their path. As the water on the surface is driven west, an upwelling of deep, cold, nutrient-rich water along the coast of South America replaces it. These wind-driven currents create the continuous stream of cold water that surrounds the islands. The water in turn cools the winds. Because the islands are on the equator, the sun continuously heats them. This combination of warm land and cool winds causes the islands to be dry. The cool sea winds warm up and absorb moisture as they pass over the land, just as the dew on a lawn dries as the sun rises and heats the air. When El Niño events occur, this flow of cold water ceases, and the waters and winds around the islands warm up. As they do, their drying effects wane, and rainfall can increase dramatically. El Niño events occur at irregular intervals, ranging from two to seven years. They are often interspersed with droughts, so, over time, the amount of rainfall fluctuates widely and irregularly.

The same ocean currents and trade winds that dictate the climate today were the conveyors of most of the plants and animals that colonized the islands. Charles Darwin surmised this when he visited these islands. His visit came after he had spent nearly four years exploring South America, so he was very familiar with the plants and animals found on the mainland. He was impressed by the plants and animals on the Galapagos because so many proved to be new species, yet they were often similar to those found on the mainland. The wildlife was also remarkable because the animals had no fear of humans, since they had lived without human contact until the Spanish landed there in the seventeenth century. Collecting birds and lizards proved as easy as picking apples from trees.

Darwin made large collections of plants, insects, mollusks, and birds from the Galapagos. Some of his shipmates made collections as well. It was only years later, after Darwin's return to England, that the eminent ornithologist John Gould discovered the diversity of bird species that were present on the islands and fully described how they differed from one another. It was telling that Darwin did not always record the island from which he collected his birds, because he did not imagine that there could be any differences between species found on islands that were sometimes within sight of one another. This was an unusual lapse in Darwin's otherwise meticulous record keeping. Luckily, Captain FitzRoy had made collections as well and kept track of the island of origin. FitzRoy's diligence enabled Gould to determine not only that many of the birds in the collection were unique to the Galapagos, but also that some were unique to individual islands.

Darwin enlisted the help of Joseph Hooker to identify the plants. Of the 193 species that Darwin collected, Hooker determined that 109 were unique to the Galapagos. Of these, 85 were found on only one island.

At this point, Darwin had important pieces of information that drove him to the idea that speciation is an ongoing process. He knew that the islands were of volcanic origin, that they were younger than South America, and that many of the species living there were found only on the islands, but had close relatives in South America.

Modern geological methods allow us to provide specificity to Darwin's observations. For example, we can now estimate the age of the lava that is the bedrock of the islands, and know that the islands range from 700,000 to 3.5 million years old. We also know that the bird fauna includes thirteen species of finches that are found only on the islands. Investigators have shown, with the use of molecular genetic methods, that all thirteen Galapagos species share a single species of finch from the mainland of South America as their common ancestor. This combination of island endemism, molecular evidence, and islands of known age provides a basis for estimating the time required for these species to have evolved, which is actually a few million years more than the oldest extant island. This is possible because some older volcanoes in the archipelago that were once islands have since eroded away, leaving behind submerged sea mounts: the speciation must have started on those older islands before they eroded away.

These thirteen species of finch are an example of an "adaptive radiation," or the diversification of a single ancestor species into an array of separate species that fill a diversity of ecological niches. Their common ancestor arrived at a habitat already colonized by plants, insects, and other organisms, but with few birds. The islands gave the original colonizing individuals and their descendants diverse, unexploited sources of food. As the finch populations expanded onto all the islands, they diversified into separate species, each of which adapted to exploit different sources of food and different types of habitat. Diet differences are reflected in the beak structure of each species. The ground finches, one group of species, specialized in eating seeds. The species in this group now have beaks of various sizes that differ in their ability to crack seeds. Shorter, narrower beaks make birds more agile in picking up and shelling small seeds, while longer, wider beaks enable them to crack the shells of larger, tougher seeds. Other species instead specialized to live in trees and feed on insects, or to exploit other combinations of habitat and diet.

One species, the medium ground finch (*Geospiza fortis*), has become a model for the study of natural selection because of the work done by Peter and Rosemary Grant and their colleagues on the small Island of Daphne Major. It and the other islands in the Galapagos are especially good natural laboratories for studying evolution because they have few species of plants and animals relative to the mainland, making it easier to characterize them all and to study their interactions.

The Grants' remarkable long-term study has provided detailed documentation of the process of natural selection. Their detailed methodology, the properties of their study site, and propitious timing came together to turn evolution by natural selection into a visible process. First, the island—an emergent tip of a volcano shaped like a cone with a hollow interior—is manageably small, with bird habitat concentrated on the slopes and plane that fill the interior of the cone. The Grants came to know almost every bird on the island as an individual. They caught adult birds with fine mist nets, then measured and recorded various dimensions of their bodies such as wing or leg length, plus the dimensions of the beak, and then gave each bird a leg band that uniquely identified it.

There was considerable variation among individual finches in the shape and size of bodies and beaks. Birds had beaks that ranged from short, narrow, and shallow to long, wide, and deep. This is the same sort of variation found among different species of ground finch, although differences between individuals within this species were smaller than differences between species. During the breeding season, the investigators visited the nests and measured and marked all the nestlings. Because they knew who the parents were, they also knew how many surviving offspring were produced by each parent. Because they measured both parents and offspring, they could evaluate how similar they were in appearance. Offspring tended to look like their parents.

The Grants also studied food availability. Most fresh seeds were produced during periods of rainfall. During the dry season, the birds mined the "seed bank," or seeds that had accumulated in the soil. The Grants cataloged all the seed-bearing plants on the island and characterized each type of seed by its hardness, or the amount of force required to crack it. They regularly sifted samples of soil to see how many and what types of seeds were available. They also quantified how effective individual finches were in harvesting the different types of seeds and found that differences in beak dimensions were associated with differences in harvesting ability.

Then in 1977, during this study of the island's finch population and food supply, the island experienced a year-long drought—an accident of timing that gave the investigators an opportunity to observe and record the process of natural selection. There was little plant growth or seed production, forcing the birds to rely entirely on the seed bank for food. Food was so scarce that none of them produced young that year. They depleted the supply of small seeds as the drought progressed, leaving just the large, thick-shelled seeds as the more abundant food source. As the seed supply dwindled, birds began to die. Only 15% of the adults had survived to reproduce by the time the drought finally ended. Those that survived were different, on average, from those that had died. Because small seeds were harder to come by, the survivors tended to be birds that were larger overall and had longer, wider, deeper beaks than those that had died. They survived because they were better able to harvest the more abundant larger seeds in the seed bank. When the rains returned and the birds were once again able to reproduce, they gave birth to babies that looked like them. The average postdrought finch was a bit larger and had a larger beak than the average predrought finch.

Because the amount of rainfall can vary so dramatically from year to year, the nature of the seed supply also varies. During the winter of 1982–3, there was an El Niño event and a surfeit of rain that was as dramatic as the scarcity of rain in 1977–8. Grasses carpeted the inside of the volcanic cone and produced an abundance of small seeds. Whereas the finches did not reproduce at all during the drought, some of them produced multiple sets of offspring during this especially wet season. Those individuals with short, narrow beaks were better at harvesting the available seeds and were much more likely to survive and reproduce. Once again, offspring tended to look like their parents, so the average bird in the population after the El Niño was smaller and had a shorter, narrower beak than seen the year before, thus reversing the change that had occurred during the drought. This reversal is telling because it says that there is not a universal "best" bird. Whether or not a given feature of an individual gives it an advantage over another depends entirely on the circumstances. As circumstances change, so do the sorts of traits that favor one individual over another.

These changes that the Grants observed in the finch population from one year to the next, and the sequence of events that caused them, are a model for the process that Darwin called "natural selection." Darwin's first goal in the *Origin* was to define this mechanism. He named it "natural selection" because he saw it as an analogy to "artificial selection." The breeding of plants and animals for desired properties was widely practiced in the England of his time, so it was a process that was well known to his intended readership. The goals of this artificial selection ranged from developing improved domesticated plants and animals as sources of food and fiber to attaining aesthetic goals, such as developing a prizewinning rose or pigeon.

Darwin's mechanism consisted of four parts. First, all organisms make many more offspring than are required to replace themselves in the next generation. If all offspring lived, they would quickly fill the world many times over. This does not happen because something intervenes to control their abundance. They may be eaten by predators, killed by disease or parasites, succumb to competition, or, like the finches, starve during a time of scarcity.

Second, individuals are almost always different from one another in how they look or how they work on the inside. Some differences are obvious. In the finch study, individual variation was quantified by measuring beaks and the size of different parts of the body and wings. Consider the variation that we see in humans. There are differences between us in hair color, skin color, eye color, and height. There are "shape" differences, such as the relative length of the legs or trunk of the body, and a host of internal differences, such as blood type. Among other organisms, the differences may not be so apparent, but they are always there if you know how to look for them.

Third, at least some of these differences between individuals are heritable, meaning that they are transmitted from parents to offspring. The Grants were able to quantify this similarity by measuring parents and offspring. We know this from practical experience as well, since children tend to look like their parents.

Finally, these differences between individuals can influence who survives and reproduces and who dies or fails to reproduce. In finches, the difference between life and death during a drought came down to the dimensions of their beaks and their ability to harvest large seeds. Life is not as harsh for humans, so the life-and-death aspects of our variations may be hard to appreciate. But for organisms exposed to the rigors of nature, factors such as height, shape, and color can make a big difference in speed, agility, endurance, or the extent to which an individual blends into its background. Such variation really can spell the difference between life and death.

A consequence of individual variation and its differential effect on survival and reproduction is that populations as a whole can change, or evolve, because individuals who are successful are, on average, different from those who are not successful. If the environment changes, then the sort of individual that is best suited for the new environment may be different from the sort that was best before the change. Those who are best suited to the new environment are more likely to survive, will have more opportunities to reproduce, and will contribute more offspring to the next generation.

Among the finches the Grants observed during the drought of 1977–8, individuals who had larger, stronger beaks were more likely to escape starvation and survive to breed the following season, so it was their offspring who dominated the population the following year. During wet season of 1982–3, there was another episode of selection that caused the population average to change, but this time in the opposite direction. This reversal of fates illustrates an important quality of evolution by natural selection. Unlike artificial selection, evolution by natural selection lacks foresight and has no goals. It does not progress in any particular direction but is, rather, a response to present conditions alone. If conditions change, then so will the kind of selection experienced by individual organisms. If conditions remain constant, it is possible that there will be no evolution.

The Galapagos finches reveal other important properties of natural selection. First, evolution by natural selection is a process that is built on the fate of individuals, but we perceive it through changes in the average properties of a population. The year-to-year genetic changes in the average finch, whether caused by droughts or El Niño events, is evolution. In fact, a more technical, modern definition of evolution is a change in the genetic composition of a population over time. The features of an organism that determine its survival and reproductive success can thus be shaped by natural selection. For example, the diversity of beak dimensions that we see between different species of finch enables each species to specialize on a different source of food. The differences between species are larger than the variation that we see between individuals within a species. All these species were derived from a single common ancestor species. We surmise that this exaggeration of differences between species in beak morphology is a consequence of selection for specialized beak dimensions over a prolonged period of time, such that the kind of changes that we see on a year-to-year basis in the medium ground finch on Daphne Major accumulate in different directions in each species as it specializes on a given habitat and diet.

Changes in beak shape, or in any other trait of an organism that affects its survival and reproductive success and hence is shaped by natural selection, are referred to as "adaptations." Adaptations are a consequence of natural selection for a specific function. In the case of the beaks of finches, the function is their ability to obtain specific types of food. To call some feature of an organism an adaptation, one must establish that there is a cause-and-effect relationship between the trait and its presumed function. The Grants can say that the differences between species in their average beak dimensions are adaptations for feeding on different types of seeds, because they characterized the ability of the finches to harvest different types of seeds. They showed that these differences in harvesting ability are attributable to the physical properties of the seeds and the force required to open them, and the ability of birds with bills of different dimensions to generate those forces.

According to Darwin's theory, the kind of evolution that we can study over short intervals of time can cause the differences that we see between species, and in fact can cause the origin of new species; the adaptive radiation of thirteen species of Galapagos finches, all derived from a single ancestor species from the mainland, is a consequence of this same mechanism. Furthermore, Darwin argued that natural selection can account for the entire history of life on earth, save for the origin of life, and explain a diversity of phenomena that previously had been thought to be unrelated. These other phenomena include the system of animal and plant classification developed before Darwin's time and still in use today, the fossil record, extinction, similarities of different species in their patterns of development or their anatomy, the geographic distribution of species, and more. He thus proposed a unity of process, or a single mechanism—natural selection—that explains adaptation, but that can also explain evolution "writ large," including the origin of new species and the history of life on earth.

References

Abbott, I., L. K. Abbott, and P. R. Grant. 1978. Comparative ecology of Galapagos ground finches (*Geospiza*, Gould): Evaluation of the importance of floristic diversity and interspecific competition. *Ecological Monographs* 47:151–84.

Darwin, C. [1860] 1962. *The voyage of the* Beagle. Annotated and with an introduction by Leonard Engel. Garden City, NY: Doubleday and Company.

Grant, P. R., and B. R. Grant. 1995. Predicting microevolutionary responses to directional selection on heritable variation. *Evolution* 49:241–51.

Weiner, J. 1994. *The beak of the finch.* New York: Alfred A. Knopf. (Good source for the complete story of the Grants' and their associates' research on the Galapagos finches.)

Chapter 2

Variation under Domestication

Darwin begins his argument for natural selection with a discourse about the domestication of plants and animals. Why begin here? One reason is that domestic animals are familiar to all of us. Everyone knows about dogs and the remarkable diversity of dog breeds that range in size from teacup Chihuahuas to giants like the Irish elkhound, Great Dane, or Saint Bernard. Their diversity goes well beyond size. Afghan hounds have an elongated, narrow snout, while pugs have a squashed face with a lower jaw that is often longer than the upper. Dachshunds are shaped like wieners, with their short legs and long bodies. Most people also know that domestic dogs are not like wild animals but are instead the products of selective breeding. Few people know the history and origins of domestic dogs, but they at least represent a familiar point of departure for talking about how the process of selection works and how it can result in organisms that are very distinctive in appearance and habits. One of Darwin's goals in his opening chapter is to develop an analogy between breeders' efforts to shape the evolution of domestic plants and animals through artificial selection and nature's constant shaping of organisms through natural selection. He has larger goals as well. He uses domestic animals to characterize how much change is possible under selection, the properties of organisms that make this change possible, and the circumstances that facilitate change. In this way, he uses variation under domestication to shape his subsequent argument for the process of evolution by natural selection.

Darwin supported all his arguments with many examples. For the sake of simplicity, I will choose only a few from those that he included in the *Origin*.

Pigeons and Artificial Selection

Darwin, always the empiricist, took up pigeon breeding to learn more about the process of domestication. He built aviaries, stocked them with exotic breeds, then committed what breeders considered to be a cardinal sin; he crossed different breeds to learn about the laws of inheritance. He joined pigeon-breeding clubs so that he could meet with local experts and corresponded with pigeon fanciers from all over the world. There are now hundreds of named breeds, each of which is distinct and "breeds true," meaning that the parents and offspring are very similar in appearance. Midway through this first chapter, Darwin regales his readers with descriptions of the fantastic diversity in plumage and body shapes seen among different breeds of pigeons.

The diversity of the breeds is something astonishing. Compare the English carrier and the short-faced tumbler, and see the wonderful difference in their beaks, entailing corresponding differences in their skulls. The carrier, more especially the male bird, is also remarkable from the wonderful development of the carunculated skin about the head, and this is accompanied by greatly elongated eyelids, very large external orifices to the nostrils, and a wide gape of mouth. The short-faced tumbler has a beak in outline almost like that of a finch; and the common tumbler has the singular and strictly inherited habit of flying at a great height in a compact flock, and tumbling in the air head over heels. The runt is a bird of great size, with long, massive beak and large feet; some of the sub-breeds of runts have very long necks, others very long wings and tails, others singularly short tails. The barb is allied to the carrier, but, instead of a very long beak, has a very short and very broad one. The pouter has a much elongated body, wings, and legs; and its enormously developed crop, which it glories in inflating, may well excite astonishment and even laughter. The turbit has a very short and conical beak, with a line of reversed feathers down the breast; and it has the habit of continually expanding slightly the upper part of the oesophagus. The Jacobin has the feathers so much reversed along the back of the neck that they form a hood, and it has, proportionally to its size, much elongated wing and tail feathers. The trumpeter and laugher, as their names express, utter a very different coo from the other breeds. The fantail has thirty or even forty tail-feathers, instead of twelve or fourteen, the normal number in all members of the great pigeon family. . . .

In the skeletons of the several breeds, the development of the bones of the face in length and breadth and curvature differs enormously. The shape, as well as the breadth and length of the ramus of the lower jaw, varies in a highly remarkable manner. The number of the caudal and sacral vertebrae vary; as does the number of the ribs. . . .The proportional width of the gape of mouth, the proportional length of the eyelids, of the orifice of the nostrils, of the tongue (not always in strict correlation with the length of beak), . . . the relative length of wing and tail to each other and to the body; the relative length of leg and of the feet; the number of scutellae on the toes, the development of skin between the toes, are all points of structure which are variable.

Figure 1 appeared in a later book by Darwin titled *Variation under Domestication*. These drawings are of four skulls, first that of the rock pigeon, which is the ancestor of all domestic pigeons, then those of three different breeds of domestic pigeons. I showed this figure to experienced ornithologists and asked them to identify the birds. They often placed the birds in different orders. ("Order" is a formal level of classification from the Linnaean hierarchy. The complete series of categories is kingdom, phylum, class, order, family, genus, and species. An order is thus four steps up the Linnaean hierarchy from species. To give you some perspective, wolves, bears, raccoons, skunks, lions, and sea lions are from different families in the order Carnivora, so "order" is a big category that contains a very diverse group of animals.) Darwin's point here is that artificial selection has so profoundly changed pigeons that a bird specialist would consider different breeds to be different, distantly related species. He does not argue that they actually are different species; his point is instead to demonstrate how changeable organisms can be under artificial selection. Figure 2 presents photographs of four breeds of pigeons to illustrate what some of these birds look like in life.

What are the origins of domestic pigeons? Darwin found that "the earliest known record of pigeons is the fifth Egyptian dynasty, about 3,000 BC, as pointed out to me by Professor Lepsius, but Mr. Birch informs me that pigeons are given in a bill of fare in the previous dynasty" (*Origin*, pp. 27–28). Current archaeological data push the earliest known appearance of pigeons associated with humans back to 6,500 years ago, when they were placed in fertility shrines. The first morphologically distinct varieties ap-

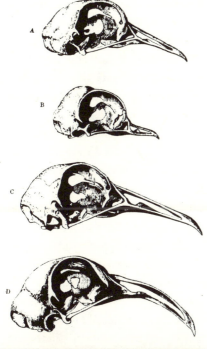

Figure 1
Caption from Darwin's *Variation of Animals and Plants under Domestication*: "Skulls of pigeons viewed laterally. A. Wild rock-pigeon, *Columba livia*. B. Short-faced tumbler. C. English Carrier. D. Bagodotten Carrier."

Figure 2 *(below)*
The courier pigeon (upper left) is very similar to the wild-type rock pigeon from which all domestic pigeons were derived. The other three birds pictured here represent only a few of the many bizarre breeds of domestic pigeons now in existence.

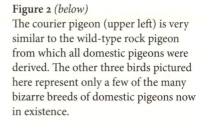

peared 3,000 years ago. Darwin argues (and we now know) that all these breeds descended from a single ancestral species. This means that the stunning differences between breeds of pigeons are the product of only a few thousand years of domestication. In contrast, the common ancestor of the various orders of birds lived tens of millions of years ago, so in just 3,000 years humans have obtained a level of morphological disparity equivalent to what took tens of millions of years by natural processes.

An alternative explanation, adhered to by some of Darwin's contemporaries, is that the diversity of pigeons actually reflects their being derived from many different ancestral species that shared this same diversity (*Origin*, p. 29). Darwin argues that this is unlikely for pigeons or any other domestic plant or animal because such ancestors are nowhere to be found among living species or in the fossil record. It strains plausibility to say that there were once several species of pigeons, dogs, horses, cattle, or any other wild animal that were ancestral to the different breeds of domesticates but also different from all living wild species. These presumed ancestors would have to have all disappeared without a trace or perhaps still exist in some quiet, unexplored corner of the world. The changeability of the living breeds argues instead that the existing diversity is a product of artificial selection. To illustrate how changeable species are under artificial selection, Darwin quotes "that most skilful breeder, Sir John Sebright" as saying that "he would produce any given feather in three years, but it would take him six years to obtain head and beak" (*Origin*, p. 31). Also, the breeds of domestic pigeons, or breeds of dogs, or cattle, or other domestic animals, will all freely interbreed. The same cannot be said for wild species that are closely related to one another, much less ones that are so different in morphology. Darwin supports this argument with breeding experiments. Rock pigeons have some color patterns that are not shared by any other living species of pigeon, nor by some of the domestic breeds, but he was able to cross different breeds, one of which was pure white and the other pure black, and recover the wild-type coloration of rock pigeons within a few generations.

We now know, based on the application of modern genetic methods, that Darwin was right in arguing that our domestic plants and animals are almost always derived from one wild ancestral species, in spite of the diversity that we now see among living breeds. Darwin made an exception for dogs, which he believed originated from more than one ancestral species. He was wrong. They all share gray wolves as a common ancestor. We can also estimate the time of origin of some of these breeds, and many fall within a time frame similar to that of the pigeon, which is when people were making the transition from being nomadic hunter-gatherers to being agriculturists. There are some exceptions: It seems that humans may have first domesticated wolves earlier, while they were still hunter-gatherers. (Dogs are the only domesticate that accompanied humans on their journey across the Bering Land Bridge from Asia to North America and across the Malay Archipelago to Australia.) And some domestic plants originated as hybrids between different species.

Darwin presses on by arguing that this process of breed formation is ongoing. His concern was that his contemporaries saw existing breeds as unchanging, or not changing enough to form new breeds. Darwin counters that change is slow relative to our day-to-day lives but can be seen with the passage of time. He relates one person's observation that the Spanish pointers in England are known to be descended from dogs imported from Spain, yet if you go to Spain you cannot find any pointers that look like those from England. The reason is that British and Spanish tastes for pointers are perhaps as different as their tastes for food, so the breeds diverged in the decades since the dogs were imported to England. To see such continuous change, we need a time capsule that allows us to compare the domesticated organism with its ancestor or to compare two populations that have been separated for a long time. Figure 3 simulates such a time capsule, in the form of a pair of photographs that illustrate the evolution of the dachshund by comparing a typical early twentieth-century dog with an early twenty-first-century dog. It appears that, over the years, dog breeders have preferred that their dachshunds look more and more like sausages, since their bodies became longer and their legs shorter.

Darwin's contemporaries also argued that the changes we see under domestication are not permanent. When animals become wild again, they revert to the appearance of their wild ancestor. It is as if domestication is like stretching a rubber band; the band springs back to its original shape once the tension is released. Darwin observes that while such reversion is often claimed, he knows of no confirmed examples. There is no question that many domesticated organisms would fare poorly if they were released back into the wild. (Try to imagine the fate of a toy poodle fending for itself in the wilds of the Canadian Rockies.) We should expect them either to be exterminated or to adapt to their new environment because conditions in nature are so different from those in domestication. Such adaptation would be a selection process similar to the one that produced the domestic variety

Figure 3
The Dachshund Then and Now
An early twentieth-century dachshund variety named "lady of the valley" (top photo)
is a mounted specimen in the British Museum of Natural History. Compare this individual with a twenty-first-century dachshund (bottom photo). Look carefully at the
length of the legs, particularly in relation to the length of the body (keeping in mind
that the lady of the valley's legs are partly obscured because it is mounted in a prone
pose). The earlier dachshund is much longer legged and proportioned more like our
current cocker spaniels than like our current dachshunds.

in the first place. The difference would be that humans selected features for
the domestic breed, but nature would select for whatever changes occurred
after the animal was released back into the wild.

Those of us who work in the tropics have seen this process in action.
Wherever you go, be it in the Caribbean, in South America, or in Southeast
Asia, there are mongrel dogs that are similar in appearance. They weigh

10–15 kilograms and have short hair, long tails, and often pointed, erect ears. We assume that life as a feral dog in the tropics is rough and that there is severe selection for this body type. However, these dogs never revert to being wolves. The changes that breeders selected for were real, and subsequent natural selection when dogs become feral resulted in the evolution of a new type of dog, not a reversion to the ancestral wolf.

Darwin also addresses the argument that species are chosen for domestication because they have special properties and hence are atypical in their capacity to change. He agrees that many domesticates are successful because they have special properties but wonders, "how could a savage possibly know, when he first tamed an animal, whether it would vary in succeeding generations and whether it would endure other climates?" (*Origin*, p. 17). Darwin argues that the only special property that is likely to be shared by ancestors of domesticated species is the ability to reproduce under their changed circumstances. The biggest hurdle to domestication is being able to successfully breed an organism that is brought in from the wild. Otherwise, species that have been domesticated are very much the same as all others. One recent form of support for Darwin's argument is the finding that the number of distinct breeds in any domestic bird species is directly related to how long it has been domesticated. Disparity under domestication is thus determined by how much time humans have had to select for new varieties, rather than any special property of the bird.

Laws of Inheritance

Artificial selection shows that it is possible for humans to look at a plant or animal, see that some individuals are different from others, then select the parents of the next generation. The selected parents produce offspring that inherit the parents' distinctive traits. When this process is repeated over many generations, differences can accumulate to the point that we might recognize some pigeon fancier's flock or some cattle breeder's herd as being a distinct breed or variety. All the individuals share traits that make them different from those kept by other breeders. In this way, the Spanish pointers of England came to differ from those of Spain, the dachshunds of today came to differ from those of early twentieth-century England, and new breeds of pigeons come into existence. An observer as skilled as Darwin might also use this process to learn something about the laws of inheritance.

Darwin freely admitted that breeders and naturalists did not then understand the laws of inheritance. We will see that his presumed mechanisms of
inheritance were wrong. We argue today that having a deep understanding
of evolution demands an understanding of genetics, yet this was clearly not
essential to Darwin when he conceived of evolution by natural selection. It
is also sometimes argued that it was a tragedy that Darwin never learned
of Mendel's work, since it was published in 1865, between the publication
of the first and sixth editions of the *Origin*. However, the history of the
rediscovery of Mendel and of the subsequent development of genetics argues that Darwin would have gained little had he read Mendel's paper. In
actuality Mendel's work lay dormant until 1900. And its rediscovery was
not like the dawning of a new day in which everyone suddenly understood
inheritance. It came at a time when there was an ongoing battle about how
inheritance worked. Mendel's laws of inheritance were claimed by one of
two competing schools of thought. The claimed ownership by one of them
meant that the rediscovery fueled a fire of conflict between the two sides,
rather than provided some form of salvation. Mendel's experiments on peas
seemed so contrived that they could be claimed as special cases that had no
generality. Three more decades passed before there was a reasonable reconciliation between these competing schools of thought and an acceptance
that Mendel's laws of inheritance applied, at least in principle, to all traits in
all organisms. I will say more about these competing camps below and in
the next chapter.

For someone who has a good understanding of genetics, reading the *Origin* shows why achieving such an understanding is so hard. The basics that
you learn in general biology and even in a first undergraduate course in
genetics fall far short of what you need to know to understand what Darwin saw when he was breeding pigeons. A consequence of the gulf between
Darwin's knowledge and what we know now is that his chapters on variation under domestication (*Origin*, chap. 1) or variation in nature (*Origin*,
chap. 2) are difficult to follow. You need to put yourself in Darwin's shoes
and imagine how the world looks when you do not know the principles of
genetics. In spite of these difficulties, Darwin proposed some rules of inheritance that are accurate and others that were good enough to enable him
to develop a theory of evolution that was correct.

To explain the difference between Darwin's concept of inheritance and
our modern perspective, I need to digress for a moment to describe the two
competing models for inheritance that prevailed shortly after the rediscov-

ery of Mendel in 1900. One model was "blending inheritance," which means that offspring are a blend of contributions from both parents. Inheritance was thought to be like the blending of white and black paint to produce gray. The key is that the mixture is inextricable; we can never recover the pure black or white paint once they are mixed. Likewise, it was thought that the contributions from each parent are inextricably blended in the offspring. "Particulate inheritance" is the alternative proposed by Mendel, based on his experiments with peas. Mendel showed that for each trait, such as plant height or pea color, each parent contributes a unit of inheritance, which we call an "allele," to its offspring. Some units are dominant in their expression over others. For example, if a tall parent was bred to a short parent, all offspring were tall, even though they inherited an allele for tallness from one parent and one for shortness from the other parent. Because these offspring still carried an allele from the short parent and could transmit it faithfully to the next generation, short plants could reappear in subsequent generations. If inheritance were like a blending of parental traits, then the recovery of the short form would never occur. The products of crosses would have intermediate heights, and all subsequent generations would be similar in appearance.

Other individuals, including Darwin, had performed hybridization studies like Mendel's, but Mendel was distinct in his methods. He was very deliberate to begin with strains of peas that "bred true," meaning that all parents and offspring looked alike for many generations. He made specific crosses between parents from different true-breeding lines that differed in one, two, or three discrete traits. He kept each family of progeny separate and classified each offspring by its appearance, and continued such crosses for additional generations to produce grandchildren and great grandchildren, each time keeping track of the appearance of the parents and their offspring and of the numbers of offspring of each type. For example, when he crossed a tall and short plant, he observed that all offspring were tall. He then "selfed" each plant, meaning that he fertilized each flower with pollen from the same individual, then counted the number of tall and short offspring in the next generation. He found that there were three tall plants to each short plant in such cohorts of grandchildren. It was this reappearance of the "recessive" (short) trait in predictable ratios in the second generation that supported his model. The original tall parent had two tall alleles for a gene that determined plant height. The short parent had two short alleles for the same gene. All offspring had one allele for tall and one for short, but the tall allele was dominant, so they all grew into tall plants. When they were selfed, the pollen and eggs combined at random: 50% of the eggs and 50% of the pollen carried the tall allele, so 25% of the offspring in the next generation (0.5 × 0.5 = 0.25) had two tall alleles, 50% had one tall and one short allele (2 × (0.5 × 0.5) = 0.5), and 25% had two short alleles (0.5 × 0.5 = 0.25). The first two categories of offspring would be tall while the third would be short, in a 3:1 ratio. This was Mendel's model, or interpretation for how inheritance works. He presented a large series of experiments that made predictions based on this model and showed that the predictions were upheld.

The problem with Mendel's experiments in the eyes of many was that he was so particular about the pureness of plant strains he used and the traits he chose to track that they questioned the generality of his results. Mendel used only true-breeding strains of plants and worked only with traits that were inherited in a discrete fashion. He knew that some traits appeared to have a blending type of inheritance, but he avoided using them because he felt that they would make it more difficult to discern how inheritance worked. He was very deliberate about choosing the simplest possible circumstances for study because he felt that these would make it easiest for him to determine the laws of inheritance. Anyone who did any kind of breeding experiment or practiced selection on domesticated animals or plants knew that most traits were inherited in a fashion that was consistent with blending inheritance. It was thus fair to argue that Mendel had studied special cases and had not revealed general laws of inheritance. Darwin's reading of Mendel would have predated a considerable amount of research on inheritance that happened between 1865 and 1900, so it is very unlikely that he would have done any better in evaluating Mendel's results.

How can we reconcile Mendelian, particulate inheritance with the appearance of blending inheritance that we so often see in practice? Mendel offered an answer with some additional, small experiments on a different species of bean plant. He tried to duplicate his results for flower color on peas (genus *Pisium*) in a bean from the genus *Phaseolus*. In peas, red flowers were dominant to white, so all offspring of the first generation were red, but red and white flowers were produced in a 3:1 ratio when the first generation was self-fertilized. In *Phaseolus*, all the flowers in the first generation were pink, as if the red and white of the parents had been blended. In the next generation he produced flowers that ranged from white to red with different shades of pink in between. This plant proved to be more difficult to breed, and Mendel was able to produce only a small number of offspring. This

Figure 4
Normal vs. Ancon Sheep
Ancon sheep were homozygous for a mutation that caused them to have limbs that
were considerably shorter than those of normal sheep. (In this photo, an ancon sheep
is on the left and a normal sheep on the right.) Darwin often refers to the ancon as an
example of discontinuous variation.

number was sufficient for him to propose a hypothesis that eventually reconciled blending and particulate inheritance. If we substitute modern genetic
terms for his original words, he proposed that flower color in these beans
was controlled by either two or three genes and that each one made an independent contribution to color, so pure red appeared only when all alleles
on all genes coded for red flowers. It was also necessary that red alleles were
not dominant to white alleles. If one allele for a given gene coded for red
and the other coded for white, then the flower would be some shade of pink.
We eventually learned that such "polygenic inheritance," or inheritance in
which more than one gene determines each feature of an organism, is very
common. Polygenic inheritance, in which the appearance of the organism
is the summed contribution of multiple genes, is the mechanism by which
Mendel's particulate inheritance can create the appearance of blending.

Darwin based his ideas about inheritance on his own observations. He
saw two distinct kinds of variation that breeders could work with. One
was the abrupt appearance of new, distinct forms that were referred to as
"sports" or "monstrosities." His favorite example of such a monstrosity was
the ancon sheep, a breed developed in Massachusetts in the late eighteenth
century (see fig. 4). The breed was based on the sudden appearance of a

short-legged lamb born to normal, long-legged parents. When the lamb matured into an adult ram, it was bred to long-legged ewes and produced fifteen offspring, two of which had short legs. When these short-legged off-spring were mated, they produced pure-breeding short-legged offspring. We can now interpret this history in terms of Mendelian inheritance. The original long-legged parents carried one normal allele and one ancon allele, and normal is dominant to ancon. When they were mated, they could pro-duce an offspring with two ancon alleles. Some of the ewes that the original ancon ram were mated to also carried one normal and one ancon allele, so two of the offspring had two ancon alleles. When the farmer was finally able to breed an ancon male to an ancon female, both parents carried two ancon alleles and all offspring had two ancon alleles, so they were all short-legged. This breed was thus the product of a mutation at a single gene that was recessive to the normal, long-leg-producing version of that gene. Ancon sheep were favored by some farmers because they were less able to jump fences and less likely to invade, trample, and eat farmers' vegetables. Sports like the ancon are generally caused by a single gene that has a large effect on an individual's appearance.

The second type of variation recognized by Darwin includes the small, indistinct differences, such as differences in size or shape, that we often see between individuals within a population. For example, within a flock of pigeons of the same breed, some might have a slightly shorter, longer, or more curved bill than others. Such differences can be extremely small. In this chapter Darwin emphasizes that being successful as a breeder demands that one be an acute observer when selecting on such variation. "Not one man in a thousand has accuracy of eye and judgment sufficient to become an eminent breeder" (*Origin*, p. 32). He describes the process of selection of me-rino sheep in Saxony: "the sheep are placed on a table and are studied, like a picture by a connoisseur; this is done three times at intervals of months, and the sheep are each time marked and classed, so that only the very best may ultimately be selected for breeding" (*Origin*, p. 31). (We assume that those not selected moved on to the dinner table, as so many animals fall prey to predators.) Subtle differences between individuals that the breeders sort out by procedures like these, Darwin argues, are the basis of virtually all the changes that become preserved in new breeds. However small such varia-tions might be, they are ubiquitous and readily exploited, at least by those with a well-trained eye to spot them. Successive generations of selecting on such variation produced the huge differences now seen between breeds.

Darwin strongly favored these small variations, rather than sports, as the real fuel for evolutionary change. Sports are rare, unpredictable, and are often accompanied by unfavorable side effects. For example, the ancon sheep grew poorly and attained smaller adult body sizes than long-legged sheep, perhaps because their mobility was so limited that they could not feed efficiently. In the long run they were not favored by farmers and have since gone extinct. In contrast, the small, subtle variations exploited by breeders were ubiquitous, readily selected upon, and could be shaped at will to produce desired traits. An important distinction between the two types of traits was that the sports bred true, meaning that they clearly showed Mendelian inheritance, while the small variations often seemed to blend when parents that differed in appearance were mated with one another. Selection requires variation but blending erases it, yet in a large enough population the variation always seemed to be present.

Where did this variation come from? Darwin observes that plants and animals that are being domesticated manifest more variation than normally seen in wild organisms. He suggests that the environment that they experience under domestication is different from what was experienced in nature. He proposes that this change in conditions induces variation among individuals and that the induced variation is incorporated among the traits that parents can transmit to offspring. Darwin also argues that such environmentally induced variation is essential for evolution to happen. Because Darwin believed that the variation between parents is lost when they mate and their traits are blended in their offspring, he also believed that there must be some process that creates new variation for selection to act upon. Darwin proposes that the environmental induction of variation is that process. Later in the *Origin*, Darwin sometimes states that he thinks such variation will only occasionally be induced, so evolution will not always be possible.

This model of the inheritance of acquired traits is not consistent with our modern understanding of inheritance. Given Darwin's strong empiricism and his accuracy as an observer, I have to wonder why he arrived at this conclusion. I take it for granted that the observation is correct: when you collect some organisms from the wild and begin to breed them in artificial conditions, you will see the appearance of heritable variation that was not observed in nature. Mendel's model of inheritance provides a likely explanation. Natural populations of organisms are known to have many rare, recessive alleles among their "pool" of genes. Because these alleles are rare, their effects are almost never seen, since it is very unlikely that two parents that each have one copy of the allele will mate with one another to produce an offspring with two copies; however, if two close relatives mate with one another, it becomes much more likely that each parent will carry one copy of a rare allele. These rare recessives can often be damaging to the offspring. It appears that many populations of humans discovered this for themselves, which is why we so often see taboos against the marriage of close relatives (yet Darwin married his first cousin). Darwin's observation of the spontaneous emergence of variation in domesticated plants and animals was likely a consequence of the breeding stock being initiated with only a few individuals, making it more likely that there would be pairings of close relatives and the emergence of offspring having two copies of some recessive alleles. The continued practice of breeding close relatives to select for some of these desired traits would then continue to reveal such formerly hidden genetic variation. Darwin argues throughout the *Origin* that natural variation is random in nature, meaning that it can be beneficial, damaging, or neutral with respect to its effect on the organism. This is an accurate description of the kind of variation that is revealed when such rare, recessive alleles become homozygous.

Our contemporary explanation for Darwin's emphasis on small variations and blending inheritance is that he was seeing that most traits are polygenic and was emphasizing differences between alleles at some of the genetic loci that contribute to the determination of each trait. Because all these genes operate under the same mechanism, particulate inheritance, they will all retain their genetic variation. It is not lost through blending, even though it appears to be, and there is no need to invoke any special mechanism for creating new variation each generation. Because many genes control most traits, it is possible for the response to selection for a desired trait to continue for a long time and to cause very large changes. This is because successive generations of selection first act on one gene, then another, and so on, so that the difference between a selected line and its ancestor can become bigger and bigger over time. The bottom line is that Darwin was wrong about how inheritance worked, but close enough to be able to arrive at a correct conceptual model of evolution.

In this chapter and throughout the *Origin*, Darwin presents other "laws of inheritance" based on his experience with animal and plant breeding. One that he invokes repeatedly is the "law of correlation," which refers to the way different parts of an organism evolve in concert with one another.

In pigeons, selection for a small beak results in birds that also have small feet. Likewise, selection for large beaks yields birds with large feet. Cats with blue eyes tended to be deaf. Domestic grazers, such as sheep, with different coat colors differ in their susceptibility to natural poisons that occur in some plants. A consequence of such correlations is that when breeders select on some particular feature of an organism, they inadvertently change others. Sometimes these changes cause problems. Birds require well-developed beaks to be able to break out of their eggs at hatching time. Artificial selection for small beak size in the short-faced tumbler pigeon resulted in offspring with beaks too small for many of them to break the eggshell, so the breeder must do this for them.

What Darwin was seeing is that the same gene can influence multiple features of an organism. For example, a single gene can affect patterns of growth in different parts of the body. Selecting for a bigger or smaller beak means selecting for faster or slower growth rates as the beak forms. Genes that affect the growth rate of the beak can also affect the growth rate of feet or other parts of the body.

Conditions That Favor the Formation of Distinct Breeds

Finally, Darwin considers the conditions that encourage the development of new breeds. His goal here is to extend his model of domestication to the origin of new races in nature. First, he observes that it is easier for breeders to develop a new breed if they begin with many individuals. Because the desired variation may be rare, having a large number of individuals gives breeders more opportunity to find what they want. Since the goal is often to produce food, the culls can proceed to the dinner table while the selected individuals can be put out to pasture to breed. Likewise, Darwin observes, plant breeders practice "rogueing"; they periodically pull up and discard undesirable individuals. There is a socioeconomic consequence to the importance of numbers: Wealthy landowners who maintain large herds have far more opportunity to improve their stocks than small landowners who maintain only a few animals. Likewise, nurserymen who keep large fields are much more likely to develop new breeds of plants than hobbyist gardeners with small plots.

Second, one must pay careful attention to whatever variation is present and practice careful selection. Darwin observes: "I have seen it gravely remarked, that it was fortunate that the strawberry began to vary just when gardeners began to attend closely to this plant" (*Origin*, p. 41). The more likely explanation is that strawberries always varied, but it was not until gardeners began to pay attention and select for certain attributes that distinct varieties were developed.

Finally, there must be limited opportunities for crossing between populations. "Wandering savages or the inhabitants of open plains rarely possess more than one breed of the same species" (*Origin*, p. 42), because they cannot keep subpopulations isolated from one another long enough to develop distinct breeds. On the other hand, in more developed regions where herds are enclosed in separate land holdings and can be kept from interbreeding, many distinct breeds are seen. Breeding habits also matter. Pigeons will mate for life, so a breeder can keep pairs from distinct breeds together in the same aviary without their interbreeding. In contrast, cats are nocturnal and promiscuous, so it is hard for a cat breeder to control paternity. As a consequence, there are many distinct varieties of pigeons, but few of cats. (Darwin added that the breeds of cats that did exist were often from islands or very isolated regions.)

In summary, we have huge diversity among domestic breeds of pigeons and other plants and animals. We can increase this diversity at will by selecting on whatever variation is present. We have modern evidence that all domesticates are descended from a single or sometimes two ancestral species and that they are of recent origin. We also know that the differences between breeds are permanent and are faithfully transmitted from parent to offspring. This was Darwin's opening salvo.

All organisms, as they exist today, share this enormous capacity to change under selection. The variation among individuals within a species that makes this change possible is ubiquitous. In the case of pigeons, the morphological differences we see now between breeds is comparable to the differences that exist between whole orders of wild birds. Orders of birds are descended from common ancestors that lived tens of millions of years ago, so humans practicing artificial selection have produced an equivalent disparity among breeds in around one thousandth to one ten-thousandth of the time required for nature to do the same. Darwin did not know the laws of inheritance that make this change possible, but now we do. We do not need to invoke anything special or outside our daily experience to explain this capacity for variation, since it has been so well documented in so many plants and animals.

This opening chapter presents the raw material that Darwin needs for his subsequent arguments about *natural* selection to work. The changeability of organisms under *artificial* selection proves that the availability of heritable variation and amount of change that is possible are easily sufficient to account for the diversity of living things that are present in the world today, given the age of the earth and the vast amount of time that has been available for natural selection to act.

References

Bradley, D. G., E. Emshwiller, and B. D. Smith, eds. 2006. *Documenting domestication*. Berkeley: University of California Press. (Source of multiple references about the origins of domestic plants and animals.)

Mendel, G. [1866] 1966. *Versuche uber Plfanzenhybriden. Verhandlungen des naturforschenden Vereines in Brunn, Bd. IV fur des Jahr 1865, Abhandlungen, 3-47.* Translated by W. Bateson, 1901, with corrections by R. Blumberg. Electronic Publishing Project, http://www.esp.org

Price, T. D. 2001. Domesticated birds as a model for the genetics of speciation by sexual selection. *Genetica* 116: 311–27. (For some updates on the history of pigeon domestication.)

Provine, W. B. 1971. *The origins of theoretical population genetics.* Chicago: University of Chicago Press. (For details about the fate of the *Origin* in the latter half of the nineteenth century, the modern synthesis era, and blending vs. particulate inheritance.)

Schwartz, K. V., and J. G. Vogel. 1994. Unraveling the yarn of the ancon sheep. *Bioscience* 44:764–68.

Wayne, R. K., J. A. Leonard, and C. Vila. 2006. Genetic analysis of dog domestication. Pp. 279–93 in *Documenting domestication*, ed. D. G. Bradley, E. Emshwiller, and B. D. Smith. Berkeley: University of California Press.

Chapter 3

Variation under Nature I

The next step in Darwin's argument is to document that the kind of variation we see in domestic plants and animals is equally observable and of the same kind as seen in nature. This part of the *Origin* may be familiar to you, so it may not seem radical; but it defied the prevailing opinions of the day—others had already dismissed variation under domestication as having no relevance to the origin of species. For example, Charles Lyell, Darwin's geological mentor, had concluded that the variation within breeds was different in character from the differences between species and was not permanent; if domestic organisms were released back into the wild, they would revert to their former state (Ruse 1999, p. 76). Of course, defying Lyell was not new to Darwin. He began his career in geology by overturning Lyell's explanation for the origin of coral reefs (see introduction). Here Darwin argues that the variation seen in domestic plants and animals is also seen in nature.

Variation is all-important to Darwin's proposed mechanism of natural selection because it is the fuel of evolution. Natural selection is the sieve through which diverse individual properties fall or are caught and retained. For this variation to be important in evolution, it must be faithfully transmitted from parents to offspring. Looking for such variation demands the mind-set that it is there and is important to observe and quantify. This mind-set did not exist prior to Darwin. The prevailing pre-Darwinian view of species was that they were the products of acts of special creation and that each species was defined by some ideal or archetype. Individual variations were just deviations from this ideal, and not important in their own right. Actually, it is not enough to say that they were not important; they were an annoyance. Variation made the task of systematists (the scientists who define and classify species) more difficult. There was little incentive to study such variation as an important property of living things.

Darwin argues in his second chapter that all organisms are like the strawberry, which seemed to suddenly vary when people took an interest in selecting for special varieties. It must have always been variable; all that had been lacking was the human effort to find and exploit the individual plants' differences. Darwin encountered such variation during the eight years that he devoted to the study of barnacles. He approached his revision of the classification of barnacles, as did all good systematists, by examining large numbers of specimens per collection and large numbers of collections for each species. But, rather than view variation as an annoyance, he often took special interest in it and reported on variation in many of his species descriptions. In his second volume on the living barnacles, he observed: "After having given up several years to the study of this class, I must express my deliberate conviction that it is hopeless to find in any species, *which has a wide range, of which numerous specimens from different districts* are presented for examination, any one part or organ . . . absolutely invariable in form or structure" (C. Darwin 1854, p. 155, his italics).

Darwin clearly appreciated the importance of variation before his barnacle years, since he invoked it in the first sketches of his theory, written in 1842 and 1844, thus predating his barnacle work; however, in his essay of 1844 he opened his chapter on variation in nature with the statement "Most organic beings in a state of nature vary exceedingly little." His earlier argument for variation was based on his conviction that variation was necessary for evolution by natural selection to occur. The evidence that he cited for such variation in 1844 was that there was a continuity of differences between populations and varieties within a species, and between closely related species. Darwin argued that these differences must have their root in what was originally variation among individuals within a population that was shaped by natural selection, in the same way that breeders create new varieties by carefully selecting who gets to be the parents of the next generation. It was his prior conviction that variation was important that caused him to make it a target of study during his barnacle years; however, the bulk of his argument for variation as presented in this second chapter of the *Origin* is the same as in his essay of 1844. His other goal in this chapter is to build a case for speciation. I will revisit the speciation part of this chapter in part 2.

The entirety of Darwin's discussion of "individual differences" in this chapter is brief and can be quoted in full:

Again, we have many slight differences which may be called individual differences, such as are known frequently to appear in the offspring from the same parents, or which may be presumed to have thus arisen, from being frequently observed in the individuals of the same species inhabiting the same confined locality. No one supposes that all the individuals of the same species are cast in the very same mould. These individual differences are highly important for us, as they afford materials for natural selection to accumulate, in the same manner as man can accumulate in any given direction individual differences in his domesticated productions. These individual differences generally affect what naturalists consider unimportant parts; but I could show by a long catalogue of facts, that parts which must be called important, whether viewed under a physiological or classificatory point of view, sometimes vary in the individuals of the same species. I am convinced that the most experienced naturalist would be surprised at the number of the cases of variability, even in important parts of structure, which he could collect on good authority, as I have collected, during a course of years. It should be remembered that systematists are far from pleased at finding variability in important characters, and that there are not many men who will laboriously examine internal and important organs, and compare them in many specimens of the same species. I should never have expected that the branching of the main nerves close to the great central ganglion of an insect would have been variable in the same species; I should have expected that changes of this nature could have been effected only by slow degrees: yet quite recently Mr Lubbock has shown a degree of variability in these main nerves in *Coccus*, which may almost be compared to the irregular branching of the stem of a tree. This philosophical naturalist, I may add, has also quite recently shown that the muscles in the larvae of certain insects are very far from uniform. Authors sometimes argue in a circle when they state that important organs never vary; for these same authors practically rank that character as important (as some few naturalists have honestly confessed) which does not vary; and, under this point of view, no instance of any important part varying will ever be found: but under any other point of view many instances assuredly can be given. (*Origin*, pp. 45–46)

Darwin acknowledges that organisms in nature seem less variable than domesticated organisms. He thought domesticates were more variable because they recently experienced a change in environment, which induced variation. As I argued in the previous chapter, the increased variation that he saw in domestic animals was likely to have been the product of inbreeding and of breeders' deliberate preservation and magnification of oddities.

For the remainder of this chapter, I will look forward in time to consider how Darwin's emphasis on variation affected future science. I do so in part to show that Darwin's emphasis on individual variation and his proposal of natural selection as the mechanism that causes evolution was at first not widely accepted. The study of natural selection began in the second half of the nineteenth century, then stalled and lay dormant for decades. It did not reemerge as a well-defined discipline until the mid–twentieth century. In between lay the rediscovery of Mendelian inheritance and its reconciliation with natural selection.

Darwin's emphasis on variation spawned the development of a new branch of science soon after the *Origin*. An early leader in this new discipline was his cousin, Francis Galton. Galton's research involved measuring large numbers of parents and children so that he could quantify the nature of variation within populations and the patterns of similarity between relatives. Summarizing and interpreting such data required statistics, so Galton and his successors, most notably Karl Pearson, applied their backgrounds in mathematics to the development of statistics to characterize and quantify variation and the similarities between relatives. Pearson's "product moment correlation" remains in common use today. Galton found that variation in traits such as height could be described with the bell-shaped curve that we call a "normal" distribution and that the appearance of the parent was a good, but imperfect, predictor of what the offspring would look like. Galton attracted a school of adherents that later became known as the "biometricians" because of their quantitative approach to measuring biological variation.

Ironically, Galton also inspired the formation of a second group of investigators who became the antagonists of the biometricians. Galton disagreed with Darwin's emphasis on small differences between individuals as the fuel of evolution and with the idea that evolution was a process of gradual change. He instead thought that "sports" or "mutations" were the real fuel for evolutionary change and that evolution happened in discrete steps. Darwin advocated the importance of small differences because of his practical experience with artificial selection. Such small differences between individuals could always be found, and breeders could always exploit them with artificial selection. The appearance of "sports" such as the ancon sheep was an unpredictable, rare event and was often associated with damaging side effects. Galton disagreed with Darwin because he found that offspring of parents with extreme characteristics, such as the tallest or shortest individuals within a population, tended to be less extreme, or to "regress" to the average of the population. Because of such regression, he thought that selection on small differences between individuals could not generate offspring whose appearance fell outside the range of variation seen in the population as a whole. Selection could never result in a future population that was taller or shorter than the tallest and shortest individuals now present. He reasoned that novelty, or individuals with appearances that fell outside the range of what was already present, had to be the product of large mutations.

Many investigators adopted Galton's perspective and promoted the idea that evolution was caused by large, mutational steps. In 1889 Hugo DeVries, who was a leader in this field, published a book in which he proposed his formal "mutation theory" for the origin of species. He recognized the sort of individual variation that was promoted by Darwin, but he viewed species as being separated by unbridgeable gaps. These gaps could be crossed and new species created only by single mutational events. Prior to 1900, there does not seem to have been a formal name for these individuals who were the antagonists to the biometricians, although I have sometimes seen them referred to as "mutationists." After the rediscovery of Mendel, they became known as "Mendelians" because they adopted Mendel's laws of inheritance as the mechanism that could cause such discontinuous changes in organisms.

W.F.R. Weldon (1860–1906) was one of the leaders in developing biometry. He was the professor of comparative anatomy and zoology at University College, London, a fellow of the Royal Society, and served as president of the British Association for the Advancement of Science. Weldon took Darwin's premises for natural selection and shaped them into a program of empirical research. He argued that a study of natural selection must be statistical in nature: it must characterize how traits varied among individuals within a population, the extent to which trait variation was transmitted from parent to offspring, and the extent to which traits influenced survival.

This was really just a restatement of Darwin's premises for the process of evolution by natural selection (see chap. 1 above), except that it was made in a Galtonian framework by focusing on the empirical measurement of traits in a large sample of individuals from a population.

Weldon collaborated with Karl Pearson and inspired Pearson to develop the necessary statistics for characterizing variation within populations, how this variation changed within and between generations, and the degree to which the appearance of parents predicted what their offspring would look like. In one group of studies, Weldon quantified the shell shape and size of crabs (*Carcinus maenus*) found near the laboratory of the Marine Biological Station on Plymouth Sound. He showed that the pattern of variation of frontal breadth was well described by a normal distribution. He found that the juveniles had more variation in shell shape than adults, which suggested that individuals with either very narrow or very wide shells were less likely to survive to adulthood than were individuals that were closer to the population average. Pearson calculated the extent of "selective death" required to cause such a narrowing of the size distribution. We now recognize this selection against extremes, or "stabilizing selection," as a common form of natural selection that does not cause evolution, in the sense that it maintains the population average over time rather than causing it to change.

Weldon argued that stabilizing selection would be the most common process seen in nature. In addition, he argued that selection that changes the average value of a trait in a population, which we now call "directional selection," would occur rapidly in response to a change in the environment. Directional selection would be seen only if one were lucky enough to be watching a population when such a change occurred, as were the Grants and their colleagues when drought hit the Galapagos Islands. Weldon may have had a similar stroke of good luck when he observed that the frontal breadth of the crabs he was studying became narrower after Plymouth Sound was isolated from the ocean by an artificial breakwater, then filled with fine clay sediment discharged by rivers emptying into the sound. He showed experimentally that individuals with broader shells were less likely to survive exposure to suspended sediments. He hypothesized, but never proved, that a narrower frontal breadth would prevent the accumulation of sediment in the gill chambers. He also never proved that this variation was heritable.

Though Weldon fell short of proof of directional selection by modern standards, our current understanding of evolution by natural selection indicates that all of Weldon's general conclusions were correct. Furthermore, his general approach to the study of natural selection is similar to that of modern investigators. Weldon and Pearson combined their empirical and statistical skills to formalize biometry into a field of study that attracted many other investigators. They initiated the scientific journal *Biometrika* as a place to publish such research. The birth of this journal was a by-product of their dispute with the Mendelians and their difficulty in publishing in existing journals. This journal is still published today.

You might think that such success so soon after the *Origin* would have made Weldon's an enduring name in evolutionary biology. In fact, references to his work disappeared without a trace in the modern scientific literature. His name appears only in publications that deal with the history of science during that period; he had no direct impact on the way we now study evolution by natural selection. He disappeared because he and others who identified themselves with the biometricians opposed not only the "mutationist" emphasis on large, sport-like variations but also Mendel's model of particulate inheritance. As the fortunes of Mendel's laws rose, those of the biometricians faded. The cause of Weldon's disappearance may also be that he died at an early age, in the midst of battles with the Mendelians.

Mendel's laws were rediscovered in 1900. The mutationists quickly claimed them as support for their theory of evolution. The characteristics of plants that Mendel used in his experiments, such as tall versus short, red versus white flowers, or green versus yellow peas, were viewed as having the same discrete, discontinuous properties as the imagined mutations that created new species; so the mutationists adopted Mendelian inheritance as their own, so much so that the they became formally known as the Mendelians after 1900. The biometricians rejected Mendelian inheritance and instead promoted blending inheritance. They did so because their formal comparisons of parents and offspring showed that the characteristics of offspring tended to lie between those of their parents. While Mendel's results were clear, it was possible to argue that they represented special cases, rather than a universal mechanism of inheritance. Ironically, Mendel had already shown that the appearance of blending inheritance was compatible with his model of particulate inheritance when he crossed *Phaseolus* bean plants with red versus white flowers and got offspring with pink flowers (see chap. 2). Nearly two decades passed between the rediscovery of Mendel's laws in 1900 and the formal proof that a polygenic model of inheritance could reconcile these opposing models of inheritance.

As evidence in favor of Mendelian inheritance accumulated, support for blending inheritance faded. More remarkably, as the Mendelians prevailed, the acceptance of natural selection as the mechanism that caused evolution temporarily died, as an innocent bystander of this battle.

A consequence of the Mendelian-biometrician battles is they divided the early success of the *Origin*. Evolution and the transmutation of species attained general acceptance fairly quickly, but there was not the same acceptance of natural selection as the cause of evolution. The scientific community accepted evolution and speciation because Darwin's arguments were based on a wealth of supporting evidence. We will see in part 3 of this book ("Theory") that Darwin also showed that evolution by natural selection presented the simplest and most coherent explanation for phenomena that ranged from embryonic development to the fossil record.

Doubts were expressed about natural selection as a mechanism for evolution long before the Mendelian-biometrician controversy came to a climax in the early 1900s. Natural selection had a heartless, materialistic quality that was difficult to accept because it explained all-important matters of life and death, of species survival or extinction, as the products of seemingly random variation among individuals. Furthermore, natural selection was based on unknown laws of inheritance and unknown sources of heritable variation. Even Thomas Huxley, Darwin's most ardent and vocal defender, expressed skepticism about natural selection and gradual change.

The growing proof of the universality of Mendel's laws rose under the aegis of the Mendelians, who argued that evolution was *not* caused by natural selection's gradual shaping of a population through the favoring or disfavoring of subtle differences between individuals. They argued instead that evolution was caused by large changes that appeared abruptly and were either retained because they were favorable or lost because they were damaging. Even new species were created at once by single mutations. This proposal differs from Darwin's mechanism because Darwin envisioned a continuum of gradual change, driven by small differences between individuals that affect their probability of survival. The initial opposition to natural selection was thus now joined by an alternative mechanism for evolution that came along with a proven law of inheritance. This was a potent combination that drove interest in variation and natural selection into the background for the next three decades.

Widespread interest in variation and natural selection did not reemerge until after the modern understanding of Mendelian inheritance grew to the point that the Mendelians' emphasis on abrupt, large changes could be reconciled with the kinds of continuous variation that Darwin saw when he contemplated domesticated plants and animals. It eventually became clear that the distinction between continuous and discontinuous variation was a false dichotomy; both could be attributed to the same mechanism of inheritance, and there was a continuum that joined the imagined extremes. This "modern synthesis" of evolutionary theory began in the late 1920s, and the empirical study of variation and adaptation began again soon after. Part of the modern synthesis involved the development of population genetics as a discipline. Population genetics consists of the synthesis of Mendelian genetics with evolution by natural selection. These new research programs evolved independently of the earlier work of Weldon and his colleagues and proceeded with little reference to that work. The biometricians were relegated to history. If we look forward in time from the inception of Galton's studies of variation, we see the empirical study of natural selection fade away between 1900 and 1910. If we look backward from the present, we find the origins of our current approaches to the study of adaptation in the late 1930s or later. In the gap in between, we see little interest in the empirical study of individual variation, natural selection, and adaptation.

After the modern synthesis, the study of variation became an enduring component of evolutionary biology. One school of modern study has properties very much like those adopted by Weldon and Pearson; we focus on measurements of the physical appearance or other properties of organisms and analyze patterns of similarity between relatives derived from designed experiments to make inferences about the genetic underpinnings of these similarities. This field is called "quantitative genetics" and traces its origins to R. A. Fisher and papers that began to appear in 1918. Fisher was certainly influenced by Pearson and the Mendelian-biometrician battles, so there is an indirect connection between the biometricians and this modern approach to variation. This discipline plays a central role in the use of artificial selection to develop improved domesticated plants and animals. One of the major endeavors in the related field of population genetics is the study of factors that maintain genetic variation in natural populations. A third major area of research involves the use modern genetic and biochemical techniques to analyze variation at the level of the genetic code or the proteins that are produced from this code. So, even though Darwin's emphasis on the importance of variation receded for a time, it reemerged and remains a major component of scientific research today.

This brief history illustrates the long reach of the *Origin*. Darwin recognized the importance of variation and the laws of inheritance. The *Origin* elevated individual variation from being a source of annoyance to being a topic of central interest in evolutionary biology. Although Mendel worked independently of Darwin, it was Darwin and the *Origin* that popularized scientific investigations of the laws of inheritance. It was in this context that Mendel's work was rediscovered and appreciated after a thirty-five-year hiatus. This rediscovery in turn spawned the subsequent development of genetics into a modern discipline. It was also in this context that we saw the invention of branches of statistics that remain in common use today, since statistics are required for a formal analysis of variation. The modern disciplines of quantitative and population genetics are also a direct outgrowth of the research spawned by the *Origin*. Individual variation and the factors that sustain such variation in natural populations are central to all these disciplines, for the same reason that they are central to Darwin's argument. Darwin's initial observation that such variation is ubiquitous has been upheld, as has his proposal of its central importance for fueling evolution by natural selection.

References

Darwin, C. 1854. *A monograph on the subclass Cirripedia.* Vol. 2: *The Balanidae.* London.

Darwin, Francis. 1912. *The foundations of the "Origin of Species": Two essays written in 1842 and 1844 by Charles Darwin.* Cambridge: Cambridge University Press.

Provine, W. B. 1971. *The origins of theoretical population genetics.* Chicago: University of Chicago Press. (Main source for all details presented here on the Mendelian-biometrician controversy and the later rise of population genetics.)

Ruse, M. 1999. *The Darwinian revolution—science red in tooth and claw,* 2nd ed. Chicago: University of Chicago Press.

Weldon, W.F.R. 1895. Attempt to measure the death-rate due to the selective destruction of *Carcinus maenus* with respect to a particular dimension. *Proceedings of the Royal Society* 57: 360–79.

Weldon, W.F.R. 1899. Presidential address. In *Report of the sixty-eighth meeting of the British Association for the Advancement of Science, held at Bristol in September, 1898.* London: John Murray.

Chapter 4

The Struggle for Existence

Imagine the English countryside that was familiar to Darwin and his contemporaries. There were meadows, woodlands, and ponds, each filled with plants and animals in abundance. All life was going about its daily task of finding whatever it needed from its surroundings, growing and reproducing. Many animals seemed to have leisure time. Birds sang, turtles sunned themselves on logs, foxes cavorted in meadows at sunrise. It would be easy to think of nature as being benign and bountiful. Darwin's perspective was quite different from this romantic vision of nature that prevailed at the time. He saw all organisms as constantly striving to increase in abundance and doing so at the expense of others, perhaps by eating them or competing with them for the same resources. These interactions were so severe as to cause the extinction of inferior competitors. In short, according to this view—which is the theme of his third chapter—nature is not benign. All living things are engaged in constant competitive, exploitative interactions to gain an advantage.

Darwin felt that these "biotic interactions," or interactions between living things, were the most important cause of evolution by natural selection. He saw these interactions as being analogous to the role that breeders play when they select for desired properties in domestic plants and animals. However, Darwin saw nature's reach as being far more pervasive than that of humans: "We have seen that man by selection can certainly produce great results, and can adapt organic beings to his own uses, through the accumulation of slight but useful variations, given to him by the hand of Nature. But Natural Selection, as we shall hereafter see, is a power incessantly ready for action, and is as immeasurably superior to man's feeble efforts, as the works of Nature are to those of Art" (*Origin*, p. 61).

The physical, or "abiotic," environment, including temperature, water availability, wind, or the seasons, can be a source of natural selection, but Darwin considers these factors to be minor in comparison with biotic interactions. He emphasizes the various competitive and exploitative interactions between organisms as the dominant cause of change. Consider some examples. The external appearance of many organisms causes them to blend in with their background so that they will not be seen, either by predators looking for a meal or by prey trying to avoid predators. The stripes of the tiger are as important a form of camouflage as is the dappled coat of a fawn. In contrast, some organisms have bright coloration that attracts attention. Flowers advertise an award of nectar for visiting bees. Skunks display a "touch me not" advertisement of their noxious defense.

We can also see adaptation in shape and performance. The antelopes and cheetahs of the Serengeti plains have evolved longer legs and more flexible spines that enhance their ability to accelerate and attain higher running speeds. This process is the root of high-speed chases, as cheetahs pursue their intended prey. Yet there are no antelopes so fast that they can invariably escape predation, nor cheetahs that no longer face the risk of starvation; each adaptation by a predator that improves its ability to capture prey is balanced by a counteradaptation in the prey. They are locked in what we call a red-queen process, with reference to the Red Queen of Lewis Carroll's *Through the Looking Glass*, who cautioned that "it takes all the running you can do, to keep in the same place." A balance is achieved, but it does not reflect the harmony of nature that was envisioned by the Victorian English. The balance is instead like the stasis that you see between two evenly matched teams engaged in a game of tug-of-war.

Interactions between organisms also shape internal anatomy. For example, an animal's diet will shape the evolution of everything associated with food processing, including the teeth, tongue, and lips and the digestive tract. The teeth of mammalian predators tear like knives and shear like scissors. The stomachs of predators produce powerful acids that help to break down muscle and bone, and their short intestines absorb the simple nutrients that are the breakdown products. Mammalian grazers instead have teeth that cut and grind so that they can mill the abrasive leaves of plants. They have lips that can function like hands as they grasp vegetation, muscular tongues for moving the bolus of food around the mouth when chewing, and guts modified into fermentation chambers where milled grasses are digested by bacteria. The bacteria become the food source digested by the grazer.

Metabolism, reflexes, and perception all evolve as parts of the same package. Any animal that is capable of high-speed, sustained running must have lungs that can absorb sufficient oxygen and a heart and circulatory system that can deliver the oxygen to its muscles. Both predators and prey must have appropriate senses, including sight, hearing, and smell, to detect each other's presence. Grazers have eyes that are oriented toward the sides of their heads so that their visual fields encompass close to a full circle. This wide field of view enhances their chances of seeing an approaching predator. Predators have both eyes pointing forward, which limits their field of view but enhances their depth perception and eye-paw coordination. All these features are thus shaped by how organisms interact with others as part of their day-to-day lives, be it to find and process food or to avoid being eaten.

Common sense may dictate that the physical environment, such as cold, heat, or the availability of moisture, would be more important than biological interactions. Darwin argues that these physical features prevail only in extreme environments, such as the Arctic or extreme deserts. In many cases where it seems that the physical environment is an important factor of selection, we can still argue that biotic interactions prevail. Deserts are dry, but plants still compete for the moisture that is available. Drought caused Galapagos finches to evolve, but variance in their ability to harvest seeds determined who died and who survived.

Darwin's emphasis on biotic interactions was certainly a product of his lifetime of experiences as a naturalist, but it was Robert Thomas Malthus's *Essay on the Principle of Population* that brought these experiences into focus. In his autobiography, Darwin recalled how reading this essay provided the missing link in his musings about evolution and inspired the concept of natural selection as the mechanism for evolution. One of the remarkable, coincidental features of Alfred Wallace's independent discovery of natural selection was that he too was inspired by Malthus's essay. Malthus's argument began with the observation that population size can increase exponentially and the assertion that agricultural production can increase only arithmetically as methods of cultivation improve. Human population growth in the American colonies during the eighteenth century provided one of his examples of exponential population growth. The population of the colonies had doubled every 25 years or so, which meant that after 100 years and four doublings it was 16 times $(2 \times 2 \times 2 \times 2)$ larger than its initial size. If all available land had already been under cultivation, then Malthus expected that improvements in farming methods could only have caused a fourfold increase in agricultural production during the same period. This inequality meant that there would have been many more people living on proportionately fewer resources.

Malthus observed that European colonies in the New World had much higher rates of population increase than nations in Europe had. In the United States or South America, there was an abundance of unutilized, fertile land. As the populations in the New World grew, fueled by early marriage and the production of large families, they simply spread to occupy new land. In contrast, humans had occupied northern Europe for much longer so that, by the end of the eighteenth century, most arable land was devoted to either agriculture or pasture. Population growth rates there had slowed to doubling times of 300–400 years. Malthus presented actuarial tables that compared the numbers of births and deaths. In Europe, the highest ratio in the best of times was 117 births to 100 deaths, while in the United States, where access to new arable land was not so limited, births could outnumber deaths by 300 to 100.

Malthus was concerned with the effects of such resource limitation on human societies. Darwin's take-home message was different. Darwin recognized that human populations cannot grow as quickly as those of almost any other species, save for something like an elephant, because humans are much older when they begin to reproduce and produce far fewer offspring during their lifetimes. Yet in spite of their low rate of multiplication, even humans have the capacity to increase population size at a remarkably rapid rate, given sufficient resources (see fig. 5).

If human populations are capable of such remarkable growth, then the same must be true for all organisms; yet we rarely see such explosive growth in nature. Instead, we tend to see rather stable populations in places like the placid English countryside that Darwin's contemporaries thought of as so benign. In Darwin's view, such stability can be sustained only at a cost: the destruction of most of the offspring that any organism produced. If there is variation among offspring and if this variation has any impact at all on their chances of surviving and contributing offspring to the next generation, then the many factors that regulate population size will become those that determine who will survive and reproduce. In so doing, those factors will shape the future appearance of the population.

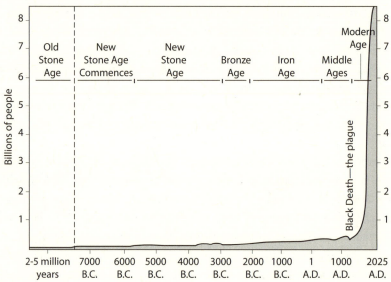

Figure 5
Human Population Explosion
Humans serve well as an example of the capacity of all organisms to increase
in population size if unchecked. Our population explosion began after the plagues
of the fourteenth century, fueled in part by the emigration of Europeans, the spread
of agriculture throughout the world, and the displacement of less numerous hunter-
gatherer populations. Later factors include the advent of mechanized agriculture
and other farming methods that improved crop production, plus increased survival
attained through sanitation and medical care.

Why did Darwin believe that all organisms have this ability to increase
so rapidly in population size, but are generally prevented from doing so?
During his travels, Darwin witnessed the fates of plants and animals that
had been transplanted from Europe to other countries, where the features
of the environment that had controlled their population growth were no
longer present. Cattle and sheep had multiplied prodigiously in Australia,
as had horses and cattle in South America. Wild artichokes that had been
transplanted from Europe to South America had gone from being relatively
scarce in their land of origin to being so dominant in the pampas of Argentina that "very many (probably several hundred) square miles are covered
by one mass of these prickly plants, and are impenetrable by man or beast"
(Darwin 1839, p. 138). In our own time, we have seen the advent of "inva-

sive species"—exotic organisms that were transplanted from one part of the world to another, either by accident or design, then experienced population explosions in their new homes. Starlings and house sparrows did so when transplanted from England to the United States, rabbits when transplanted from England to Australia, and zebra mussels when transplanted from the Caspian Sea to the inland waterways of North America. In each case, an organism that had been part of a stable community in its homeland became abundant to the point of being a pestilence in its new home. Darwin argues in this chapter that the biotic interactions that normally control the abundance of such organisms in their home countries are no longer present in their new homes, so their potential for rapid population growth is realized.

What normally checks population growth? Darwin had observed that organisms are most vulnerable in their early life stages, so he proposes here that population regulation is often attained through the destruction of very young individuals. He cites an experiment in which he marked every seedling that emerged on a square yard of bare ground. Out of 357 marked seedlings, 295 were destroyed shortly after germination, often because they were eaten by slugs. He reasons that the explosive growth of species introduced to new environments is fueled largely by the increased survival of newborn young because of an absence of such agents of control.

He then considers different kinds of controlling agents. Darwin first characterizes single factors that are important in population regulation. One factor is competition between organisms for limited resources. He cites an experiment in which he monitored plants growing on a few square yards of turf. The community began with twenty species, but in time nine of them were eliminated because they were crowded out by the more vigorous growth of others.

Predation is a second regulator of populations. Darwin observes that the popular sport of hunting grouse, partridge, and hare depends on the control of foxes. Foxes are natural predators on game birds and hence competitors with human hunters. By killing foxes, game managers make more of their prey available to hunters. Darwin estimates that hundreds of thousands of game birds and rabbits are shot every year by hunters. If the natural predators (foxes) were no longer eliminated and all hunting ceased, then Darwin predicts that the game species would probably become less abundant because foxes are more effective in controlling their numbers than all the hunting practiced by humans. We see the equivalent process today in response to the elimination of predators. In the northeastern United States, seeing deer was once a special event because they were scarce and alert, so they saw you first and ran before you could see them. In the absence of population regulation by predators, but in the presence of hunters, they have become a pestilence. They are so abundant that they have altered the structure in natural plant communities and in the gardens of suburbia.

Disease can also be an effective regulator of population size. In western Europe, plagues repeatedly decimated human populations during the fourteenth century. Even in Darwin's day, lethal diseases were far more common than they are today; three of his ten children died before reaching adulthood. However, Darwin argues that disease is a less prominent feature of natural ecosystems and will become important only when the potential victims of disease organisms are present in unusually high numbers.

Finally, Darwin acknowledges that climatic extremes can cause periodic regulation. The winter of 1854–5 destroyed four-fifths of the birds on his estate. However, he argues that such effects are generally mediated through the effects of climate on food availability. The immediate cause of death was starvation, as was the case for the medium ground finch studied by the Grants on the Galapagos. Darwin argues that the physical climate most often acts as an indirect factor that modifies biotic interactions, such as modifying food availability, rather than as a direct factor that causes the death of some individuals.

Biotic interactions can select for adaptations that may not be so obviously caused by such interactions. For example, dandelions produce plumed seeds that are carried away from the mother plant by the wind, so we can interpret them simply as an adaptation for dispersal. But why disperse? Darwin argues that such mechanisms for dispersing seeds evolved as a consequence of competition for space. If the space around the mother plant is likely to be fully occupied and if the availability of empty space is scattered and unpredictable, then casting seeds to the fate of the winds is a means of finding an empty patch of soil where they can germinate and grow. Each seed that drifts away from the mother plant is like a lottery ticket, since its fate is determined by whether or not it is lucky enough to fall on a bare piece of land where it can grow and reproduce. Other plants have evolved an alternative strategy of producing large seeds that are rich in nutrients that support early growth. Being large means that the seed cannot be dispersed by the wind and will come to rest close to the mother plant. In this case, maternal provisions support the early growth of the seedling so it can compete effectively with others.

Darwin also argues that the Victorians' habit of importing exotic plants for their gardens and animals for their zoos shows that the physical environment is not often a limiting factor. Every garden contains flowers from different climates, and the London zoo is full of animals from warmer climates, yet all the species survive, and most of them reproduce when protected from competition with local species. Thus their natural distribution is not prescribed by their tolerance for certain physical environments per se but, rather, by their ability to disperse and the organisms that they interact with.

Darwin next considers population regulation via far more complex ecological interactions, rather than just single factors like competition or predation. In this regard, many ecologists recognize Darwin as a pioneer in defining their discipline as well. First he describes a tract of barren, grazed land on the estate of one of his relatives. Twenty-five years earlier, one portion of the heath had been fenced to exclude grazing cattle, then planted with scotch fir. The community that developed in the young woodland was dramatically different from its surroundings. It had twelve species of plants and six species of insectivorous birds not seen in the heath. The heath had two or three species of insectivorous birds not seen in the tree plantation. These differences in the bird community implied that the insects in the woodland versus the field were also different, since the birds fed on insects but differed in the types of insects that they preferred.

During his travels in South America, Darwin saw that feral cattle and horses, both of which had been imported from Europe, had become very abundant in many areas and had similarly modified the structure of local communities by suppressing tree growth and hence indirectly modifying the remainder of the plant and animal community. However, the grazers were excluded from parts of Paraguay because of a fly that lays its eggs in the navels of newborn grazing mammals and kills most of them before they mature. Elsewhere, insectivorous birds suppressed the abundance of these flies. We thus have birds controlling the abundance of the parasitic fly, and so allowing the multiplication of grazing mammals, which in turn suppress tree growth, which in turn results in a modification of other elements of the community, "and so onwards in ever-increasing circles of complexity" (*Origin*, p. 73). When the birds are absent, insects control the abundance of grazing mammals, which prevents the cascading effects those mammals would unleash on the rest of the community.

As a third example, Darwin had shown in earlier work that pollination by bumblebees greatly increased seed production by red clover and heartsease. He infers from this and many other observations that insect pollinators play a major role in controlling the abundance and distribution of flowering plants. The abundance of bees is in turn controlled by field mice, which often destroy their colonies. Bees tend to be much more common near villages where people keep cats, because the cats in turn regulate mouse abundance. Darwin imagines that there is a complex interaction of cats controlling the abundance of mice that in turn control the abundance of bees that in turn control the abundance of flowering plants, which presumably would also affect other elements of the community, such as the insects that feed on the plants and the birds that subsist on the insects. Nature is defined by such "battles within battles," but they are usually so well balanced that we perceive stability. Disruptions can readily tip the balance and result in population outbreaks for some organisms, perhaps resulting in the extirpation of others.

Darwin ends with the introduction of two important concepts that we will revisit later in the book. One pertains to which organisms will be each other's most severe competitors. The struggle for existence is often pictured as a contest between a host and parasite or between predator and prey, which are very different and distantly related organisms. However, "the struggle will almost invariably be most severe between the individuals of the same species, for they frequent the same districts, require the same food, and are exposed to the same dangers" (*Origin*, p. 75). Likewise, the struggle between varieties, or distinct populations, of the same species will be almost as severe as that between individuals within the same population. Darwin considers the practical experiences of those who cultivate varieties of wheat in the same field or flowers in the same garden. If all seeds that are produced are harvested and replanted, then after a few years the variety that is best suited for that particular patch of soil prevails. Gardeners who want to maintain varieties of the same plant together usually have to deliberately and separately harvest and replant each of them to sustain this diversity. The same loss of diversity happens if different varieties of sheep are grazed on the same land or different varieties of medicinal leech are kept in the same culture. Likewise, a competition between species in the same genus will be more severe than one between species from different genera, again because the more closely related organisms will be more similar to one another in their requirements. Darwin predicts that for similar varieties or species to coexist, natural selection will strongly favor those individuals from each that are most divergent in how they earn a living. Over time, such selection will cause coexisting varieties or species to become more different from one another. He named this concept "divergence of character" or the "principle of divergence." The different-size bills in Darwin's finches and the differences in the seeds that they eat (chap. 1) represent such a divergence of character.

The second concept Darwin introduces here is extinction, which is a potential consequence of these biotic interactions, particularly between species that are closely related. The reality of extinction had become generally accepted only in the few decades before the *Origin* was published, and the cause of extinctions remained a mystery. The discovery of so many fossils of organisms that were not seem among the living had, by the early nineteenth century, led to the conclusion that there really were many animals that had once existed and had since disappeared. Why? Darwin observes that his contemporaries often invoked unseen causes: "so profound is our ignorance, and so high our presumption, that we marvel when we hear of the extinction of an organic being; and as we do not see the cause, we invoke cataclysms to desolate the world or invent laws on the duration of forms of life!" (*Origin*, p. 73).

He argues instead that, to understand why extinction happens, we need not look any farther than the day-to-day interactions between species that we can so easily observe and quantify. All organisms are constantly striving to increase in numbers, often at the expense of others. Although we think of nature as being finely balanced, the advantage that one species has over another, however small, will enable it to multiply more quickly and eventually displace a competitor, one population at a time, until the loser in the struggle disappears in entirety. Extinction is thus just the collateral damage inflicted by the greater success of one organism over another in this day in, day out struggle for existence. (Our current concept of extinction is broader than this; I will expand on this topic in part 3 of the book.)

Darwin is acting again as Charles Lyell's apostle (see introduction) with this proposed cause of extinctions. Lyell argued that day-to-day processes that we can see and quantify—like the wind, rain, and tides—have worked over vast intervals of time to shape the geological features of the earth. We need make no recourse to cataclysms that lie outside our common experience. Darwin argues that the same logic applies to biological processes and, in this case, to the causes of extinction. Extinction is just the product of the biotic interactions that we see constantly at work in nature.

Darwin was well aware that his "nature red in tooth and claw" view of the world would shock the sensibilities of many of his contemporaries, so he ended the chapter with a note of consolation: "When we reflect on this struggle, we may console ourselves with the full belief, that the war of nature is not incessant, that no fear is felt, that death is generally prompt, and that the vigorous, the healthy, and the happy survive and multiply" (*Origin*, p. 79).

If only this were true.

References

Darwin, C. 1839. *Journal of researches into the geology and natural history of the various countries visited by H.M.S. Beagle.* London.

Malthus, T. 1978. *An essay on the principle of population, as it affects the future improvement of society with remarks on the speculations of Mr. Goodwin, M. Condorcet, and other writers.* London: Printed for J. Johnson, in St. Paul's Church-Yard.

Chapter 5

Natural Selection I

How will the struggle for existence ... act in regard to variation? Can the principle of selection, which we have seen is so potent in the hands of man, apply in nature? I think we shall see that it can act most effectually. Let it be borne in mind in what an endless number of strange peculiarities our domestic productions, and, in a lesser degree, those under nature, vary; and how strong the hereditary tendency is. ... Let it be borne in mind how infinitely complex and close-fitting are the mutual relations of all organic beings to each other and to their physical conditions of life. Can it, then, be thought improbable, seeing that variations useful to man have undoubtedly occurred, that other variations useful in some way to each being in the great and complex battle of life, should sometimes occur in the course of thousands of generations? If such do occur, can we doubt (remembering that many more individuals are born than can possibly survive) that individuals having any advantage, however slight, over others, would have the best chance of surviving and of procreating their kind? On the other hand, we may feel sure that any variation in the least degree injurious would be rigidly destroyed. This preservation of favourable variations and the rejection of injurious variations, I call Natural Selection. (*Origin*, pp. 80–81)

Here, at the start of his fourth chapter, Darwin completes his analogy between natural and artificial selection. Artificial selection is possible because differences between individuals can be found for all traits. These individual differences are faithfully transmitted from parents to offspring. Breeders can exploit this variation by selecting those individuals that possess desired traits to be the parents of the next generation. Doing so for successive generations makes it possible to magnify small differences between individuals into big differences between breeds. The same selection process occurs in nature because wild organisms also vary and because so many more offspring are produced than are required to replace their parents. Those who survive do so because their individual variations give them some advantage in the struggle for existence. As with artificial selection, the continued action of natural selection can convert individual differences into distinct, new varieties.

You might imagine a plant or animal evolving in response to some change in its physical environment. Darwin argues that evolution by natural selection can happen even without such a change. The complexity of interrelationships between organisms in their struggle for existence means that any change in one species can cause a cascade of changes in the other species that interact with it. Remember that cats eat mice, mice eat bees, bees pollinate flowers, and flowers are part of a plant community that sustains insects, which are in turn eaten by birds. Darwin thus envisions nature as being like a complex jigsaw puzzle in which each piece interlocks with many others, so a change in the shape of any one piece inevitably leads to changes in others. Because all organisms constantly overproduce offspring that strive to find a place for themselves in the environment, some individual variant may find a new way of supporting itself, thus initiating a change that will affect others.

In spite of these tight interactions, the environment will never be fully occupied, because there is always a potential to utilize it in a different way. We see possible evidence for unoccupied spaces in the success of invading species. Invaders have often been released from the biotic factors that controlled their abundance in their native land, be they competitors, predators, or disease, so their populations expand rapidly in their new home. Their success as invaders may be attributable to their being superior competitors, once they have escaped some factors that control their abundance in their native land, but may also suggest that they find some unoccupied space in the community they are introduced to. The success of invaders is just a more dramatic manifestation of what must be happening all the time, since all species constantly produce excess offspring that vary in characteristics and probe the environment for available resources.

Darwin observes that there are limits to the analogy between artificial and natural selection. "Man can act only on external and visible characters: nature cares nothing for appearances, except so far as they may be useful to any being. She can act on every internal organ, on every shade of constitutional difference, on the whole machinery of life" (*Origin*, p. 83). Recall my comparison of a cheetah and an antelope in the previous chapter and how each diet specialization is associated with a complete reorganization of the body and metabolism. We do not see the same scope of integrated adaptations in the products of artificial selection.

A second difference between artificial and natural selection is that humans sustain all the products of artificial breeding with special care, so many traits that would adversely affect survival in nature, such as the specialized beaks and feathers of domestic pigeons or the stubby legs of a dachshund and of ancon sheep, are shielded from natural processes. In nature, any aspect of an organism that contributes to its ability to obtain and process food or in other ways helps to ensure survival will be under constant "scrutiny" by natural selection. Domestic animals rapidly evolve, or die out, when they are released back into the wild because they have been released from the shielding influences of humans' care and are once again subject to natural selection. Furthermore, a human life span is very short, so the duration of any one person's efforts is limited. Nature has all time at its disposal. Darwin also argues that natural selection will be slower than artificial selection, so we cannot expect to see it in action except when we get glimpses of its affects over longer intervals of time, such as in the fossil record, where we can see that organisms of the past are different from those alive today. Nevertheless, natural selection is always active and influences everything, so its long-term effects will be larger and more pervasive than those of artificial selection.

Darwin argues that one measure of the potency of natural selection is the way it has "seen" and modified seemingly trifling details of organisms. Consider the way the coloration of so many animals conceals them so well in their chosen background. Leaf-eating insects are green, while those that feed on bark are mottled brown. The ptarmigan, a bird found in northern latitudes, is pure white during the winter to blend in with the snow, while the black grouse is dark to blend in with peaty soil. Some plants have fruit or leaves that are smooth, while others are covered with a fine down. The down deters certain types of insects that feed on leaves or fruit, so it appears to be an adaptation that deters herbivores. A comparison of domestic plums that have purple versus yellow flesh reveals that each variety is susceptible to different types of diseases. We can imagine that any traits such as these will become predominant in a population as a function of the type of organisms that might consume the species, the type of diseases that it is exposed to, or any other feature of the biotic environment that might influence an individual's ability to survive, grow, or reproduce.

A corollary of observations like the relationship between the color of fruit flesh and susceptibility to disease is that many traits of organisms are correlated (see chap. 2). Darwin learned about these "laws of correlation" as a by-product of his experience breeding pigeons. Breeders of pigeons who selected for short beaks also inadvertently selected for birds with small feet. Darwin considers whether such correlations can limit natural selection, especially if the evolution of one trait favorable to an organism is accompanied by a correlated change that is injurious.

Darwin argues that natural selection can overcome such correlations. For example, many organisms have a larval life stage that is quite different from the adult stage; the two stages are separated by a discrete metamorphosis that transforms larvae into adults, just as tadpoles becomes frogs. Such organisms often have larvae that are exquisitely adapted to a specific environment that is very different from the environment utilized by the adult. Darwin observes that it is possible for larvae to evolve new adaptations without their having any influence on adult morphology or lifestyle. We can see such independence of life stages in sea urchins. Some species in the genus *Heliocidaris* produce larvae that live, feed, and grow in the open water column, then settle and metamorphose into adults. Other species in this genus instead produce young that bypass the free-swimming larval stage and instead develop directly into miniature adults. These direct-developing larvae lack all the adaptations that are required for an active, self-sustaining existence. In spite of these differences in the larvae, the adults of all species in this genus are very much alike in morphology and ecology.

Even animals without a metamorphosis can have specializations that characterize juveniles but not adults. Young deer often lie still on the forest floor while their parents are grazing. They have dappled coats that cause them to blend in with the background of spots of daylight shining through a leafy forest canopy. One of my colleagues described crawling through the forest, intent on finding nesting birds, then realizing that he was close enough to a fawn to reach out and touch it. Adults are instead countershaded, so that they are darker on the back and white on the belly. When illuminated by sunlight from above, the back appears lighter while the belly is in shadow. When seen from the side, this combination of countershading with the effects of sunlight causes the deer to appear to be a uniform light brown that blends in well with the background of open meadows, where they frequently graze.

The ability to evolve specializations that characterize just one life stage without incurring damaging, correlated changes in other stages represents an important distinction between natural and artificial selection. Darwin expands on this distinction with an example from pigeon breeding. Recall from chapter 2 that the short-faced tumbler pigeon was bred for a shorter beak. A consequence of such selection on adults was that the chicks were often not able to break out of the egg to hatch. Breeders had to break the eggs for them. Darwin argues that natural selection must balance how it modifies one life stage to eliminate such deleterious correlated changes in other life stages. If there were strong natural selection for adult pigeons with short beaks, then there would also have to be simultaneous selection for chicks developing in the egg to retain the ability to hatch. More generally, natural selection must in some way balance the positive and negative consequences of change. Such a balancing of costs and benefits would tend to slow down the process of evolution.

Sexual Selection

The males and females of a species often occupy the same ecological niche yet differ in appearance (fig. 6). Male cardinals are bright red, while the females are a more cryptic brown. Many species of birds have males that display similar, bright coloration that is not seen in females. Male baboons are much larger than females and have enlarged canine teeth. Male lions are also larger than females and have a conspicuous mane that females lack. If natural selection were the only factor that shaped the appearance of organisms, then the sexes should look the same. Darwin argues that some process other than natural selection must cause these differences between the sexes. He names this process "sexual selection." He proposes that such differences between the sexes are products of selection that acts on the success of males in obtaining mates and siring offspring. He attributes sexual selection to three possible mechanisms. The first is male-male combat for the opportunity to mate. The victors in such combat are not necessarily the most vigorous individuals; victory may instead be a function of their having some specialized weapon, such as the leg spurs found on roosters or the antlers of deer. The differences between sexes may also lie in male defenses against the assaults associated with male combat, such as the mane of a male lion or the shoulder pad of the boar, both of which protect the

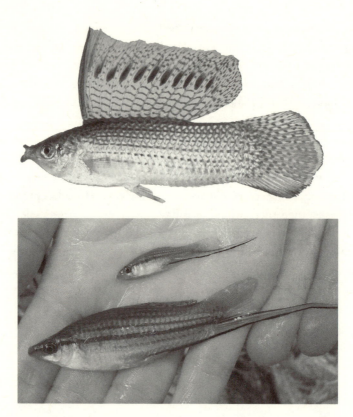

Figure 6
Examples of Sexually Selected Traits
The enlarged dorsal fin of sailfin mollies (*Poecilia latipinna*, top) and the swords of
swordtails (*Xiphophorus helleri*, bottom) are found only on males and are incorporated
in elaborate courtship displays. They serve no known function other than to attract
the attention of females. The males of both species can vary considerably in size, as
illustrated here for swordtails.

individual from attack, "for the shield may be as important for victory as
the sword or spear" (*Origin*, p. 88). Sexually selected attributes may also
reflect the tastes of females when choosing mates. The males of some bird
species gather at arenas where they display and compete for the attention of
females. Females visit the arena and choose mates. The preferred males will
sire the most offspring and will be best represented in the next generation.
"I can see no good reason to doubt that female birds, by selecting, during
thousands of generations, the most melodious or beautiful males . . . might
produce a marked effect" (*Origin*, p. 89).

Darwin predicts that natural selection will be far more potent than sexual selection because sexual selection is caused by differences in reproductive success while natural selection is caused by differences in survival. "Sexual selection is, therefore, less rigorous than natural selection" (*Origin*, p. 88). This distinction, repeated throughout the *Origin*, reveals what Darwin thought natural selection selects for, or how he would define "fitness." Today, we define fitness as the success of an individual in contributing offspring to the next generation relative to the average number of offspring contributed by an individual in the population. A fitness value greater than 1 means that the individual is better than average, while a value of less than one means that it is worse. This definition long postdates the *Origin*, so my argument here is based on the definition that I think is implicit in Darwin's presentation. The predominant theme in the *Origin* is that fitness is defined more by survival than by reproduction. Natural selection acts through a modification of risks of mortality and increases life span. Living longer means having more opportunities to reproduce. Our current definition is subtly different, because we now argue that natural selection acts on lifetime reproductive success, which is a composite of survival and reproduction. One difference between Darwin's definition and the one we use now is that we now allow for the possibility that natural selection can indirectly cause the evolution of shorter life spans or the production of fewer offspring, providing that there are compensations in other parts of the life history that cause an increase in the number of offspring that survive to adulthood. I will expand on this theme in chapter 7.

Some might argue with my interpretation of how Darwin defines fitness. Darwin hints at a broader definition once or twice in the *Origin*, but his repeated distinction between sexual and natural selection makes it clear that he thinks that the bottom-line consequence of natural selection and all the adaptations it produces is increased life span and that life span figures more prominently than reproduction in determining fitness. Using more current definitions for how natural selection increases fitness means that the distinction between sexual and natural selection is no longer as easily made as it is in the *Origin*, yet the reality of sexual selection is clearly defined by the many features of organisms that serve only to assure reproductive success, sometimes at great cost. The tail of a peacock or the bright coloration of a male guppy are both great lures for attracting the attention of females, but are a burden to males when they are not courting. What is visually attractive to a female may also attract the unwanted attention of a predator. Some balance must be struck between the effects of sexual and natural selection.

These few pages of the *Origin* represent the first formal exposition of the concept of sexual selection. Darwin expanded on this idea in *The Descent of Man, and Selection in Relation to Sex* (1871), where he interpreted the differences between men and women as being products of sexual, rather than natural, selection. The study of sexual selection has since blossomed into a major subdiscipline of evolutionary biology.

The Descent of Man is, in turn, one of the two books in which Darwin took on a topic that he studiously avoided in the *Origin*, which is the implications of his theory for human origins. Darwin published the second book, titled *On the Expression of Emotions in Man and Animals*, a year later. He hints clearly in the *Origin* at his intention to write both books, but felt this current "abstract," the *Origin*, would be sufficiently incendiary without dethroning humanity from its supposed uniqueness as the final masterwork among all of God's other special creations. It was actually Thomas Huxley who took the heat for applying Darwin's theory to humankind. He defined humanity's place among the great apes—with chimps and gorillas as our closest living relatives and Africa as our land of origin—first in a series of lectures given in 1860, then in his book *Man's Place in Nature* in 1863.

Illustrations of the Action of Natural Selection

Darwin offers imaginary examples of natural selection to show how it works. He envisions a population of wolves that experiences a time when food is scarce, such that the individuals who are the slimmest and fastest will be more likely to survive, as long as their slender build retains the strength to subdue prey. Even without such a period of scarcity, there could still be selection for different varieties of wolves. One might have attributes that make it more skilled in catching a particular type of prey that is not so well utilized by other members of the population. He cites the observations of a Mr. St. John, who found that some of his cats would only bring home rabbits and hares while others specialized on birds; these individual differences in preference and perhaps ability may represent such variation. Darwin imagines similar variation on a geographic scale. Wolf packs in a mountainous region might differ from those found in the lowland by hunting different prey in a different fashion. In time, we might see the evolution of distinct highland and lowland wolf populations.

Darwin proposes more complex scenarios that involve interactions between plants and the insects that pollinate them. One starting point for the interaction might be a plant that secretes a sugary substance consumed as food by some insects. As an insect moves from plant to plant while feeding, it may inadvertently carry pollen, fertilize ova, and hence increase the pollen donor's production of offspring. If plants vary in their ability to attract insects, then those that are more attractive may also be more successful in fathering seeds than those that attract fewer pollinators and rely more on the wind. Wind-pollinated plants produce an abundance of pollen that is cast to the fates, most to be lost, in order to sire offspring. Plants that attract the services of insect pollinators can save on this large investment in pollen.

Darwin presents his own observations on the efficiency of insect pollination of holly trees on his property. The flowers of most plants have both male (pollen producing) and female (seed producing) organs, but in this species trees are either male or female; some individuals have flowers that produce only pollen while others have flowers that produce only seeds. Darwin had a male and female tree in his garden that were growing 60 yards apart. After an interval of time when there was no wind, he examined twenty flowers from the female tree and found that all of them had been recently pollinated; some had an abundance of pollen. The weather had recently been cold and not favorable for flying insects, yet bees had still managed to fly from the male to the female tree and deposit pollen on every flower. Darwin proposes that plants can evolve an increase in the production of nectar and, at the same time, flowers that are visually more attractive to insects, in order to be more effective in exploiting insects to move pollen from plant to plant. The trade-off is that they can then invest much less in pollen production.

Darwin also proposes the joint evolution of plants and insects. As plants evolve to exploit insects as pollinators, insects evolve to exploit those plants as a source of food. The first book he published after the *Origin* was titled *On the Various Contrivances By Which British and Foreign Orchids Are Fertilized By Insects, and On the Good Effects of Intercrossing* (1862). Here he presents the extraordinary example of the orchid *Angraecum sesquipedale* from Madagascar. A general property of orchids is that they have a broad lower petal called the labium, which tapers into a narrow tube called a

Figure 7
This is the first known photograph of the long-tongued hawk moth *Xanthopan morgani praedicta* extracting nectar from the long-spurred orchid *Angraecum sesquipedale*. The nectar spur descends from the orchid along the right-hand margin of the photograph. The photograph was taken in 1992 by L. T. Wasserthal, who also documented the moth's removing pollinia and pollinating orchids while feeding, as predicted by Darwin.

nectary. The nectaries fill with nectar to attract insects that serve as pollinators (fig. 7). The orchid labium often has ridges that guide the mouthparts of the insects, which are like extendable soda straws, into the nectary and center them under the anther. The anther produces pollen and packages it in a stalked structure called a pollinium. The base of the pollinium is sticky and adheres to the mouthpart of the insect as it feeds. As the pollinium dries, the packets of pollen rotate sideways so that when the insect visits the next orchid flower, it is in a position to transfer pollen to the stigma, the female reproductive organ, and hence fertilize ovules and produce seeds.

The special feature of *Angraecum sesquipedale* is that it has a "whip-like green nectary of astonishing length" (Darwin [1862] 1979, p. 98). The nectaries on Darwin's specimens were 11.5 inches long, but had nectar only in the terminal 1 to 1.5 inches. He postulated that the flower coevolved with an insect pollinator that had a broad-based mouthpart long enough to reach the tip of the nectary, but also that the flower had to force the mouthpart to the base of the nectary to obtain the last bit of nectar in order to be in position for pollinia to adhere to it. The problem with this proposal was that no such insect was known to science. Darwin was vindicated in 1903 when a pollinator with all the predicted properties, a hawk moth named *Xanthopan morgani praedicta*, was discovered. The word *praedicta* was added to the scientific name in honor of Darwin's having predicted its existence from the structure of the orchid it pollinates. Darwin's orchid book and other works suggest many similar scenarios in which there was a "hand in glove" coevolution of flowers and the insects that pollinate them.

When the *Origin* was published, one might have been able to dismiss such scenarios as storytelling. We are now in a much stronger position than Darwin was to illustrate the process of natural selection, since we now have many well-studied examples like those given above, or the Grants' study of the Galapagos finches. Nevertheless, even well-documented examples of change within a species fall short of Darwin's claims for what natural selection can ultimately accomplish. Darwin argues later in the *Origin* that this same process can, given sufficient time, cause the origin of a new species or the evolution of complex structures like our eyes. Here, he foreshadows this argument by drawing an analogy with geology. He argues that before Charles Lyell's *Principles of Geology* (1830–1833), it was popularly thought that the major features of the earth were products of cataclysmic events, such as a single, massive flood carving out the Grand Canyon. No one would have accepted that processes that we can see as part of our daily lives could have shaped such large-scale features of the surface of the earth. By Darwin's time, it had become accepted that the ordinary action of wind, waves, or running water could sculpt all these features, given sufficient time. Likewise, Darwin argues, the same process of natural selection that we can observe making changes over the course of our lives is capable of causing the much larger events that we associate with the history of life.

On the Intercrossing of Individuals

> I have collected so large a body of facts, showing, in accordance with the almost universal belief of breeders, that with animals and plants a cross between different varieties, or between individuals of the same variety but of another strain, gives vigour and fertility to the offspring; and on the other hand, that *close* interbreeding diminishes vigour and fertility; that these facts alone incline me to believe that it is a general law of nature . . . that no organic being self-fertilises itself for an eternity of generations; but that a cross with another individual is occasionally— perhaps at very long intervals—indispensable. (*Origin*, pp. 96–97)

Why are there separate sexes? Not all organisms have them. In Darwin's time (and even today) many people did not realize this. One of Darwin's insights, derived from his barnacle work and his early observations on plants, was that many organisms are hermaphroditic (meaning that every individual has both male and female sexual organs) and that species with separate sexes evolved from hermaphroditic ancestors. He argues here that the advantage of separate sexes is to ensure that the offspring of every generation will be produced by crosses between different individuals, rather than self-fertilization.

Darwin's ideas about sex were inspired by his work on barnacles. Before Darwin, it was thought that all barnacles were hermaphrodites. Adults are permanently attached to a solid substrate, like a rock, the hull of a ship, or the nose of a whale, so they cannot venture forth in search of mates. They compensate for their immobility by being the best endowed of all animals, with penises several times their body length that snake around their surroundings in search of mates, but even such prowess limits them to the hermaphrodite next door. In the absence of available mates, some can fertilize themselves. During his barnacle years, Darwin discovered diversity in barnacle mating systems. Some species in the genera *Ilba* and *Scapellum* have both hermaphrodites and males (fig. 8). The males start life like the hermaphrodites: as larvae that live, grow, and disperse in the open water. Hermaphrodites metamorphose into typical adults that attach themselves to a solid substrate with a stalk, or peduncle. Males instead invade hermaphrodites' protective shells, then attach themselves to the hermaphrodite's body, much as a parasite becomes embedded in its host. Most of their larval organs degenerate so that they become little more than a testis and source of sperm. Males continually rain down on established hermaphrodites. Darwin found multiple males attached to some of them. Dispersing males means having a wider range of mates to choose from than would be the case for those species that rely on roving penises. Some species of *Ilba* and *Scapellum* have gone a step further to enforce outcrossing by replacing hermaphrodites with females. Darwin suggests that what he discovered in barnacles is true for all organisms—separate sexes evolved from hermaphroditic ancestors and did so to ensure that mating is always between individuals.

Most flowering plants produce flowers that have both male and female organs, yet have mechanisms that foster mating between individuals rather than self-fertilization. Darwin found that the flowers of *Lobelia fulgens* re-

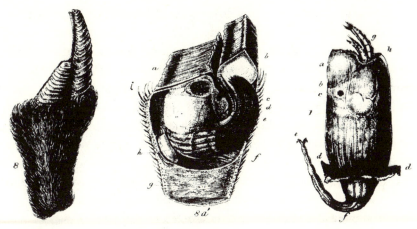

Figure 8
Illustrations of *Ilba cumingii* **from Darwin's**
Monograph on the sub-class Cirripedia **(1851)**
This species of barnacle has separate males and females, rather than males and hermaphrodites (see text). On the left is a female, magnified 4 times in Darwin's original
figure, with an internal view of the female's body to its right showing the attached male
(*h*). The male of *Ilba cumingii* (right) was magnified 32 times in Darwin's original figure. The appendage that swings to the left on the bottom of this drawing (labeled *f*) is
what the male uses to anchor himself to the female—you can see the same appendage
in the internal view of the female's body. In females, this same appendage becomes the
peduncle, or stalk, that attaches the female to a solid substrate.

lease all their pollen from the anthers before the stigma is ready to receive
pollen. Even among flowers that have stigmas and anthers close together
and have pollen ripening at the same time the stigma is ready to receive
pollen, there is a strong tendency for ova to be fertilized by some other
individual.

Darwin describes an experiment in which he reared several varieties of
cabbage together, then reared 233 plants from the seed crop they produced.
In spite of all plants having many flowers that produced a profusion of pollen at the same time the stigmas were receptive, he found that only 78 of the
233 seedlings were "true to kind"; the remainder were hybrids and hence had
to be products of crosses between individuals rather than self-fertilization.
In other experiments, Darwin found that the pollen from a different individual or variety had greater success in fertilizing ova than did pollen from
the same individual, so some mechanism must favor the success of pollen
from different individuals and hence reinforce outcrossing.

Trees present a special challenge to mechanisms ensuring cross-pollination: each tree can produce such a profusion of flowers that a pollinator is much more likely to continue feeding at the same tree rather than move on and carry pollen to a different tree. Some species, like the holly trees in Darwin's garden, have distinct male and female individuals, which ensures that seeds will receive pollen from other trees. Darwin surveyed British plants and found that trees were much more likely than other types of plants to have separate sexes. Joseph Hooker confirmed that this same trend was seen in New Zealand, as did Asa Gray for the United States. Hooker found that, for some unknown reason, this pattern did not seem to prevail in Australia. Our more complete knowledge of plants supports Darwin's observation that trees are much more likely to have separate sexes than other types of plants.

It is this diversity of mechanisms that organisms have evolved to ensure such outcrossing that causes Darwin to infer that "an occasional intercross with a distinct individual is a law of nature" (*Origin*, p. 101) because it improves the quality of the offspring that are produced.

Darwin saw connections between the evidence for the importance of such outbreeding and his own life. He had married his first cousin, which was a common practice in those times, and saw that some of his children suffered from health problems similar to his own. If crosses between unrelated individuals improved the fitness of domesticated plants and animals, and if natural selection had caused the evolution of elaborate mechanisms that ensured crosses between individuals because it improved the "vigor" of the offspring, then the same principle should apply to humans—who were, after all, just another member of the animal kingdom. Brooding about this possibility in the late 1860s as he was writing *The Descent of Man*, Darwin saw an opportunity to obtain an answer to the question of whether or not marriages between cousins resulted in the production of less healthy offspring. He campaigned to add a question to the census of 1871 that asked, simply, whether or not the head of the household was married to his cousin. The answer to this question could then be correlated with the number of children in the family, which would be a combined measure of the joint fertility of the parents and the health of the offspring. He envisioned including these results in his upcoming book. He made use of his connections, which included friends who were members of Parliament, and of his good reputation as a scientist to put the adding of such a census question before Parliament. The request was denied.

Like so many other issues raised in the *Origin*, the evolution and maintenance of sexual reproduction has become a subdiscipline of its own. In the century and a half since the publication of the *Origin*, Darwin's insight has proven to be almost always true, and it is true for the reason posed by Darwin, which is that outcrossing increases the fitness of offspring. There are competing ideas for why the crossing of individuals is important for increasing fitness, but the bottom line that outcrossing is important has prevailed. There are some possible exceptions to Darwin's assertion that all organisms must at least occasionally intercross. The best case has been made for the bdelloid rotifers, which are microscopic multicellular organisms mostly found in freshwater environments. There is genetic evidence that these organisms have reproduced without sex for tens of millions of years. We have yet to figure out how they persist without sex.

Circumstances Favorable to Natural Selection

In "Variation under Domestication" (see chap. 2 above), Darwin summarized the conditions that favor the development of distinct breeds of domestic organisms. Here he extends the analogy between artificial and natural selection by applying those same conditions to the evolution of locally adapted varieties, then species via natural selection.

The first factor that favors the evolution of new varieties is population size. Natural selection acts on heritable variation among individuals. If more individuals are present in a population, there will also be more of this raw material for selection to act upon. Finding a new place in an already packed ecosystem may well require some unusual and rare variation. Any rarity becomes more likely to appear if the population is large. If only one in 1,000 individuals displays some rare trait and there are only 100 individuals in a population, then it is likely that no one will have that trait. If there are 10,000 or 100,000 individuals in the population, then it is likely that multiple individuals will have it.

The second factor that favors the evolution of local varieties is restricted crossing. If a species is distributed over a wide area that incorporates considerable variation in habitat and if individuals move freely between populations, then such movement and interbreeding would limit the degree to which a population could adapt to its local conditions.

It is easy to envision a balance between the effects of the distribution of different types of habitat, the tendency of individuals to move between habitats, and the mode of reproduction in determining the extent to which a population will adapt to local conditions. An organism like a barnacle or a plant may have limited means of dispersal. A population of such a species consists of individuals that are hermaphroditic and capable of self-fertilization, so they do not necessarily cross with a different individual every generation. When individuals do breed, they tend to pair with another from the same vicinity. This combination of traits can lead to populations that are well adapted to a given location. Populations of an organism like a bird will instead often consist of individuals that are long-lived, range widely, and always reproduce by mating with another individual, rather than self-fertilizing. This kind of biology can result in more mixing between populations dispersed over a wide area. A large area of land will in turn often contain a diversity of habitats that grade into one another. Such an organism is less likely to form races that are adapted to a narrow range of local conditions.

Darwin considers it possible for organisms like birds to evolve local races if interbreeding with other members of the species is restricted by some means. A local population might breed at a time of year slightly different from that of other populations, occupy a specific type of habitat, prefer to mate with another of its own variety, or in some other way be restricted from breeding with members of a different variety.

The physical isolation of a population can also promote local adaptation. Because Darwin envisions the process of adaptation as being very slow, being isolated means that a new variety can gradually evolve and improve in its ability to exploit the local environment without being exposed to constant interbreeding with immigrants from a different environment. However, if the isolated population is very small (e.g., on a small island), then its rate of evolution will be limited, because there will be little variation for selection to act upon.

The ability of organisms to become adapted to a local environment is thus determined by a combination of population size, physical isolation, the mobility of the species, and its mode of reproduction.

Darwin closes this section with a reconsideration of the very slow rate at which he expected natural selection to occur. If evolution is caused by changes in the physical environment, then it will be slow because those changes happen slowly. Its rate will also be governed by the availability of variation in the population experiencing such change and by the degree to which the population is isolated from immigration. A continuing influx of migrants from different environments could halt adaptation. Selection can instead be caused by changes in any species participating in a complex ecological interaction. Such change will cause selection on all other parties to that interaction, but adaptation will again be governed by the availability of variation and intercrossing with migrants. "Many will exclaim," Darwin acknowledges, "that these several causes are amply sufficient wholly to stop the action of natural selection. I do not believe so. On the other hand, I do believe that natural selection will always act very slowly, often only at long intervals of time. . . . I further believe that this very slow, intermittent action of natural selection accords perfectly well with what geology tells us of the rate and manner at which the inhabitants of this world have changed" (*Origin*, pp. 108–9).

Darwin is arguing for an ongoing process, yet also one that has never been directly observed. He reconciles the apparent contradiction between pervasiveness and invisibility by once again drawing an analogy with Lyell's view of geology, which is that imperceptibly slow processes can cause large changes if given sufficient time to act. The Grants' work on the Galapagos finches has already shown us that Darwin seriously underestimated the potential speed of evolution by natural selection. I will present additional examples later.

In the remainder of this chapter, Darwin further develops his argument about extinction and the divergence of character. He introduced these concepts at the end of the previous chapter. He also considers the longer-term consequences of evolution by natural selection, which include the origin of new species and then the origin of higher divisions in the taxonomic hierarchy, such as genera and families. Since these topics are central to the second ("Speciation") and third ("Theory") parts of this book, I will return to his fourth chapter later.

Some Properties of Natural Selection

Here I borrow material from chapter 6 of the *Origin* ("Difficulties on Theory," pp. 200–203) that I think is particularly appropriate for this first exposition of the principle of natural selection.

First, I consider what natural selection cannot do. Natural selection cannot modify a species purely for the benefit of some other species. We often see in nature that one species profits from others. Some birds and insects feed on the nectar produced by plants. This does not mean that plants produce nectar to provide for those that consume it. They instead evolved flowers with nectar to attract birds and insects to serve as pollinators. Plants that had these properties gained more efficient sexual reproduction, which in turn gave them a selective advantage over ancestors that lacked flowers and nectar. Darwin adds: "If it could be proved that any part of the structure of any one species had been formed for the exclusive good of another species, it would annihilate my theory, for such could not have been produced through natural selection" (*Origin*, p. 201).

Natural selection can act only on differences in fitness between individuals and hence can benefit only individuals and their offspring. We will see later that there is some possibility for variation on this theme of individual advantage. Traits can evolve to benefit relatives, but never other species.

Natural selection can never cause the evolution of a trait that is injurious to its possessor. Some traits might be associated with a mixture of costs and benefits, but Darwin argues that an appropriate analysis would reveal that the benefits outweigh the costs and that natural selection will, over time, reduce the costs. One example of such a balance of costs and benefits cited by Darwin is artificial selection for short beaks in short-faced tumbler pigeons. Breeders of these birds found that many soon-to-hatch young did not have sufficient beak development to be able to break out of the egg. The breeders had to break the eggshells for them in order to continue selection for shorter beaks in the adults. Under natural selection, there would be a balance between the benefits associated with the evolution of short beaks in adults and the costs of reduced beak development before hatching. Short-beaked adults would evolve only if the costs were less than the benefits incurred over the lifetime of the individual.

Natural selection will not produce perfection. It will only result in descendants that are more successful in the "struggle for existence" than were their immediate ancestors. This evolution occurs in the context of a community of biological interactions, so it will be appropriate only to that community at that time and will not necessarily translate into success in any other context. Recall the evolution of beak sizes in Galapagos finches. In a dry year, natural selection favored the evolution of larger beak sizes. During the El Niño event that followed, it favored the opposite, or the evolution of shorter, narrower beaks. Most of evolution by natural selection is a tracking of these more ephemeral features of the environment and will not lead to continuing change in a particular direction or to the formation of new species. While this is a more modern perspective on evolution by natural selection that recognizes its operation on a finer scale than recognized by Darwin, his conclusions are the same; most evolution leads to local adaptation but goes no further. It is only in special circumstances that this process leads to the bigger events that we associate with evolution, such as the evolution of new species or the evolution of complex, highly specialized structures like the eye.

There is a second reason why natural selection will not produce perfection. François Jacob described evolution as a process of tinkering rather than engineering. An engineer designs something from scratch according to some ideal, then assembles it from new materials. A tinkerer instead makes do with whatever material is at hand and modifies it to suit current needs. Evolution is tinkering because it involves modifying an existing organism for some new set of circumstances. Each such adaptation has the property of being better than what was present in its ancestor and "good enough" for its present function.

Darwin presents many examples of this process of tinkering and recycling. In many cases an organ serves one function in one type of animal but a different function in others. Fish have a gas-filled bladder in their body that, in some, serves as an auxiliary to the gills for obtaining oxygen. In others, this same bladder helps the fish maintain neutral buoyancy as it moves up and down in the water column: gas can be added or expelled to either increase or decrease the buoyancy of the fish. Darwin postulated that an organ that had originally evolved to maintain buoyancy was co-opted to supplement gas exchange, but we now think it happened the other way around. Either way, a gas-filled bladder has served different functions in different fishes. This kind of pervasive recycling, or co-option of organs for different functions in different organisms, is seen throughout the animal and plant kingdoms, prompting Milne Edwards to say "nature is prodigal in variety, but niggard in innovation" (*Origin*, p. 194).

Elsewhere in the *Origin*, Darwin emphasizes the imperfection of the products of natural selection to differentiate them from what we would expect from special creation. If organisms were specially created, they should be perfectly suited to their environment. A host of phenomena tells us, however, that they are only "good enough."

Evolution Today: Natural Selection after the Modern Synthesis

Darwin defined evolution as adaptive change through time and defined natural selection as the cause of evolution. The modern synthesis, which began in the 1920s with the invention of population genetics, primarily by R. A. Fisher, J.B.S. Haldane, and Sewall Wright, gave us a different definition of evolution and made natural selection one of four causes of evolution. These mathematical biologists reconciled Mendelian inheritance with evolution by natural selection by representing what Darwin called "individual variations" as differences between individuals in genotype, or the two alleles that each individual possesses at each genetic locus. They defined evolution as a change in the frequency of an allele at a given locus over time and defined mutation, migration, and genetic drift as the three additional causes of evolution. Scientists today see evolution as happening in the same way that Darwin did, which is at the level of the population. Like Darwin, they envision a species as being an aggregate of populations that are connected by the migration of individuals from one population to another.

Mutation refers to a change in the sequence of the nucleotides (building blocks) that make up DNA molecules. It is this sequence that defines the genetic "code." Mutations in the DNA sequence are most often caused by errors made when these complex molecules are replicated during the process of cell division. When such mutations occur in the cells that produce eggs or sperm, they can be transmitted to offspring. Since mutations create new alleles, they cause a change in allele frequency when they are transmitted to offspring, so they are a mechanism of evolution. More importantly, mutations are the ultimate source of the new alleles and hence the individual variations that are the fuel for evolution caused by migration, drift, and natural selection.

Migration is a factor in evolution if the individuals that migrate into a population cause a change in allele frequency. This is especially likely if immigrants come from a population that is different in genetic composition. Darwin saw migration as important because he thought that adaptations would first appear in a local population, then spread as individuals migrated to new areas. He also thought that highly mobile species would be less likely to adapt to local environments because individuals from different regions would constantly mix and interbreed. Both of these ideas appear in the *Origin*, and both became formalized in the later development of population genetics. Researchers often refer to this phenomenon as "gene flow" rather than migration. The distinction is that an individual must first migrate to a new population, then must successfully reproduce so that its genes become incorporated into the population.

The final mechanism of evolution is genetic drift, which is a change in allele frequency that occurs by chance, or sampling error, in small populations. I teach this concept in my evolution class by bringing in a bag that contains an equal number of kidney and pinto beans. I pick out beans at random, then give four beans to each student and ask him or her to imagine that the beans represent the male and female founders of a new population. Each founder will have two alleles at each locus, so the two founders together carry four alleles, represented by the four beans. We then tabulate what the genetic composition of each new population will be. In the big bag of beans, or the original population, the two alleles were equal in frequency. If the beans are taken at random from the bag, then approximately three-eighths of the new populations will also have two each of the different types of beans. Around half of the new populations will have three of one type of bean and one of the other. Around one-eighth of the new populations will have only one color of bean. If these beans represented a male and female emigrating from some large population to start a new population elsewhere, then many of the new populations would have a genetic composition very different from that of the parent population. Such differences between new populations and the parent population happen by chance alone, but they fulfill the formal definition of evolution because there has been a change in allele frequency.

The change in allele frequencies caused by the establishment of a new population is a special case of genetic drift called the "founder effect." Such random variation in allele frequencies also happens in established populations if they have few individuals. If our new populations continued to replace themselves, generation after generation, with just one male and one female offspring, then they would soon all have only one allele at the locus in question, but around 50% would have just the allele represented by the kidney bean and 50% would have just the allele represented by the pinto bean. This form of evolution is not necessarily adaptive. Whether or not an allele is retained or lost by this process has nothing to do with whether it benefits or harms the individual.

The influence of genetic drift declines as population size increases. We can visualize this relationship with computer simulations of populations, each represented by one locus with two alleles, A1 and A2. The two alleles are equally common (each comprises 50% of the population) at the beginning of the simulation, but they vary their frequency, or relative abundance, over time through the same sort of random sampling process as my blindly pulling beans out of the bag. One simulation (fig. 9, upper panel) models eight populations that each consist of 4 individuals. After 20 generations, all the populations contain only one allele—five contain only allele A1, while three contain only allele A2. Another simulation (middle panel) models the same process for eight populations of 40 individuals. After 100 generations, four of the populations still have both alleles, three contain only allele A2, and one contains only allele A1. A third simulation (bottom panel) repeats the process for eight populations of 400 individuals. After 100 generations, both alleles are still present in all eight populations, although some of them have drifted far from a 50:50 distribution.

Population geneticists then considered how natural selection, mutation, migration, and drift interact with one another. One question was to ask what the relative importance of natural selection and genetic drift might be in populations of different sizes. Drift dominates natural selection when populations are very small. These populations will evolve, but evolution may actually be damaging because beneficial alleles are lost. We see such damaging evolution in some endangered species that persist only as small, isolated populations. In larger populations, the influence of drift wanes, and natural selection will prevail, so evolution will be adaptive.

A second question was to ask what the fate of a new mutation would be. Most new mutations are soon lost regardless of whether they harm or benefit the individual that carries them. The reason is that mutations first appear as a single copy. Whether or not that first copy is transmitted to offspring is much more a matter of chance than of the harm or benefit it causes. If, by chance, the original mutation becomes more common over the next few generations, it then becomes more likely that it can be "seen" and acted upon by natural selection. Mutations tend to accumulate over time; their abundance will be a function of how they affect the quality of the individual. Even potentially lethal mutations can be found in natural populations. It is estimated that all of us have a few of them. They can persist as long as they are recessive and very rare. Their presence is why breeding with close relatives can be a dangerous practice, since close relatives are much more likely to share such damaging mutations and hence to produce homozygous offspring that will suffer the ill effects. Seeing such unhealthy offspring is probably the source of the taboo against mating between close

(a) Population size = 4

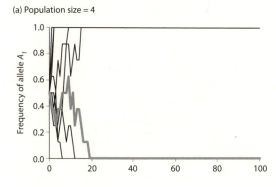

(b) Population size = 40

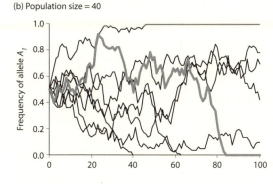

(c) Population size = 400

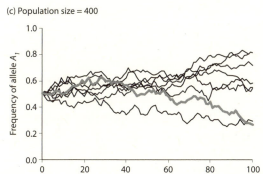

Figure 9
Genetic Drift
The three panels illustrate the fate of allele A1 in simulations of randomly mating populations of different size. The y axis represents the proportion of allele A1 in the populations. The x axis represents the number of generations of random mating. The frequency of A1 is 50% at the beginning of all simulations. See text for details.

relatives in many human societies. Darwin recognized this hazard as well, even if he could not identify the cause, which is why he tried to get the Parliament to add a question about marriages between cousins to the 1871 census. Damaging mutations will remain uncommon unless they occur in small populations and are given a boost by genetic drift. Beneficial mutations in large populations will tend to come under the influence of natural selection and as a result increase in frequency.

We are able to find multiple alleles segregating at most loci in large populations because of the gradual accumulation of new mutations. This is one important source of the variation that is available for natural selection to act upon. The way an allele affects the fitness of the individual that bears it depends on the environment. An allele that is damaging or neutral in one environment could become beneficial in a different environment, so a change in conditions could initiate an episode of adaptive evolution as a once rare allele increased in frequency. The "individual variations" in domestic organisms described by Darwin were very likely to be rare, recessive alleles that became homozygous as a consequence of inbreeding. Even if Darwin misinterpreted the cause of the variation (he thought that it was environmentally induced), his conclusion that the variation was random in nature was correct. The constant presence of such genetic variation in natural populations decouples mutation from evolution. Evolution by natural selection is not simply a matter of waiting for the right mutation to come along and then capturing it. It instead is a process that capitalizes on genetic variation already present in the population. This interpretation of evolution is quite consistent with Darwin's vision that evolution capitalizes on individual variations and that such variation is more likely to be found in large rather than small populations.

References

Browne, Janet. 2002. *Charles Darwin: The power of place*. Princeton, NJ: Princeton University Press. (Pp. 165–80, details of Darwin's orchid studies; pp. 326–69, details on the writing of *Descent of Man* and *Expression of Emotions*; pp. 326–28, Darwin's attempt to add a question to the census of 1871.)

Darwin, C. 1851–1854. *A monograph on the subclass Cirripedia*. Vol. 1: *The Lepadidae*. London, 1851. Vol. 2: *The Balanidae*. London, 1854.

Darwin, C. [1862] 1979. *Fertilization of orchids by insects*. Reprint of the first edition published in 1862 by J. Murray, London, under the title *On the Various Contrivances By Which British and Foreign Orchids Are Fertilized By Insects, and On the Good Effects of Intercrossing*. Earl M. Coleman, publisher, Conklin Hill Road, Stanforville, NY 12581.

Huxley, T. H. [1863] 2001. *Man's place in nature*. New York: The Modern Library.

Jacob, F. 1977. Evolution and tinkering. *Science* 196:1161–66.

Villinski, Jeffrey T., Jennifer C. Villinski, Maria Byrne, and Rudolf A. Raff. 2002. Convergent maternal provisioning and life-history evolution in echinoderms. *Evolution* 56:1764–75.

Welch, D. M., and M. Meselson. 2000. Evidence for the evolution of bdelloid rotifer without sexual reproduction or genetic exchange. *Science* 288:1211–15.

Chapter 6

Laws of Variation

Variation among individuals is the fuel for evolution. If there are no differences between individuals in traits that affect their survival and that are inherited by their offspring, then evolution cannot happen. For this reason, Darwin was interested in knowing where variation came from, how it was maintained, and if there were any patterns to be found in the occurrence of variation in nature. It is here that we see him confront some of the greatest difficulties caused by his ignorance of how inheritance works. We can also appreciate how difficult it is to see Mendelian inheritance when manifested in real organisms. Even though the basic rules of inheritance in peas are universal, many steps can lie between the action of a single gene and its effect on the phenotype (observable traits) of real organisms, even when the traits being observed are as simple as flower color or stem height. These added steps between genes and phenotypes are what made it so difficult to define the laws of inheritance. Darwin tried as hard as anyone before him to understand the laws of inheritance as he bred pigeons or discussed animal breeding with experts. His persistent difficulties, combined with his excellence as an observer and scientist, illustrate how hard genetics really is and why Mendel's work lay unappreciated for thirty-five years. Even after the rediscovery of Mendel's laws, an additional thirty years or so passed before we could reconcile the laws of inheritance with evolution by natural selection.

It is also in this chapter that we can see some of the fuel for what would become an epic conflict between Darwin and one of his former mentors. Richard Owen was only five years older than Darwin, but was already emerging as England's preeminent anatomist by the time of Darwin's return from the voyage of the *Beagle*. He had just been appointed the Hunterian Professor at the Royal College of Physicians while still in his early thirties. He was introduced to Darwin at a party organized by Charles Lyell shortly after Darwin's return from the *Beagle* voyage. Owen and Darwin became close friends and scientific colleagues. Owen described all of Darwin's fossil mammals and authored the fossil mammal contributions to the zoology volumes published after the voyage of the *Beagle*. He also played a key role in inspiring Darwin to undertake his epic study of the systematics of the barnacles. Darwin had set out to describe a single, aberrant species that he found on the voyage, but Owen suggested that he expand his study into a broad comparison of barnacles and a consideration of their common structure. Owen had adopted the concept of an archetype, or ideal body plan, that was common to all organisms in a taxonomic category. He argued that evidence for such an archetype should be apparent in the barnacles. Owen had also proposed many subsidiary laws of anatomy. In this chapter, Darwin likely put a few small bugs up Owen's nose by arguing that natural selection can explain some of these subsidiary laws. The full scope of their disagreements goes far beyond this chapter.

Here we also see Darwin's application of William Whewell's philosophy of science. Whewell argued that the true test of the worth of a theory was its ability to unite a diversity of phenomena under a single explanatory framework (his "consilience of inductions"—see introduction). Here and throughout the *Origin*, Darwin argues that evolution by natural selection is such a theory. In this chapter, Darwin argues that patterns of variation in nature that were reported by others can all be understood as being products of natural selection. Darwin's more general message here and throughout the *Origin* is that there is only one true law of the life sciences. That law is evolution by natural selection.

Because of Darwin's lack of understanding of inheritance, however, many portions of this chapter are archaic. For the sake of brevity, I will present synopses of only some parts of the chapter. I have chosen those topics that illustrate some of his correct inferences in spite of his not knowing how inheritance works, and also those that give me an opportunity to show how Darwin's puzzles led to later solutions.

Effects of External Conditions

First, Darwin addresses where variation comes from and repeats the argument from his first chapter: variation (he believed) is induced by changes in the environment. He argues that the reason domestic organisms are so much more variable than wild organisms is that variation is induced by the conditions of domestication: "the much greater variability, as well as the greater frequency of monstrosities, under domestication or cultivation, than under nature, leads me to believe that deviations of structure are in some way due to the nature of conditions of life" (*Origin*, p. 131).

Darwin's observations are accurate—domestic animals are often more variable than wild animals. However, the greater variation under domestication is caused by factors that could not have occurred to him. As we argued in chapter 1, domesticated organisms are often inbred. Such inbreeding causes rare, recessive alleles that would rarely or never be expressed in a natural population to be expressed when close relatives are mated with one another in domestic populations. A second possible reason for seeing more variation under domestication is that a wild individual with a harmful trait will quickly die, but might have persisted and been seen in the more sheltered arena of domestication.

Variation among individuals within a population can lead to differences between populations. Some trends of differentiation between populations had been reported by naturalists of Darwin's time. The great ornithologist John Gould "believes that birds of the same species are more brightly coloured under a clear atmosphere, than when living on islands or near the coast" (*Origin*, p. 132). When widespread species that show such differences between populations overlap with species that have narrower ranges, they tend to have the same character as the species with narrower ranges. So, for example, species of shelled organisms that are found only in more equatorial regions or shallower water tend to be more brightly colored than species found exclusively in deeper water or higher latitudes. E. Forbes reported that, in species that are found across a range of depths and latitudes, the shells of populations found closer to the equator or in shallow water are more brightly than those found in higher latitudes or deeper water. This parallel involving differences between populations of a widespread species and differences between species with narrow ranges suggests that the differences between local populations of the widespread species represent local adaptations.

Darwin speculates that it is possible for such matches between the appearance of a local population and the local environment to be caused by plasticity; the constitution of the species could be the same everywhere, but the local environment induces some change in its appearance. Darwin notes that many species are known to vary, but some remain the same in very different climates or vary considerably in the same climate. Because variation may or may not be correlated with the environment experienced by each population, Darwin reasons that the importance of such direct action of the climate is minor. He argues instead that these differences between populations are likely to be heritable and the product of natural selection, rather than being nonheritable and induced by the environment.

Effects of Use and Disuse

Richard Owen observed that "there is no greater anomaly in nature than a bird that cannot fly" (*Origin*, p. 134). Yet birds with reduced wings or no wings at all are often found on islands where there is no risk of predation. Since many birds use their wings primarily to escape predators, Owen and others attributed such losses to disuse. Likewise, many animals that live in caves or underground either have reduced eyes or have completely lost their eyes. Again, disuse was an attractive explanation for the loss of eyes.

Darwin takes issue with Owen and others and argues that more than a lack of use is required to cause the degeneration or loss of a structure. He argues instead that such losses are a direct consequence of natural selection. A structure that is not used can become a liability. He cites his experience with tuco-tucos during the voyage of the *Beagle*. Tuco-tucos are burrowing rodents with reduced eyes that are found in Patagonia. Darwin found that his pet tuco-tuco was blind. When he dissected it, he found that the nictitating membrane, a movable membrane found under the eyelid of some animals, was inflamed. Any such inflammation, and possible infection, would be injurious to an animal, so Darwin postulated that natural selection might first favor individuals with reduced eyes, then those whose eyes were protected by "the adhesion of the eye-lids and growth of fur over them" (*Origin*, p. 137), since such changes would make them progressively less susceptible to infections. If having an unused eye is a liability, then any individual that for some reason has less of an eye will have a selective advantage. The difference between Darwin's argument versus prior arguments concerning the role of disuse is that his predecessors felt that disuse alone would cause a character to deteriorate. Darwin instead thought that natural selection caused this deterioration because individuals with reduced development of the organ would have a selective advantage over those without such reductions.

We see a similar loss of eyes in organisms that adapt to living in caves in different parts of the world. The depths of limestone caverns that are the home to blind cave organisms in Europe and America are very similar habitats. If these organisms were the product of special creation, then we might expect the blind cave animals of Europe to be the same as those found in North America. We instead see that the blind cave animals in Europe are most closely related to animals found in the surrounding countryside than they are to other blind cave organisms in North America. Likewise, the cave animals of North America are most similar to species found nearby. Furthermore we find, as we descend from the surface into the depths of caverns on either continent, that they contain organisms that display a similar gradient of adaptation to darkness. Darwin cites Schiadöte as saying that "animals not far remote from ordinary forms, prepare the transition from light to darkness. Next follow those that are constructed for twilight; and, last of all, those destined for total darkness" (*Origin*, p. 138). Such patterns suggest that surface-dwelling animals on each continent invaded the caverns and adapted first to life in the twilight near cave openings, then to the total darkness deeper inside.

Modern research carries the argument for the role of natural selection in the evolution of cave organisms a step further than was possible for Darwin. The animals found in individual caves can represent independent occurrences of local adaptation. Kane, Culver, and Jones (Jones et al. 1992; Kane et al. 1992) used molecular genetic markers to compare the relatedness of populations of the amphipod (a type of small crustacean) *Gammarus minus* found in springs on the surface and in nearby caves in North America. Verovnik, Sket, and Trontelj (2004) used similar methods to study surface and cave populations of a small isopod crustacean (*Asellus aquaticus*) in cave networks found in northern Italy and Albania (fig. 10). In both cases, the cave populations had reduced eyes and larger antennae in comparison to those found on the surface, and had a variety of other adaptations to life underground. Antennae are a substitute for eyes in a perpetually dark habitat because they enable animals to probe and sense their surrounding environment. In both species, the investigators found cave populations that were genetically more similar to nearby surface populations than they were to other cave populations. This pattern suggests that individuals from the nearby surface populations invaded each cave system. Each time this happened, natural selection favored the evolution of reduced eyes and increased antenna length as adaptations to the perpetual darkness. Darwin saw that adaptation to the caves of Europe and North America represented independent events; we can now see that such adaptations can happen in-

Figure 10
Surface vs. Cave Populations
Two individuals from a cave-dwelling population of the isopod crustacean *Asellus aquaticus* are head to head on the right side of this photograph, and one individual from a nearby surface population is on the left. The differences in pigmentation are obvious; less obvious but visible differences are the reduced eyes and the elongated antennae and other appendages in the cave-dwelling individuals. All these appendages provide sensory information in place of lost vision.

dependently much more often than Darwin suspected and on a much finer geographic scale. Even caves that are only a few miles apart can contain animals that are adapted to cave life but whose adaptations were independently derived from traits of the surface populations close to each cave.

Acclimatization

Here Darwin presses his argument that biotic interactions, or interactions with other organisms, are the most important agents of natural selection, even if casual observation suggests instead that the physical environment is more important.

Closely related species that share a common ancestor are often found in different climates and frequently show what seem to be adaptations to their respective habitats. Darwin's colleague Joseph Hooker collected seeds from pines and rhododendrons at different elevations in the Himalayas, grew some of those seeds at Kew Gardens, and found the resulting plants to "possess different constitutional powers of resisting cold" (*Origin*, p. 140) that were well correlated with how cold their climate of origin was; however, Darwin argues that "the degree of adaptation of species to the climates under which they live is often overrated" (*Origin*, p. 139). In Darwin's view, even though some influence of climate can be found, the biotic environment plays a far more important role in shaping how organisms evolve. He repeats his observation that it is possible to transport plants and animals to very different climates and then successfully breed them. All domestic plants and animals can be found growing and reproducing in a wide range of environments. They almost certainly were originally chosen for properties other than their ability to live in those climates. Their tolerance of a wide range of climates is instead a constitutive property of most organisms. Rather than being an adaptation to climate, the ranges of most species are instead shaped by the other species with which they interact.

Darwin notes that gardening manuals often instruct that certain varieties will do best in specific climates, which seems to argue for local adaptation to the physical climate. He remains skeptical and suggests that, if one really wanted to prove the source of such apparent adaptations, then a selection experiment would be in order. "Until some one will sow, during a score of generations, his kidney-beans so early that a very large proportion are destroyed by frost, and then collect seed from the few survivors, with care to prevent accidental crosses, and then again get seed from these seedlings, with the same precautions, the experiment cannot be said to have been even tried" (*Origin*, p. 142).

Darwin concedes that various features of "habit," by which I assume he is referring to an organism's ability to be plastic and adjust to different climates, may play some role in enabling organisms to survive in a given area. However, he concludes that all such environmental influences "have often been largely combined with, and sometimes overmastered by, the natural selection of innate differences" (*Origin*, p. 143).

Correlation of Growth

Referring to "correlation of growth" in an organism, Darwin explains: "I mean by this expression that the whole organization is so tied together during its growth and development, that when slight variations in any one part occur, and are accumulated through natural selection, other parts become modified" (*Origin*, p. 143). He wrote this section as a preemptive strike against those who would argue that such correlations mean that these traits are obligately linked and hence represent a law of variation. Darwin argues instead that each trait can be independently shaped by natural selection, in spite of its correlation with other traits.

First Darwin considers cases where it seems that correlations of traits might be obligatory. One such class of correlations is adaptations in the larval life stage that affect the structure of the adult. A second class might be when different parts of the body vary in a similar fashion, such as in the symmetry of the right and left sides of the body. However, Darwin argues that even bilateral symmetry is a correlation that can be modified by natural selection. He reports on a family of stags that had an antler on only one side. If this were favored by natural selection, then he would expect such asymmetry to become characteristic of whole populations or species. Asymmetry rarely evolves, but not because it cannot evolve. One example from my own research is a fish called *Xenodexia ctenolepis*, which is a live-bearing relative of guppies that is found in remote parts of Guatemala. Males of this species have a normal left pectoral fin and a right pectoral fin that is longer, more muscular, and has a notch at the base of the fin. The anal fin is modified to function as an intromittent organ that inserts packets of sperm into the female. The anal fin is also asymmetrically twisted to the right side. It has been suggested that the pectoral fin functions as a clasper that holds the female in place on the male's right side; then the anal fin swings to the right to inseminate her.

Darwin argues that correlations may represent traits that functioned together but evolved independently. For example, Alph. de Candolle observed that winged seeds are found only in plants that have fruits that open to disperse seeds. Darwin notes that having fruits that open is a necessary precursor to evolving mechanisms for the dispersal of independent seeds. If fruits did not first open to release seeds, then there could be no selection for seeds with wings that enhance dispersal. These two traits are found together because one must be present before natural selection can cause the evolution of the second trait.

J. W. Goethe, who is better known as a poet, novelist, and philosopher, suggested that characters are correlated because of an economy of nature: "in order to spend on one side nature is forced to economize on the other side" (*Origin*, p. 147). Darwin observes that there are many apparent examples of such trade-offs, but this does not necessarily mean that the reduction of one trait was causally linked to the elaboration of another. He suggests instead that "natural selection will always succeed in the long run in reducing and saving every part of the organization, as soon as it is rendered superfluous, without by any means causing some other part to be largely developed in a corresponding degree" (*Origin*, p. 148).

Darwin illustrates his argument with observations from his work on *Proteolepas*, a parasitic form of barnacle. The heads of nonparasitic barnacles contain three anterior segments with well-developed nerves and muscles that perform vital functions, since they include antennae and organs associated with obtaining and processing food. In this parasitic species "the whole anterior part of the head is reduced to the merest rudiments" (*Origin*, p. 148) that serve as organs of attachment to the host without any compensatory elaboration of some other part of the body. Conversely, he argues that natural selection can cause the evolution of increased development in any organ should it be advantageous to the individual "without requiring as a necessary compensation the reduction of some adjoining part" (*Origin*, p. 148). Darwin thus argues that such apparent correlations are again products of selection acting independently on each of the traits in question rather than being a consequence of a trade-off in resource allocation, such that devoting more resources to one trait is gained at the expense of another.

Darwin's take-home message is not that true correlations of traits do not exist, but rather that they should not be accepted uncritically as a measure of a causal relationship. The more important take-home message is that, when Darwin argued for each trait being independently honed by natural selection, he was arguing for the predominance of natural selection in shaping all features of organisms.

Why Some Characters Are Highly Variable

Darwin's original subheading here is "A part developed in any species in an extraordinary degree or manner, in comparison with the same part in allied species, tends to be highly variable." All the elaborate examples in this section share, in Darwin's view, a single explanation that draws on his continuing comparison between artificial and natural selection. I present that explanation first, before describing select examples. Darwin observes that when a breeder begins to select for the elaboration of a particular trait in some domestic organism, there is initially considerable variation in the expression of that trait. If the breeder persists in selecting on the trait for several generations, then the amount of variation declines and individuals of the developing breed became more uniform in appearance; however, active selection is required to maintain this uniformity. If the breeder fails to maintain selection, then the breed will once again become variable. Likewise, Darwin argues that traits that are highly variable in nature are subject to weak selection or have only recently become subject to strong selection. Selection has not yet persisted long enough to render the population uniform in character. Conversely, traits that vary little in nature have been subject to intense selection for a prolonged period of time.

Owen observed that the extremely long arms of the orangutan were quite variable in length. Darwin agrees with Owen's observation and considers it to be generally true that when individual species have exaggerated structures in comparison to their close relatives, those structures tend to be unusually variable. However, the wing of the bat is also highly aberrant relative to forelimbs of other vertebrates, yet it is quite uniform in structure among the many species of bats. Darwin argues that the difference between the forelimbs of orangs versus bats lies in how recently each became subject to intense natural selection. The arms of orangutans are variable because they have not been under selection for a prolonged period of time. The wings of bats are not variable, because they are the product of selection that began a long time ago and has persisted long enough for the common ancestor of all bats to diversify into a large number of species that share the trait.

Secondary sexual characters also tend to be highly variable. These are features that characterize only one sex, such as the elaborate tail that is seen in male peacocks but not females. In the previous chapter, Darwin argued that sexual selection is weaker than natural selection. He proposes here that the greater variation in these traits is a product of their being subject to less rigorous selection.

Darwin sees an extension of this principle to the characters that are used for classifying species. Characters that are used to distinguish closely related species from one another are more variable than those that are used to distinguish genera. The traditional classification of organisms is hierarchical, so that species that are closely related to one another are grouped into the same genus. Closely related genera are in turn contained within the same family, families are contained within orders, and so on. We shall see later that Darwin interprets this organization as being a product of how natural selection causes the formation of new species. For now, it is important to accept that all species in the same genus are derived from a single common ancestor while all genera in the same family are also derived from a single common ancestor, but one that long predates the common ancestor of species within a genus. Because a species has a more recent origin than the genus that it is part of, characters that distinguish one species from another have been subject to natural selection for a shorter interval of time than those that distinguish genera from each other. The tendency of specific characters to be more variable than generic characters is thus a signature of their having been under selection for a shorter interval of time.

In summary, the degree to which characters vary will be a product of the intensity of selection that they are subject to and to the duration of that selection. Weak selection, as in the case of sexually selected traits, will always result in traits being more variable. Intense selection will initially cause an increase in variation but will, in the long term, cause a reduction in variation. Traits that vary little may be ones that inherently lack variation for some reason, but are also likely to be ones that have been under intense selection for a prolonged period and are still subject to strong selection.

Darwin argues that all these patterns, as diverse and unconnected as they may seem, share a common explanation:

> Finally, then, I conclude that the greater variability of specific characters, or those which distinguish species from species, than of generic characters, or those which the species possess in common;—that the frequent extreme variability of any part which is developed in a species in an extraordinary manner in comparison with the same part in its congeners; and the not great degree of variability in a part, however extraordinarily it may be developed, if it be common to a whole group of species;—that the great variability of secondary sexual characters, and the great amount of difference in these same characters between closely allied species . . . are all principles closely connected together. All being mainly due to the species of the same group having descended from a common progenitor, from whom they have inherited much in common,—to parts which have recently and largely varied being more likely still to go on varying than parts which have long been inherited and have not varied,—to natural selection having more or less completely, according to the lapse of time, overmastered the tendency to reversion and to further variability,—to sexual selec-

tion being less rigid than ordinary selection,—and to variations in the same parts having been accumulated by natural and sexual selection, and thus adapted for secondary sexual, and for ordinary specific purposes. (*Origin*, p. 158)

Conversely, under the alternative theory that all species are independently created, no such patterns should exist. (Incidentally, whenever I or others quote Darwin, we are inclined to cite his clearest and most eloquent passages. I thought that it would be constructive to present some of his thornier prose, since it helps to illustrate some of the difficulties of the *Origin*.)

Atavisms and Analogous Variations

Darwin's original subheading here is "Distinct species present analogous variations; and a variety of one species often assumes some of the characters of an allied species, or reverts to some of the characters of an early progenitor." This very elaborate subheading refers to two alternative interpretations of peculiar patterns of inheritance. One is what we now call "atavisms," or "the reappearance in an individual of characteristics of some remote ancestor that have been absent in intervening generations" (*Webster's Encyclopedic Unabridged Dictionary of the English Language*, 1989). The second, which Darwin calls "analogous variations," refers to the independent appearance of similar traits in closely related species. Darwin argues that both of these phenomena can be explained by his theory of evolution by natural selection.

Darwin first draws examples from domestication. Within all breeds of pigeons we see "the occasional appearance . . . of slaty-blue birds with two black bars on the wings, a white rump, a bar at the end of the tail, with the outer feathers externally edged near their bases with white. As all of these marks are characteristic of the parent rock-pigeon, I presume that no one will doubt that this is a case of reversion" (*Origin*, pp. 159–60).

Such apparent reversions are also seen in the offspring of crosses between two very different breeds of pigeons, neither of which has any of the apparently atavistic characters observed in the offspring. In chapter 1, I explained that the prevailing view in Darwin's time was that inheritance was a blending of the genetic contributions of each parent. Blending means that the contributions of each parent are inextricably mixed. If two breeds have been propagated separately from one another for twelve generations and then a male from one breed is mated to a female from the other, Darwin calculates that they will only have one part in 2,048 (2^{12}) of their blood inherited from their common ancestor, "and yet, as we see, it is generally believed that a tendency to reversion is retained by this very small proportion of foreign blood" (*Origin*, p. 160). Darwin argues instead that "when a character which has been lost in a breed, reappears after a great number of generations, the most probable hypothesis is, not that the offspring suddenly takes after an ancestor some hundred generations distant, but that in each successive generation there has been a tendency to reproduce the character in question" (*Origin*, pp. 160–61). He thus appears to argue that something other than blending inheritance must account for such reversion, since the latent ability to express some characters is retained and faithfully transmitted from generation to generation, even if it is not expressed.

He offers a second explanation for the appearance of the same trait in two closely related species, which is that the trait is analogous, or arose independently in each lineage because they inherited similar constitutions from a common ancestor. It is often not possible to distinguish between these two explanations, because the appearance of the ancestor is not known. Each explanation is consistent with Darwin's ultimate interpretation of the phenomenon, which is that each is a product of the shared ancestry of the breeds or species in question.

Although Darwin could not offer a specific mechanism of inheritance to explain these phenomena, we now can. In the early twentieth century, W. Bateson and his student R. C. Punnett provided a potential explanation for atavisms with their discovery of "epistasis." Epistasis refers to an interaction between different genes such that the alleles that are present at one gene can mask or modify the expression of alleles present at another gene. In its simplest form, we can picture epistasis as a single molecule that is modified by two genes that act in succession, such that the first gene produces an enzyme that converts some precursor molecule A into molecule B and then a second gene produces an enzyme that converts molecule B into molecule C.

$$A + (\text{gene 1 enzyme}) \rightarrow B + (\text{gene 2 enzyme}) \rightarrow C$$

If an individual has functional copies of both genes, then the phenotype of the individual is C. If gene 1 is functional (converts A to B) but gene 2 for some reason cannot convert B to C, then the phenotype is B. If gene 1 cannot convert A to B but gene 2 can convert B to C, then the phenotype is A;

gene 2 never encounters any B to convert to C, so gene 2's presence cannot affect the phenotype. The phenotype would also be A if both genes were not functional. Gene 1 is said to be epistatic to gene 2 because it can preempt the effects of gene 2 on the phenotype; the absence of a functional gene 1 prevents gene 2 from ever having access to molecule B.

Imagine that developing a given phenotype (C) requires the presence of functional copies of genes 1 and 2. The first breed of pigeon might lack a functional copy of gene 2 because it is homozygous (has two identical alleles) for a mutant allele at that locus, which can readily happen as a by-product of inbreeding; so it has phenotype B. The second breed of pigeon lacks a functional copy of gene 1 because it is homozygous for a mutant allele at that locus, but its gene 2 is functional, so it retains the ability to convert molecule B into molecule C. It has phenotype A. When the two breeds are mated, at least some of the offspring will inherit functional copies of both gene 1 and gene 2, so their ability to convert molecule A into molecule B and then B into C will be restored. In this case the offspring will have phenotype C. This means that they have a phenotype that was not present in either breed and may not have been seen in either of them for many generations. Epistasis thus offers a plausible explanation for the spontaneous appearance of an atavism when different breeds are crossed: it could be due to the restoration of a metabolic pathway by a mechanism similar to the one outlined here.

More generally, the genetic basis for any given feature of an organism is usually not so simple as the "one gene–one trait" pattern that we see in Mendel's classic experiments. In Mendel's experiments, one gene determined whether a plant was tall or short or a pea was smooth or wrinkled. If this were true for all traits, then the laws of inheritance would have been more apparent, and we would not have had to wait so long or worked so hard to understand them. We instead find that any given feature of an organism tends to be the product of actions involving a number of genes that can either make independent contributions to the trait or can act in concert with each other via pathways like the simple one illustrated above. Individual genes can also influence multiple traits. When we combine the two phenomena of many genes contributing to each trait and individual genes contributing to multiple traits, we see that the action of an individual gene can be (and, in fact, most often *is*) embedded in a network of interacting genes rather than a "one gene–one trait" relationship. When we add such complexities to inheritance, the basic rules of inheritance can become all but impossible to discern from the kinds of pigeon crosses performed by

Darwin or his contemporaries. Now that we know the fundamental mechanism of inheritance, we can appreciate Mendel's genius in developing such an elegant means for defining the laws of inheritance and also appreciate why it took so long for the value of his work to be realized. He described laws of inheritance that are universal, but are obvious only in the special circumstances that he chose for his experiments.

Darwin then extends his argument to an example that includes natural differences between species, as opposed to differences between breeds of a domestic species. He uses the example of striping in wild and domestic horses to show that atavisms and analogous variations are not just an artifact of domestication. Everyone knows that zebras have stripes; to be more specific, they are striped "all over"—on their bodies and legs. The now-extinct quagga was also striped, but to a variable degree among its different subspecies. Some subspecies were striped on the head, neck, and shoulder region but not elsewhere, while others had stripes that extended from the head all the way to the hips. They had little or no striping on the legs. The Mongolian wild ass ("hemionus" in the *Origin*) may have no or only traces of shoulder stripes as an adult but can have faint shoulder and leg strips as a juvenile. The domestic donkey can have bold transverse bars on its legs and either one or two shoulder stripes. Most domestic horses are not striped but some show traces of the stripes seen in wild species. Duns, which are a breed of domestic horse, may have a faint shoulder stripe and bars on the legs. The Kattywar breed from India had a stripe down the middle of the back, two or three shoulder stripes, and often bars on the leg and stripes on the face. Hybrids between some of these species and domestic horses can exhibit striping that is not seen in either parent: "In Lord Moreton's famous hybrid from a chestnut mare and a male quagga, the hybrid and even the pure offspring subsequently produced from the mare by a black Arabian sire, were much more plainly barred across the legs than is even the pure quagga" (*Origin*, p. 165).

The cross of a domestic donkey with a Mongolian wild ass yielded an offspring with barring on the legs, three shoulder stripes, and zebra-like stripes on the face. Neither parent had either leg or facial stripes or three shoulder stripes. Darwin hints that he has many similar examples but does not have the room to present them, since the *Origin* is just an abstract of his theory. Many did appear later in his book-length treatment of variation under domestication, which was published between the appearance of the first and sixth editions of the *Origin*.

Darwin sees strong parallels between domestic pigeons and horses. The wild ancestor of all domestic pigeons is a slaty blue with other distinct markings. This same background color and these same markings can appear spontaneously in even the oldest breeds and can be generated with crosses between breeds. Striping appears to various degrees in many of the wild equids, but also appears occasionally in domestic breeds. It can also reappear in hybrids in a fashion that is not present in either parent. Darwin argues that we see these patterns in domestic pigeons because they are all descended from the rock pigeon so they all share common properties, even if they are not seen in every generation. Likewise, he argues that we see these patterns of striping in equids because they are all descended from a common ancestor that was striped. This interpretation represents an argument against prevailing doctrine at the time, which was that all species were products of individual acts of special creation:

> He who believes that each equine species was independently created, will, I presume, assert that each species has been created with a tendency to vary, both under nature and under domestication, in this particular manner, so as often to become striped like other species of the genus; and that each has been created with a strong tendency, when crossed with species inhabiting distant quarters of the world, to produce hybrids resembling in their stripes, not their own parents, but other species of the genus. To admit this view is, as it seems to me, to reject a real for an unreal, or at least for an unknown, cause. It makes the works of God a mere mockery and deception; I would almost as soon believe with the old and ignorant cosmologists, that fossil shells had never lived, but had been created in stone so as to mock the shells now living on the sea-shore. (*Origin*, p. 167)

While Darwin's few small disagreements with Owen in this chapter would have likely been minor irritants, this last paragraph would have made Owen's, and many others', blood boil, since Owen and the scientific community as a whole argued that species were the product of individual acts of special creation.

Darwin summarizes the chapter by admitting that "our ignorance of the laws of variation is profound" (*Origin*, p. 167). Yet he was still able to discern some basic patterns of variation that are recognized today, given our far more complete understanding of what genetic variation is, where it comes from, and how it is maintained. One correct inference is that traits that are under weak selection will tend to be more variable. Conversely, persistent strong selection will cause a reduction in heritable variation. While he could not explain the sorts of atavisms or analogous variation that he reported for pigeons and horses, he still correctly interpreted them as an index of common ancestry. More generally, this chapter is a more direct demonstration of a quality that prevails throughout the *Origin*. Darwin constantly strove to meet Whewell's criterion of consilience by testing the limits of his theory to explain a diversity of phenomena. He argued that every phenomenon presented here could be explained either by natural selection or by different species having descended from a common ancestor, as opposed to having been independently created. These are steps in building the argument that evolution by natural selection represents a simple and universal explanation for the history of life. Not knowing the laws of inheritance created difficulties for Darwin, but his observations were accurate, and his interpretations were far more often right than wrong.

References

Browne, J. 1995. *Charles Darwin: Voyaging*. Princeton, NJ: Princeton University Press. (Pp. 475–76, Richard Owen's influence on Darwin's barnacle studies.)

Hubbs, C. L. 1950. Studies of Cyprinodont fishes. XX. A new subfamily from Guatemala, with ctenoid scales and a unilateral pectoral clasper. Miscellaneous Publications of the Museum of Zoology, University of Michigan, no. 78, pp. 1–28. (Xenodexia biology.)

Jones, R. T., D. C. Culver, and T. C. Cane. 1992. Are parallel morphologies of cave organisms the result of similar selection pressures? *Evolution* 46:353–65.

Kane, T. C., D. C. Culver, and R. T. Jones. 1992. Genetic-structure of morphologically differentiated populations of the amphipod *Gammarus minus*. *Evolution* 46:272–78.

Ruse, M. 1999. *The Darwinian revolution—science red in tooth and claw*, 2nd ed. Chicago: University of Chicago Press. (Pp. 174–80, details on Whewell's philosophy of science.)

Strickberger, M. W. 1976. *Genetics*, 2nd ed. New York: MacMillan. (Pp. 203–10, Bateson, Punnett, and epistasis.)

Verovnik, R., B. Sket, and B. Trontelj. 2004. Phylogeography of subterranean and surface populations of water lice *Asellus aquaticus* (Crustacea: Isopoda). *Molecular Ecology* 13:1519–32.

Chapter 7

Evolution Today: A Modern Perspective
on Natural Selection

I close the first and second parts of this book with a brief essay titled "Evolution Today." Each essay describes the present study of evolution and, by its chosen example, the growth of evolutionary biology into a mature science. Both essays have something in common: I have sought examples of the contemporary study of evolution that demonstrate how quickly evolution can occur. Darwin always emphasized that evolution was so slow as to be essentially invisible unless one could look for it with the benefit of evidence that somehow makes visible the changes that take place over long intervals of time. The fossil record was an excellent place to look for evolution, and so were the Galapagos Islands, since they were of volcanic origin and younger than South America—but still old enough to provide evidence for evolution. Because the islands were initially lifeless and then were colonized from the mainland, the species that were found to be endemic to the islands but different from those on the mainland must have evolved after the islands were formed. We have developed new ways to study evolution and have found that it is far more amenable to direct observation than Darwin could have imagined. In my first example, I will describe some of my own research.

Guppies are small fish with the technical Latin name *Poecilia reticulata*. They are native to northeastern South America and some Caribbean islands. The females have a uniform tan color, while the males have a diversity of colorful spots and stripes. I have studied guppies for thirty years. My research is divided between studies of natural populations on the island of

Trinidad and guppies kept in a laboratory at the University of California, Riverside. The lab has around eight hundred aquariums that contain, altogether, thousands of fish at any one time. Guppies live up to four years, and each adult female can produce dozens of live-born babies every three to four weeks, so hundreds of babies are born in the lab every day. Each baby can have babies of its own after only twelve weeks. At the other end, only a few fish die per day, so the lab's population could grow to many millions of guppies within a year if only there was a place to keep them all. Their constantly burgeoning numbers are a testament to Malthus's insight and Darwin's preoccupation with the struggle for existence. In their natural environment in Trinidad, we rarely see such evidence of the guppies' ability to propagate. They are found in small, scattered populations living in clear-running mountain streams.

Because guppies are easy to care for and easy to propagate, as well as colorful, they make ideal pets for household aquariums. They turn out to be ideal subjects for the study of evolution for similar reasons—not only because they are so well suited for laboratory experiments that emphasize propagation, but also because their natural environment is so much like a lab, with each separate branch and segment of a stream giving natural selection an opportunity to launch separate "experiments."

Charles Darwin often said that science is much more than simply looking at nature. It is always a matter of looking at nature with a hypothesis in mind. I followed this advice when I set out to study guppies.

Natural Selection as a Cause of Evolution

My goal at the outset of my research career in the mid-1970s was to study the process of natural selection and to experimentally test aspects of the theory of evolution in a natural setting. I began at a time when there was abundant precedent for doing selection experiments in the laboratory on model organisms like fruit flies, but our knowledge about evolution in the natural world relied more on indirect inferences. One such inference involved the famous peppered moths of England, which could be either black or "peppered" (which means a patchy mix of black, gray, and white). The black form was a rare collector's item in the 1850s but became more and more abundant over time, until nearly all that you could find were black moths in and around some cities. This change took about eighty years. The history of the change could be seen in retrospective in the dated collections of thousands of moths accumulated by moth fanciers. You might wonder that such a person could ever exist, but there were ample numbers of them in Victorian England. They were a part of Charles Darwin's audience.

This change in moth populations over time was interpreted as evolution in action because genes caused moths to be either black or peppered. It was a good guess that the shift from a predominance of peppered to a predominance of black moths was caused by the industrialization of some parts of England, since it was around the cities where the black moths became common. One idea was that air pollution killed the lichens that normally coated the trunks and branches of trees, turning the tree bark from a mottled gray and black to a sooty, solid black. Moths fly at night but spend the day resting on tree surfaces. The peppered moths were hard to see when they came to rest on lichens, but were easily seen on a blackened tree trunk. The reverse was true for the black moths. Birds ate these moths during the day, while they were at rest on tree trunks, so perhaps the birds were more successful in seeing and then eating the ones that did not blend in well with their background. If so, the black moths would be more and more likely to escape predation as more trees became darker from pollution. This idea was tested, first by H. Kettlewell in the 1950s, then again and again by others over the decades that followed, was shown to be true, and certainly was at least partly responsible for the changes in moth coloring that occurred between 1850 and the 1920s.

The moth story and others were bona fide studies of evolution by natural selection. But I was after something different. I wanted to find some modern manifestation of evolutionary theory that made predictions about how organisms should evolve in response to some feature of their environment. Then I wanted to do experiments on natural populations that would allow me to test this prediction—that is, I wanted to manipulate or control some aspect of the environment, then sit and watch evolution happen, and in this way see if the prediction was fulfilled. This sort of endeavor had not been attempted because we all thought, as Darwin had originally said over and over in the *Origin*, that evolution was a slow process that could not be seen while we sat and watched. It could only be seen in retrospect, such as in the dated collections of moths that accumulated over decades. Yet, in spite of its slow pace, evolution had caused big changes, so big that it could account for the origin and diversity of all the life-forms that we see in the world today. The insight that resolved this paradox in Darwin's time was provided by Lyell's geology: the world is billions of years old, and the process of evolution had this enormous interval of time to work with.

If you have the preconceived notion that evolution operates too slowly to be seen in your lifetime, you have little incentive to look for it. Studying changes that have already happened, as in the peppered moth story, is a safer route. The difference, though, is that doing a direct experiment with appropriate controls is the only way to formally test a hypothesis. It also yields the potential to define cause-and-effect relationships more precisely than is possible by trying to reconstruct the history of past events. Part of the virtue of doing experiments is to have the opportunity to measure the strength of natural selection and the rate at which evolution occurs.

This is the logic that brought me to guppies. I had been studying a related fish, the mosquito fish *Gambusia affinis*, for two years and had been thinking about a topic called "life history evolution," which I will soon define. But driving across the Walt Whitman Bridge from Philadelphia to New Jersey and smelling the sickening, yeasty odor from the beer brewery on the New Jersey side of the bridge while en route to mosquito-fish habitat in Cape May County, New Jersey, did not conform to my romantic vision of fieldwork. Then I attended a lecture given by John Endler, who spoke about guppies on the island of Trinidad, their predators, and how the predators selected for changes in male coloration. What Endler described was a perfect fit for my desire to test theories of life history evolution in nature. Predators clearly could select for changes in guppies. Endler's work, even by 1977 when I attended his seminar, showed that predators could have a big effect in a short time. I knew that guppies were easy fish to work with because I kept them as pets, starting at a very early and not so responsible age. They were also perfect because working on them could be my ticket to the tropics, which was one of my ulterior motives for going into research as opposed to going to veterinary school. I had dreamed of the tropics since I was a youngster watching National Geographic shows on TV while ice built up outside the windows during northeastern winters. Guppies might make that dream come true.

The Setting

The key environmental factor that makes guppies in Trinidad ideal for an experimental study of evolution is that the mountains forming the northern edge of the island are mostly uplifted limestone. Limestone dissolves and erodes in the heavy tropical rainfall, forming steep, forested ravines filled with clear running rivers that have many waterfalls (fig. 11). Rivers are like a tree, being large and wide at the base, then dividing into progressively smaller branches as you move up into the headwater streams. There are dozens of species of fish found downstream, but each set of waterfalls is a barrier to some of them. The headwater streams, which are like the small twigs at the ends of branches, may contain only a single fish species or none at all.

Figure 11
Male and female guppies, seen here in one of their native Trinidadian streams, live in habitats separated by features such as this barrier waterfall.

Guppies are found in a wider range of habitats than most fish. They can be found in the large, lowland rivers and all the way up into some of the smallest headwater streams. In the larger, lowland rivers they live with many species of predators. Many of these predators are relatively large fish that eat large guppies. Guppies can be rare in such big rivers. They lead a fugitive existence, traveling in schools, quickly being eaten when they appear in open water but multiplying quickly when they find a refuge, as in the submerged branches of a fallen tree, where they are safe from predators. In the headwater streams, there is usually only one other species of fish present. This other fish is the killifish *Rivulus hartii*. It is small and will eat almost anything, including guppies. Because it is small, it tends to eat small guppies.

There are many rivers draining the well-watered slopes of the Northern Range mountains. Because Trinidad is an island, the actual diversity of species is small relative to what is found nearby on mainland South America. The types of species found in each river are very much the same, so each stream is like a replicate of the others. If one is interested in studying how guppies adapt to the presence or absence of predators, then each river can be thought of as an independent experiment in which guppies evolve when they invade headwater streams and leave the predators behind.

The Theory

Life history theory is a branch of evolutionary theory that one can find, with a bit of imagination, foreshadowed in the *Origin*. Its more formal roots trace to the 1930s, but it did not really begin to develop as a quantitative branch of evolutionary theory until the 1950s. It deals with the evolution of how organisms reproduce themselves. The connection to the *Origin* and evolution is easy to imagine, even if the formal theory long postdates the *Origin*. Darwin postulated that evolution happens because individuals differ in how successful they are in contributing offspring to the next generation. Life history theory predicts how natural selection should shape the way organisms parcel their resources into making babies. Important features of their life history include the age at maturity and first reproduction, how often organisms reproduce, and how they allocate their resources to reproduction, either by making few offspring, each of which gets lots of resources, or many offspring, with little being invested in each of them. A human will produce and rear only a small number of offspring over a very long lifetime and lavish extended care on each of them. A large ocean fish such as a cod will produce hundreds of thousands of small eggs every year and cast them to the fates of ocean tides.

The theory predicts how these different parts of an organism's life history evolve in response to factors like predation. Predators are a major source of death in nature, but the risk of death can vary a great deal in different places. Also, some predators specialize on eating prey that are small, young, and immature, while others may instead prey only on large adults. The risk of predation and who the predators eat turns out to make a big difference in how they shape the life history of their prey.

To understand how the theory works, you need to think of your life as being like a pie, divided into slices that represent how you allocate your resources. One big slice is maintenance costs, or the cost of replacing all your parts that are constantly wearing out or in need of repair. Skin, blood cells, and the lining of your gut are some of the parts that wear out quickly and are constantly being replaced. Another slice is fat storage. It may be hard for you to think of fat reserves as adaptation, but that is because so many of us make so many more deposits than withdrawals. For many organisms, food is not always abundant, so fat is a way of storing energy in times of plenty for use in times when food is scarce. Another slice is for growth. Finally, there is the slice that is devoted to reproduction. The pie is finite, so increasing the size of one slice means making another slice smaller. If the slice that is devoted to making babies is to be larger, then it must take resources away from growth, maintenance, or fat storage.

Life history theory predicts the best way to divide the pie in the face of an organism's day-to-day risks of dying. Natural selection can make the size of the slice that is devoted to reproduction bigger or smaller by changing the age you begin to have babies, how often you have them, how many you have, or the amount of resources you devote to caring for them. A human does not begin to have babies until the teen years or later, and produces only a few over a lifetime. A dog can begin having babies in a year and can produce a half dozen or more at a time. A mouse can begin having babies after a few months and produce a dozen or more at a time. These differences in age at maturity and baby production are associated with differences in life span. Humans often live past eighty, dogs rarely past twenty, and mice rarely past two years of age.

If the risk of being killed by a predator is high, then natural selection will favor those individuals that devote a larger slice of the pie to reproduction by beginning to have babies when they are younger and making more babies. The increase in the size of the reproduction slice makes other slices smaller. A smaller investment in maintenance can mean being at greater risk of dying in the future because you may become more susceptible to disease or because your body deteriorates more quickly. Investing less in fat storage may put you at greater risk in the future if food becomes scarce. Investing less in growth can reduce the number of babies you can produce in the future (many organisms continue to grow throughout their lives, and bigger individuals make more babies). But, if the odds of living from one day to the next are low because of the constant risk of being eaten by a predator, then the benefit of producing babies early and producing more of them is predicted to outweigh these costs. There is little use in storing fat and little gain in growing if you are not likely to live much longer. Conversely, if predators are scarce or absent and the risk of dying is low, then this long life expectancy shifts the balance in how to best invest resources. The theory then predicts that natural selection will favor those individuals who devote a smaller slice of pie to reproduction and a bigger slice of pie to their own maintenance. Theory also predicts that they will produce fewer babies and devote more resources to each of them.

I liked this theory because it expanded on Darwin's original concept of fitness. In the *Origin*, natural selection is seen as favoring whatever traits make it more likely that an individual will survive. Those who survive longer will have more opportunities to reproduce and will contribute more offspring to the next generation than will those who die young. Yet Darwin did not elaborate. Virtually all he says in the *Origin* about life history evolution is contained in a single paragraph. In his chapter on the "struggle for existence" (*Origin*, p. 66) he mentions the likelihood of natural selection playing some role in shaping the number of young that are produced, since this number varies so radically (from ones to millions) among different types of organisms. He argues that the number of young produced is perhaps a function of their risk of dying and is perhaps an adaptation to a fluctuating environment. If the risk of dying when young is very high, then an organism must produce more to compensate for this loss. If the environment alternates between good and bad years, then it pays to be able to capitalize on the good years by producing a large number of babies.

In the theory of life history evolution, fitness is gauged by a composite of an organism's survival and its successful production of offspring. This composite includes trade-offs between what an organism does now versus what it is capable of doing in the future because it involves starting with a pie of finite size, then modifying the size of the slice that is devoted to reproduction and, as a consequence, changing the size of the remaining three slices. According to this theory, natural selection can do seemingly counterintuitive things. It can favor the evolution of a shorter life span or the production of fewer young, providing that there are appropriate trade-offs at other stages of the life history. I found this added complexity attractive because it seemed flexible enough to account for the huge diversity in life span that we see among organisms, and more generally in how they allocate resources to reproduction. More to the point, this theory makes predictions that are testable in nature if the right circumstances can be found.

The Research

The theory deals with how the risk of death alters how the pie is divided, but does not address the cause of death. The role of predators in causing death is where the natural history of guppies meets theory and where it became possible to use the theory to make predictions that can be tested on guppies. Guppies that live in the lower reaches of streams with many predators should suffer higher mortality rates than guppies that live in the headwater streams, where most predators are absent. If this is true, then life history theory predicts that, in sections of streams with predators, natural selection will favor those individuals that attain maturity at an earlier age and devote a bigger slice of their pie to making babies. In the headwater streams, the theory predicts that natural selection will instead favor guppies with delayed maturity that devote a smaller slice of the pie to making babies and larger slices to their own growth and maintenance.

Is it really true that guppies that live with predators have a higher death rate? I measured death rates with census-like studies involving the capturing and marking of individual guppies. I worked in streams that consisted of pools of water that were bounded up- and downstream by a riffle (a place where the stream is narrower and its descent is steeper, so the water flows more quickly). Guppies like to congregate in pools and are disinclined to swim the riffles to go to neighboring pools. It is possible to collect every guppy in a pool. I found that I could create a "guppy magnet" by sitting at the upstream side of a pool and gently stirring up some sediment. Most species of fish would hide, but guppies come to check out the sediment to see if there is anything good to eat. By holding a butterfly net in each hand, I could chase them from one net to the other. In that way, I could catch them all without disturbing the habitat. I measured each fish, marked it with dots of paint injected under the skin, then released it back into its home pool. I re-collected them two weeks later. By comparing the number released with the number recaught, it was possible to estimate the number that had died. It was also possible to see how well those that lived were growing, to quantify the birth of new youngsters, and to study the movement of individual guppies from one pool to the next.

I found that the death rate over a two-week period was about 15% higher in localities with high predation. Over a seven- to eight-month period, this higher death rate meant that a guppy that lived without predators was twenty to thirty times more likely to still be alive than one that lived with predators. Living with predators really does mean that guppy life expectancy is much lower than when predators are absent. This result brought me one of my proudest moments in science, which was to be featured in the *National Enquirer* under the headline "Uncle Sam wastes $97,000 dollars to learn how old guppies are when they die." I actually learned a great deal more.

I then used a method championed by Darwin, which is to let nature do the experiment. There were already many populations of guppies that lived with the larger, more dangerous predators plus other populations of guppies found in the headwaters of the same streams that did not live with those predators. Does theory accurately predict differences in their life histories? Comparisons of populations from these two types of localities showed that guppies in nature do indeed differ from one another as theory predicts. The guppies from localities with high predation were smaller and presumably younger when they began to reproduce. They produced many more babies in each litter. The individual babies were smaller. The proportion of the mother's body weight that consisted of developing babies was larger.

This was just a beginning. I showed that theory accurately predicts differences in wild-caught guppies that live with or without predators, but are these differences controlled by their genes? I did multigeneration laboratory experiments that showed that what I had seen in nature did indeed have a genetic basis. The more controlled nature of the laboratory environment also meant that I could measure things that were difficult or impossible to measure in nature. I showed that the guppies in high-predation localities were younger at maturity. They devoted more of their available food resources to each litter of babies. They produced litters of babies more often. In a different kind of study, I also showed that there was genetic variation within each population of guppies. So, while the average age at maturity was younger in guppies from a high-predation locality, there were also genetic differences in the age at maturity between families within each population. This variation is the raw material that Darwin's mechanism of natural selection calls for, and its presence in natural populations is what the theory of population genetics predicts (see chap. 6). It means that all these populations have a continuing capacity to evolve, should their environments change.

So far, so good, but all this was still just the preliminaries. I wanted to make predictions from theory and test them by doing experiments not only in the lab but also in the guppies' natural habitat. John Endler had already shown that the distribution of guppies and predators makes such experiments possible because their range is often punctuated by barrier waterfalls. The first barrier waterfall often stops the large predators but not guppies or some other smaller fish. Each successive barrier stops more fish. Some waterfalls exclude all but guppies and *Rivulus*, the small killifish that is the only other species to invade these headwater streams and that occasionally eats small guppies. Other barriers exclude even guppies, so *Rivulus* is the only fish present. *Rivulus* is able to breach barriers more effectively than other fish because it will actually jump out of the stream on rainy nights, hop around through the forest, and invade aquatic habitats that no other fish can reach.

These discontinuities created by waterfalls make it possible to think of a stream as being like a giant test tube. John Endler realized this when he studied the effects of predators on the evolution of male coloration. He took guppies that lived with predators below some barrier waterfall and introduced them into the previously guppy-free habitat above the waterfall. Doing so enabled him to study how guppies evolved when they no longer lived under the threat of predation. Change could be perceived by comparing the descendants of the introduced guppies with the population of guppies found below the barrier waterfall. At the start of the experiment, the introduced guppies had been just like those in the population found downstream from the barrier. This meant that any differences between the introduced guppies' descendants and those of their "downstream control" had evolved after the introduction.

I adopted Endler's introduction experiment, which had been initiated two years before I began working on guppies, and then replicated it by transplanting guppies from a high- to a low-predation site in another stream. Just as in Endler's experiment, life history theory predicts that natural selection should favor transplanted individuals that devote fewer resources to reproduction. This means that, over time, the population of guppies in the introduction site should evolve to have a later age at maturity and a lower rate of investment in reproduction relative to those from the downstream control.

I also did a different type of experiment, in which I added predators to a stream that had previously contained guppies and *Rivulus* but not the larger predators. Doing so increased, rather than decreased, guppy mortality rates. In this case, life history theory predicts that an increase in mortality rate will cause the affected guppy population to evolve earlier maturity and a higher rate of investment in reproduction. Their control was a population of guppies found above a waterfall farther upstream that blocked the upstream migration of the predators.

To actually see whether or not guppies had evolved, I collected adult females from the experimental and control sites and transported them to my laboratory in the United States. I put each female in her own aquarium and took advantage of a neat feature of guppy reproductive biology. Female guppies store sperm and reproduce year-round. This means that every isolated female produced litters of young. The young from each female then became a numbered pedigree in a multigeneration genetic study like the one I had done years earlier when I quantified genetic differences between guppies from high- and low-predation localities. The difference was that I was now comparing control and experimental populations. I knew how long ago I had changed the environment, and I had been able to measure the effects of the change by measuring mortality rates in nature; so now I could see how much the guppies had evolved relative to their controls.

In both types of introduction experiments, guppy life histories evolved as predicted by theory. When guppies were transplanted from a high-predation site to a previously guppy-free low-predation site, they evolved a later age at maturity and a lower rate of investment of resources into making babies, relative to the guppies from the control site found below the barrier. Likewise, when predators were introduced, guppies evolved earlier ages at maturity. This test goes beyond the "natural experiment" because what we see is the product of our own manipulation and change is judged relative to an experimental control.

The combination of comparative studies done on guppies from many rivers, genetic studies on a subset of these rivers, introduction experiments, and mark-recapture studies argued that predators and higher mortality rates have played an important role in shaping life history evolution in guppies.

This was all very good, but there were still some surprises in the results. When I described this experiment to other biologists before I did it, I often saw sympathetic smiles and was told that they hoped I would live long enough to see something happen. It turns out that I did not have to wait long. Males changed completely within four years, which is six to eight generations. Females began to show significant change in seven years. Our traditional concept of how fast evolution can occur had been similar to Darwin's in the *Origin*. Like Darwin, we too had looked at the fossil record as the primary source for understanding the history of evolution. Many scientists had quantified how fast evolution occurs by measuring the rate of change in traits like body size they found in fossils from successive geological strata. Doing the same calculations for guppies showed that the fish in my experiments had evolved at a rate that was 10,000 to 10 million times faster than what had been considered to be rapid evolution in the fossil record. It may be that guppies evolve unnaturally quickly, but I am inclined to think otherwise. Similar high rates of evolution, such as those in the Galapagos finch example discussed in chapter 1, have now been seen in many other studies. In fact, such rapid evolution has been documented so many times between the inception of my work in 1976 and now (2008) that the attitude of evolutionary biologists has shifted from considering such experiments as a quixotic search for something that cannot be seen to seeing rapid evolution as an ordinary phenomenon. The difference between rates estimated from fossils versus rates estimated in contemporary studies probably occurs because the fossil record seriously underestimates how fast organisms can evolve and how strong natural selection is. This is the entry to a different story, but it will be relevant later to consider that the potential rate of evolution is in fact far greater than the rate imagined by Darwin.

This scenario brings us back to the mechanism of natural selection and how it works. Darwin's first premise was that organisms produce more offspring than are required to replace themselves in the next generation. Anyone who has kept guppies knows how quickly they can multiply. A few guppies could become millions within a year if all were well fed and survived. When the guppies in a high-predation locality find a refuge, such as in the branches of a fallen tree, they quickly fill it with babies. We generally see such productivity only in these small bursts because so few individuals ordinarily survive predation. Our laboratory studies showed that there are genetic differences between guppy populations but also that there is genetic variation within populations. In each population, the family averages for traits like the age at maturity were different, so guppies have a continuing capacity to evolve if the environment changes. It was this genetic variation and the overproduction of babies that enabled the guppy populations to evolve when their environment was changed by their being transplanted from a high-predation site below a barrier waterfall to a low-predation site above the waterfall. The same variation enabled the residents of low-predation sites to adapt when a predator was added to their community. The presence of the variation meant that no matter what the level of predation was, some individuals were more successful than others in producing offspring in the time that was available to them. If predators were present, then life was short, and those who matured quickly and produced many babies were more successful. If predators were absent, then life expectancy was long, and natural selection favored those who invested more in their own maintenance and in producing a smaller number of higher-quality young.

This story illustrates, by example, one view of what evolution has become since Darwin. We have developed Darwin's original concept of fitness into quantifiable features of organisms. We have also developed quantitative theory that can predict how organisms will evolve in response to features of their environment. Theory has enabled us to design experiments that show that natural selection is an observable, quantifiable, contemporary process, rather than just being something that we can see only through the lens of history.

The research on guppies in Trinidad also illustrates what we need to know in order to really understand natural selection. We need to know about the variation that selection acts on, its source, how much of it is there, and how it is transmitted from parents to offspring. We need to know about the capacity of organisms to change from generation to generation under selection and how much change is possible. We need to know what in the environment controls population size and how it is controlled. Is it predators, other organisms that compete for resources, disease, or some combination of these and other factors? Whatever regulates populations is what causes evolution by natural selection to occur, and is what will determine how an organism evolves in response to that regulation. Darwin addressed all these issues in the opening chapters of the *Origin* as part of his first goal, which was to define the process of natural selection. We are now in a position to quantify them all in natural environments, and to do so in the context of testing the predictions of new, quantitative versions of the theory of evolution.

References

Reznick, D. N., H. Bryga, and J. Endler. 1990. Experimentally induced lifehistory in a natural population. *Nature* 346:357–59.

Reznick, D., M. J. Butler IV, and H. Rodd. 2001. Life history evolution in guppies 7: The comparative ecology of high and low predation environments. *American Naturalist* 157:126–40.

Reznick, D. N., M. J. Butler IV, F. H. Rodd, and P. Ross. 1996. Life history evolution in guppies (*Poecilia reticulata*) 6: Differential mortality as a mechanism for natural selection. *Evolution* 50:1651–60.

Reznick, D. N., H. F. Rodd, and M. Cardenas. 1996. Life history evolution in guppies (*Poecilia reticulata*: Poeciliidae) 4: Convergence in life history phenotypes. *American Naturalist* 147:319–38.

Reznick, D. N., F. H. Shaw, F. H. Rodd, and R. G. Shaw. 1997. Evaluation of the rate of evolution in natural populations of guppies (*Poecilia reticulata*). *Science* 275:1934–37.

Part Two

Speciation

Chapter 8

Preamble to Speciation

Charles Darwin titled his book *On the Origin of Species . . .* , yet not everyone agrees that Darwin really wrote about the origin of species. Ernst Mayr, who proposed the first modern definition of a species and presented the first well-documented argument for how speciation occurs, said the following in the opening of chapter 7 of his book *Systematics and the Origin of Species* (1942):

> Darwin entitled his epoch-making work not "The Principles of Evolution," or "The Origin and Development of Organisms," or by some other title which would stress the general problems of evolution. Apparently he considered these titles too speculative and therefore chose the more concrete one, "The Origin of Species." To him this was apparently more or less synonymous with these other titles, which is not surprising if we remember that Darwin drew no line between varieties and species. Any pronounced evolutionary change of a group of organisms was, to him, the origin of a new species. He was only mildly interested in the spatial relationships of his incipient species and paid very little attention to the origin of the discontinuities between them. It is quite true, as several recent authors have indicated, that Darwin's book was misnamed, because it is a book on evolutionary changes in general and the factors that control them (selection and so forth), but not a treatise on the origin of species. (p. 147)

So did Darwin write about the origin of species or not? I will argue that Darwin did, in fact, write a book about speciation, but with goals that were tailored to an audience different from Mayr and his contemporaries and with a definition of species that was different from Mayr's. Mayr and Darwin were separated by a gulf in time during which occurred the discovery of genetics, its reconciliation with the idea of evolution by natural selection, and the growth of evolution as a scientific discipline. This growth included the development of concepts that did not exist in Darwin's time: the modern definition of a species and proposed mechanisms for the formation of new species. The discrepancy between Darwin's intention and Mayr's interpretation is defined by this gulf.

Darwin was not nearly so concerned with defining what a species is as he was with proving what it is not. His goal was to prove that species are not the products of acts of special creation. Today we think of the advocates of special creation as representing nonscientific, religious opponents to evolution. In Darwin's day, they were the scientific establishment. Virtually everyone, ranging from his professors at Cambridge to all those who had the greatest influence on Darwin's intellectual development, advocated some form of special creation. The Cambridge dons, including John Henslow, Darwin's botany professor and most significant mentor, Adam Sedgwick, who introduced Darwin to field studies of geology, and John Herschel and William Whewell, each of whom promoted a philosophy of science that shaped how Darwin wrote the *Origin*, were all ordained ministers of the Church of England; this was required to become a professor at Cambridge or Oxford. Others who contributed to Darwin's development, including the geologist Charles Lyell and the anatomist Richard Owen, held the same views. These individuals were among those whom Darwin was most trying to convert. As a consequence, Darwin's emphasis was on the process of "transmutation," or the origin of new species from others already in existence. He constantly emphasized the indistinctness of species, or the continuity between what he referred to as varieties and species, as evidence for speciation being an ongoing process. Recall that "variety" usually refers to a locally adapted population that is in some way distinct from other populations of the same species. Thus Darwin, as Mayr correctly states, was not concerned with defining what a species was. Darwin's chief hurdle was to convince the reading public that if speciation is an ongoing process, then the boundaries between species should sometimes be indistinct. The indistinctness of boundaries between species is evidence of "species in the making." If species were always distinct and there was no evidence for the ongoing formation of new species, then those who advocated special creation would have a point and Darwin would have a hard time getting them to accept his theory of species transmutation.

The nature of Darwin's goals is implicit in the first paragraph of his second chapter: "No one definition has as yet satisfied all naturalists; yet every naturalist knows vaguely what he means when he speaks of a species. Generally the term includes the unknown element of a distinct act of creation" (*Origin*, p. 44). As we shall see, the remainder of this chapter ("Variation Under Nature") is devoted almost exclusively to an argument for the continuity between varieties and species and against the distinctness of species and hence against distinct acts of creation.

The Modern Protagonists

Before we deal with the *Origin*, I will outline the modern perspective of species and speciation presented by Theodosius Dobzhansky in *Genetics and the Origin of Species* (1937) and Ernst Mayr in *Systematics and the Origin of Species* (1942), as updated by Jerry Coyne and Alan Orr (2004). Dobzhansky's and Mayr's books are cornerstones of what we now refer to as the "modern synthesis" of evolutionary biology because they represent the recasting of Darwin's ideas after the reconciliation of Mendelian genetics with evolution. Coyne and Orr are leaders in the current study of the genetics of speciation. Beginning in the present and then turning back to the *Origin* will give you a perspective for understanding how Darwin bridged the gap between special creation and our current understanding of species and speciation.

New York City was the epicenter of the modern synthesis, since Dobzhansky was a professor at Columbia and Rockefeller universities for much of his career, while Mayr was a curator of ornithology at the American Museum of Natural History. A third modern synthesis classic, *Tempo and Mode in Evolution* (1945) by G. G. Simpson, then a curator of paleontology at the American Museum of Natural History, will be discussed in the third part of this book. These seminal books were based on each author's Jessup Lectures, which were an endowed series of lectures given at Colombia University.

These books followed on the work of population geneticists, including R. A. Fisher, J.B.S. Haldane, and Sewall Wright, mathematical biologists who ushered in the era of modern synthesis in the 1920s and 1930s by developing formal mathematical models of evolution that reconciled the rift between the Mendelians and the biometricians (see "Evolution Today" in chap. 5). Recall that these two opposing schools of thought disagreed about the mechanisms driving evolution, with the Mendelian camp ascribing it to the laws of inheritance in the form of single mutation events and the biometricians favoring the idea of continuous variation and blended inheritance (see chap. 3 above). The population geneticists' mathematical models showed the compatibility between Mendelian inheritance and Darwin's original idea of evolution by natural selection, which helped reinstate natural selection as the primary mechanism that causes evolution; population geneticists also defined the roles of mutation, migration, and genetic drift in evolution (see chap. 5).

In fact, Mendelian inheritance provided an elegant solution to all the details of inheritance that had troubled Darwin and his contemporaries. Mendel's laws showed that all the complexity of inheritance could be resolved through the simple rules of each parent contributing one allele, or one copy of each gene, to its offspring, as follows: Each parent possesses two alleles but transmits only one to each offspring, and there is no mechanism controlling which one is transmitted; it is like the random tossing of a coin. Parental traits can apparently disappear in offspring but reappear in later generations because the expression of some alleles dominates others, but also because of phenomena like epistasis (see chap. 6), or the way multiple alleles act together to determine the appearance of an individual. The source of variation is mutations, which permanently alter an existing allele; the new allele is in turn faithfully transmitted from parent to offspring as were all the preexisting alleles. By the 1930s, mutations had been well characterized and were shown to be abundant and easily documented in natural populations. Importantly, it was also shown that mutations often have only small effects on the individual, some of which are beneficial. This ran counter to the misconception that mutations could only cause the deleterious monstrosities that represent the popular image of the word.

The particulate nature of inheritance solved another problem associated with the alternative theory of blending inheritance. Whereas the blending theory predicted a constant loss of variation as the contributions of each parent became inextricably mixed in each offspring, particulate inheritance did not. Each parent's contribution remained distinct in each offspring, even if it was recessive and was not expressed in offspring, and would continue to be faithfully transmitted to subsequent generations.

Mendel demonstrated that his laws of inheritance were easily described in simple, mathematical terms. Theoretical population geneticists codified

Mendel's laws into mathematical models that represented the transmission of alleles from parents to offspring, but did so at the level of populations of individuals rather than crosses between individual parents. The important point with regard to the modern synthesis is that Fisher, Haldane, Wright, and others used Mendelian inheritance to formulate models for the process of evolution by natural selection. They presented a formal definition of such evolution as a change in the relative abundance of alleles at individual loci. The rift between the Mendelians and biometricians was fully healed, and natural selection was rehabilitated as a mechanism that causes evolution.

Although population genetics served well to describe local adaptation, it did not address speciation. This is where Dobzhansky and Mayr came in.

Dobzhansky began his scientific career in Kiev, where he studied natural history and systematics. He came to the United States in 1927 to study classical, laboratory-based genetics with Thomas Hunt Morgan at Colombia University. Morgan pioneered the use of the fruit fly, *Drosophila melanogaster*, as a model organism for studying genetics and was a major contributor to the development of Mendelian genetics and the empirical side of the reconciliation of the biometrician-Mendelian rift. Dobzhansky wove together the experimental skills of a geneticist with the observational skills of a field naturalist to create a career devoted to the study of genetic variation and adaptation in natural populations.

Mayr began his academic career in 1923 studying medicine at the University of Greifswald in his native Germany, more to fulfill his family's ambitions than his own. His real passion was ornithology. His escape from medicine began when he spotted a pair of red-crested pochards in Germany. The bird had not been seen in Europe for over seventy years. After Mayr convinced Eric Stressman, the curator of ornithology at the Berlin Zoological Museum, that the sighting was correct, Stressman offered him a position at the museum and an opportunity to do fieldwork in the tropics, provided he could complete a PhD in sixteen months. Mayr succeeded and joined field expeditions to New Guinea and the Solomon Islands, then became a curator of ornithology at the American Museum of Natural History in 1931. The insights that he contributed to the modern synthesis were those of a field-oriented systematist with a particularly strong background in biogeography, or the study of species distributions. His analyses of species distributions were the foundation for his ideas about the causes of speciation.

Jerry Coyne and Alan Orr are part of Dobzhansky's academic pedigree and are in the middle of active careers devoted to the study of the genetics of speciation. Coyne did his PhD studies at Harvard under the direction of Richard Lewontin, who had received his PhD studying with Dobzhansky. Alan Orr did his PhD studies with Coyne at the University of Chicago. One hallmark of Coyne's career was to modify some of Dobzhansky's experimental designs and use them to rekindle the long-dormant study of the genetics of speciation. Coyne and Orr have played a major role in the development of our current approaches to the study of the genetics of speciation.

The Modern-Synthesis Perspective on Species

Dobzhansky and Mayr dealt separately with the definition of the species concept and the definition of mechanisms that cause speciation to occur. Darwin's statement at the opening of his second chapter, quoted earlier in this chapter, reflects his nonchalance about the term species. As Mayr correctly stated, Darwin saw species as only a point in a continuum, so he never presented a clear definition for what a species is. He argued simply that natural selection is a single process that causes single species to diverge into multiple species over time. The alternative promoted by Dobzhansky (1937) is that natural selection attains two ends. It causes diversity as populations within a species adapt to different environments, but it also causes discontinuity, or a break in this continuum of variation, with the formation of distinct species. This break in the continuum between diversity among populations within a species and the existence of distinct species can be seen in the coexistence of closely related but distinct species. If species were just a point in a continuum, then whenever different populations of a species were reunited, they would freely interbreed, and the differences between them would disappear. However, in nature we often find that closely related species can occupy the same area, encounter each other on a day-to-day basis, and compete for resources, yet remain distinct from one another. The fact that their differences persist even when they physically mingle means that some threshold has been crossed that keeps them distinct and allows them to be recognized as different species.

Dobzhansky argued that for species to remain discrete, there must be barriers that keep them apart. Coyne and Orr (2004) offer an updated list of barriers between species (their pp. 28–29) that divides them into three main categories. First are "premating isolating barriers" that prevent individuals from different species from mating. Individuals may fail to mate because of differences in behavior. Often species have elaborate courtship displays; differences between populations in how courtship is executed or in signals associated with courtship, such as scents used to attract mates, can often prevent members from two different populations from interbreeding. A variety of ecological factors, such as preferred habitat or the timing of reproduction, can also enforce separation. Genitalia can work as a "lock and key" system, so differences in the morphology of genitalia can impede mating between species.

Second are "postmating, prezygotic isolating barriers." A zygote is the product of a successful union of male and female gametes (sperm and eggs in animals; pollen and ova in plants) that results in a developing embryo. If mating occurs, several barriers can prevent the union of gametes and the formation of a zygote. For example, sperm (pollen) from different species may be less effective than those of the same species in reaching or penetrating the egg (ovum).

Third are the "postzygotic isolating barriers" (hybrid sterility and inviability). If the egg (ovum) is fertilized, then there can be various forms of incompatibility between the genes of different species that prevent the egg from developing normally, resulting in an inviable embryo. If the embryo is viable, then it may be infertile because of some incompatibility between the genes of the two parents. Alternatively, if it is born it could suffer in competition with the parental species, either in obtaining resources for survival or in mating to produce the next generation. Any one factor could serve as an effective barrier that prevents individuals from different species from successfully interbreeding and maintains the distinctness between them. Very often, multiple factors come into play in maintaining the distinctness of closely related species.

Having the capacity to interbreed and produce viable, fertile offspring is not by itself a criterion for declaring two populations as belonging to the same species. We sometimes find that different species can be induced to interbreed and can produce perfectly healthy, fertile offspring; however, they would rarely or never interbreed under natural circumstances because of premating isolating mechanisms. In addition, the reduced ability of their offspring to survive and reproduce in a natural setting can reinforce their separation. We do sometimes see different species interbreed in nature or find individuals that are the product of such interspecific crosses, so the barriers that separate species are not absolute, but the presence of reproductive isolating mechanisms still maintains the differences between them.

Dobzhansky then argued that the prominent role reproductive barriers play in defining the boundary between species, and hence in the enforcement of discontinuity, leads directly to a definition of species. Mayr (1942, 1995) formalized Dobzhansky's definition as the biological species concept: "Species are groups of interbreeding natural populations, which are reproductively isolated from other such groups" (1995, p. 5).

Coyne and Orr summarize all the above by stating: "Groups of populations thus constitute different species under two conditions: (1) their genetic differences preclude them from living in the same area, or (2) they inhabit the same area but their genetic differences make them unable to produce fertile hybrids" (2004, p. 30).

The biological species concept defines species by whether or not they freely interbreed in nature and produce viable, fertile offspring. It is either the absence of the ability to interbreed or a reduction in the fitness of the offspring that the species produce if they do interbreed that creates discontinuity, or distinct species, in nature. This species concept has its limitations, and dozens of alternatives have been proposed. One prevailing ambiguity is that it the biological concept is defined by sexual reproduction, but some species are mostly or entirely asexual. We have no single definition that can accommodate all the known variations in mode of reproduction. Other important ambiguities arise when we are not able to determine whether organisms are capable of interbreeding. For example, when we apply the term "species" to the fossil record, we have to instead define species by anatomical features. Also, it is often not practical to apply the reproduction criterion when classifying living species, so we again use either anatomy or, more and more frequently, genetic criteria, such as comparisons of DNA sequences that code for specific genes. Using these other criteria to distinguish between species implies a definition of species that is based on physical or genetic similarity rather than the ability to interbreed.

How is it that a process that causes a continuum of change—from a population that contains individuals that differ from one another, to different populations that are adapted to different environments, to well-defined races within species, to subspecies—can then cause a break, or discontinuity in the ability to interbreed, resulting in a new species? Answering this question brings us to the second half of Dobzhansky's (1937) and Mayr's (1942) landmark contributions, which is the articulation of mechanisms that can cause speciation to occur. Dobzhansky and Mayr, like Darwin before them, were impressed with the tension that would exist between populations becoming different through adaptation to their local environment, yet also eliminating such differences by migrating and interbreeding with individuals in other populations found elsewhere. Migration and interbreeding would establish some measure of genetic continuity between populations, yet species are defined by the discontinuity of being reproductively isolated from one another. How can this barrier between species arise?

Richard Goldschmidt was a contemporary of Dobzhansky and Mayr who proposed an alternative to natural selection as the source of discontinuity (1940). Goldschmidt did superior work for his time in characterizing local adaptation in gypsy moths and had no difficulty with the idea that natural selection could cause such adaptation, but felt that since species were separated by unbridgeable gaps that could not be breached by natural selection, some other mechanism must come into play. He proposed that some sort of "macromutation," or big change in the genetic material that is transmitted from parent to offspring, must occur in a single step that makes all the descendants from that individual so different from the rest of the species that interbreeding is no longer possible. Goldschmidt's proposal is a restatement of de Vries's 1889 mutation theory (chap. 3 above). This dilemma of discontinuity and the difficulty of understanding how it can be caused by natural selection, which is a continuous process, is a recurrent theme in the *Origin* and remains with us today as a source of contention.

Dobzhansky (1937) instead argued that species were separated by many small genetic changes that were likely to have accumulated over a long interval of time. Dobzhansky, like Darwin, was a hard-core empiricist and based his ideas on a wealth of observations and experiments. He presented one example from his own research (1937, pp. 238–41) in which he evaluated mating success between female *Drosophila miranda* and male *D. pseudoobscura* from different populations. These are closely related species of fruit flies that have partly overlapping distributions in the western United States. There are places where the two species can be found together but others where only one of the two species is found. Dobzhansky set up artificial populations in milk bottles, then scored the percentages of females that had been fertilized. When such populations consisted of males and females of the same species, nearly 100% of the females mated. When crosses were made between male *D. pseudoobscura* and female *D. miranda*, the percentages of females who mated was considerably smaller, ranging from 18% to 52%, even after nine days of forced confinement with males of a different but closely related species. There clearly was some degree of premating isolation between these species, but the degree of isolation varied among populations of *D. pseudoobscura*. This pattern of variation in premating isolation suggests that isolation is not the product of a discrete "on-off" switch, as would be the case if Goldschmidt were correct and speciation were caused by a single, large mutation. The pattern suggests instead that isolation is a product of multiple genetic changes that have accumulated over time.

Mayr argued that the small genetic differences that cause reduced interbreeding could accumulate if a population that was once distributed over a large area was subdivided into geographically isolated subpopulations. Geographic separation creates a barrier to the movement of individuals between populations, so the isolated subpopulations can evolve independently. It is this geographically enforced isolation that creates the opportunity for the diversity that always exists within a species to become discontinuity between species.

Over geological time, meaning over timescales large enough to encompass large changes in climate or geography, there are many ways for populations of a species that were once continuous to become isolated from one another. For example, our climate has become warmer and in many places drier since the most recent recession of the glaciers, around 10,000 years ago. During the peak of this glaciation, so much water was tied up in the polar ice caps that sea levels were more than 50 meters lower than they are today. Islands like Trinidad (chap. 7) were part of the mainland but became isolated as glaciers melted and sea levels rose. Continuous, forested habitat retreated in many areas and became broken up into "sky islands" on the top of mountains as parts of the earth warmed and dried out. New land for colonization appeared as the glaciers retreated northward. This land was punctuated with new freshwater environments, including rivers, lakes, and marshes that were initially lifeless and represented "blank canvases" for the creative forces of invasion and adaptation as aquatic organisms migrated north to occupy them. On longer timescales, landmasses either drift apart or fuse together, mountain ranges develop, then erode away, and new landmasses, like the Galapagos archipelago, can emerge from the ocean and provide an opportunity for colonization. All such changes can cause populations of plants and animals spread over large regions to become broken up into isolated subpopulations or allow new populations to be established as new habitat becomes available.

These scenarios encompass the potential for two different modes of speciation. Both modes are now formally recognized. One mode applies when a large population is subdivided into two or more isolated populations, such as when rising sea levels subdivide landmasses into islands. A second mode follows the colonization of a new habitat by a small number of individuals from some source population, as when a volcanic eruption creates a new oceanic island that is later colonized by a few individuals.

Mayr postulated that if any population was so well isolated from others of the same species that migration between populations was either impossible or exceedingly rare, then reproductive isolation could evolve as a by-product of this isolation as each of these populations adapted to its respective location. These differences could be ones that play an important role in local adaptation, but they could also be incidental to the independent evolution of the separate populations. For example, evolution by sexual selection could follow different paths in different populations, according to the prevailing whims of the females in each of them. Such differences could conceivably result in populations that were ecologically very similar to one another, but reproductively isolated because they evolved differences in male secondary sexual characters, courtship displays, and female preferences.

The true test of whether or not these isolated populations have become different species will come if they are brought back into contact with one another, perhaps caused by some later change in geography or climate that dissolves the geographic divide between them. When such secondary contact occurs, the once-isolated populations may not have attained reproductive isolation, in which case their "gene pools" may merge as they interbreed. Any differences between them would disappear. It may also be that reproductive isolation is complete and the two species remain distinct from one another without interbreeding. A third possibility, proposed by Dobzhansky (1937), is that there is only partial reproductive isolation when the populations are reunited, such that members of the two populations can interbreed but the hybrid offspring suffer some sort of selective disadvantage. This disadvantage may be a consequence of the hybrid offspring being less able to compete for resources or less able to obtain mates than the offspring produced by crosses between parents from the same population. Dobzhansky postulated that natural selection will act to increase the separation of the two incipient species when such hybrid inferiority occurs. Selection will favor those individuals that prefer to mate with others from their own population and will disfavor those individuals who for some genetic reason are more likely to make the mistake of mating with an individual from the other population. The reason natural selection will have this effect is that individuals that prefer to mate with members of their own population will produce offspring that are superior to those produced by crosses between parents from different populations.

Dobzhansky named this process "reinforcement," since he was arguing that natural selection would act to reinforce or strengthen the reproductive isolation of the incipient species. Reinforcement thus represents the direct hand of natural selection in sealing the process of speciation. The concept of reinforcement later fell out of favor, based on theoretical and verbal arguments rather than data. Coyne and Orr played a major role in rehabilitating the idea, based on a comprehensive review of hundreds of studies of patterns of reproductive isolation in *Drosophila* flies. We now at least accept the possibility that reinforcement can occur and play a role in sealing the process of speciation, but cannot say how prevalent it is.

Speciation that is caused by the spatial separation of populations has traditionally been referred to as "geographic" or "allopatric" speciation because of the presumed importance of geographical isolation in causing the evolution of reproductive isolation. Allopatric means, literally, "from a different fatherland." Coyne and Orr added the word "vicariant" to allopatric speciation to capture the idea that reproductive isolation evolved vicariously, or as an incidental by-product of geographic isolation, as opposed to its evolving as a consequence of natural selection acting directly on the evolution of reproductive isolation. Natural selection would play a direct role in speciation only if reinforcement occurred on secondary contact.

Coyne and Orr (2004) summarized the considerable evidence that supports the role of vicariant allopatric speciation in the formation of new species. For example, nineteen different experimental tests of the feasibility of such speciation have been performed on *Drosophila* and other flies (their table 3.1, pp. 88–89). In each experiment, investigators created artificial populations of flies, then selected for their ability to survive in different environments. Ten of the nineteen experiments yielded evidence for the incidental evolution of reproductive isolation, some after as few as five generations. Collectively, these experiments argue that reproductive isolation can evolve as a consequence of the combination of geographic isolation followed by adaptation to different environments. Such isolation represents incipient vicariant allopatric speciation because it arose as a by-product of populations being isolated from one another and adapting to different environments while they were isolated. It is important to realize that it is not necessary for all nineteen studies to produce reproductive isolation to prove the point; success in half of the studies shows that isolated populations adapting to different environments can evolve reproductive isolation from one another at least some of the time.

The biogeography of closely related species offers strong support for vicariant allopatric speciation having occurred in nature. Coyne and Orr cite investigators dating to Wagner (1873) who have observed that pairs of closely related species often inhabit geographic ranges that are separated by barriers like rivers, mountain ranges, or seas, implying that such geographic separation is essential to species formation. Our best examples of vicariant allopatric speciation include those for which we have the benefit of applying a modern "molecular clock" to the problem of estimating the time since sister species descended from a common ancestor, and comparing this time with the age of the geographical barrier that separates them. One such example is based on thirteen pairs of sister species of snapping shrimp in the genus *Alpheus* that are found in the Caribbean Sea and the Pacific Ocean. Their ranges were potentially contiguous prior to the most recent elevation of the Isthmus of Panama, which fully separated the Caribbean from the Pacific approximately 3 million years ago. Comparisons of DNA sequences and the application of the molecular clock show that individual species pairs were separated from one another for intervals that range from 3 to 10 million years. Those found in deeper water tend to have been separated for longer intervals of time, which makes sense when you consider that the isthmus was formed by a ridge of progressively shallower ocean before it emerged as dry land. It could well have been a barrier to species that inhabit deep water long before it was a barrier to those that occupy shallow water. We consider vicariant allopatric speciation to be the best explanation for this pattern because we can associate the estimated age of the species with the age of the barrier that separates them.

Peripatric (literally "around or beyond the same fatherland") speciation is a variation on the theme of geographic speciation because the process is generally initiated when a small number of migrants colonize a new habitat. Such new habitats, like the Galapagos Islands, tend to be sufficiently isolated for repeated colonization to be unlikely, so the movement of individuals between the source population and the new population is a very rare event. The new habitat may have a very different environment from that experienced by the original colonists before they emigrated, so the natural selection experienced by the colonists can be particularly intense. Such circumstances can promote rapid evolution and speciation.

There is also evidence that species can form without geographical isolation. Such "sympatric" (same fatherland) speciation occurs when the new species form within the "cruising range" of an individual, meaning that whatever geographical separation exists does not lie beyond the possible regular movements of individuals, so that geography alone is not a barrier to reproduction. A special property of all such modes of speciation is that natural selection must play a role in actively selecting for reproductive isolation as part of the process. Such an active role is a necessary component of vicariant allopatric speciation only if reinforcement comes into play when once-separated populations come back into contact with one another.

Coyne and Orr present the best available arguments that species really are objective realities, rather than simply points on a continuum as argued by Darwin, and that species can be defined by reproductive isolating mechanisms. For example, there is a close correspondence between the distinct species of birds, frogs, and reptiles recognized by scientists and those recognized by the indigenous people in different regions of New Guinea. The fact that two different groups of people using very different criteria come up with very similar classifications suggests that the distinctions are real. These categories, species, can also be distinguished with formal statistical analyses of morphology and on the basis of DNA sequences. However, at the same time, among the millions of species that are known today, many are described as "problematic" because traces of continuity between some of them remain. For example, we sometimes find regions in which different species seem to blend with one another via a bridge of populations that have properties that lie between the two, making it impossible to pigeonhole all organisms into discrete species. These problematic species are the source of Darwin's argument that speciation is an ongoing process. If new species can form, then we should expect to see living examples of this ongoing process that defy objective classification.

None of our current species concepts or research on the causes of speciation could have happened until it was generally accepted that speciation is an ongoing process as opposed to a supernatural occurrence, with species being the products of acts of special creation. Until the scientific community accepted the concept of the "transmutation" of species, it was not possible to develop our modern definitions of species nor consider how and why speciation occurs. My argument for Darwin's primary role in establishing the study of speciation as a legitimate scientific endeavor is analogous to the one that I made in chapter 3 for Darwin's role in stimulating interest in variation. Before Darwin, variation was an annoyance. After Darwin, variation became biologically important and a target for study. Likewise, before Darwin, very few individuals even considered speciation to be possible. There was precedent for the idea of species transmutation, and proposals had been made before Darwin for how new species could form, but none of them remain part of science today. After Darwin, speciation was widely accepted and became a legitimate area for research. The post-Darwinian approach to the subject was fundamentally different from the pre-Darwinian approach because of Darwin's proposed mechanism for evolution. Dobzhansky's (1937), Mayr's (1942), and Coyne and Orr's (2004) books can be viewed as progress reports in the ongoing endeavor to understand species and speciation. If one were to rephrase Mayr's (1942) opening paragraph as an appraisal of the *Origin* in light of the historical context in which Darwin wrote it, then one would have to begin by accepting that the *Origin* really is about the origin of species. One could fault Darwin for being so single-minded in promoting speciation only as a process and in describing species as just points in a continuum. But, in the end, criticism of this shortcoming is an unreasonable remonstration. It was necessary to dislodge the doctrine of special creation before it was possible to define species or characterize the process of speciation. The latter became tasks for those who followed Darwin.

References

Coyne, J. A., and H. A. Orr. 2004. *Speciation.* Sunderland, MA: Sinauer Associates.

Dobzhansky, T. 1937. *Genetics and the origin of species.* New York: Columbia University Press.

Goldschmidt, R. 1940. *The material basis of evolution.* New Haven, CT: Yale University Press.

Mayr, E. 1942. *Systematics and the origin of species.* New York: Columbia University Press.

Mayr, E. 1995. Species, classification and evolution. Pp. 3–12 in *Biodiversity and evolution*, ed. R. Arai, M. Kato, and Y. Doi. Tokyo: National Science Museum Foundation.

Provine, W. B. 1971. *The origins of theoretical population genetics.* Chicago: University of Chicago Press.

Chapter 9

Variation under Nature II

Here I revisit Darwin's second chapter to discuss it again in the light of speciation. Darwin says little in this chapter about individual variation and its role in natural selection. He deals only briefly with variation as we study it now, so I devoted most of my previous discussion (chap. 3) to addressing the consequences of Darwin's success in convincing the world that "individual differences" are not noise, but are rather the notes that could comprise a symphony, if properly assembled: "These individual differences are highly important for us, as they afford materials for natural selection to accumulate, in the same manner as man can accumulate in any given direction individual differences in his domesticated productions" (*Origin*, p. 45).

Darwin devotes most of this chapter to documenting the continuum between individual variations and species. His argument, as accurately described by Mayr (1942; see chap. 8), is that species simply represent points in a continuum of change. His goal is not to define what a species is, but rather to document the consequences of the process of evolution by natural selection. If species are truly the products of acts of special creation, then they should always be distinct and easily defined. If they are instead part of an ongoing process of change, then the boundaries that separate species should often be indistinct.

Darwin first offers an expanded lexicon of variation. A key term in his lexicon is "variety," which he never formally defines. It is implicit that he is usually talking about a population of individuals that share some distinctive traits that distinguish their population from other populations of that species. Varieties were distinct enough to be given their own names in the descriptions of flora and fauna of the day. The traits that defined a variety presumably accumulated through the process of natural selection, in the same way that the traits that defined a breed of domestic animal accumulated through artificial selection. Another term in Darwin's lexicon is "monstrosities," which are injurious forms of individual variation. The short-legged ancon sheep is an example of such a monstrosity. It appeared suddenly but was then cultivated by some farmers who sought to develop herds of short-legged sheep that could be more easily confined. "Polymorphisms" are discrete phenotypes that are sometimes found within populations. The snail *Cepea nemoralis* can be brown, yellow, or pink with or without bands, and can have from one to five bands. These different-appearing individuals are often found together, but the relative abundance of different types of morphs varies from place to place. Darwin does not know what to make of such variation, but suspects that it is unimportant. (He was wrong, but it took over eighty years for polymorphisms to become the targets of research and for their significance to be revealed.) Darwin then addresses the continuity that he perceives from variety to subspecies to species. He adopts the generic term "form" to describe populations that are distinct from one another. But the question remains regarding how best to classify these forms. Are different forms simply varieties of the same species, different subspecies, or different species?

A usual criterion, inconsistently adhered to, is that if different forms are geographically separated and if there is no sign of intermediates between them, then they are named as different species. If there are linking populations that have individuals that are intermediate in character, then a systematist will be inclined to call one of them, either the first described or the most widespread and abundant, a species and the others varieties of that species. However, systematists themselves vary in how they interpret variation: "In many cases . . . one form is ranked as a variety of another, not because the intermediate links have actually been found, but because analogy leads the observer to suppose either that they do now somewhere exist, or may formerly have existed; and here a wide door for the entry of doubt and conjecture is opened" (*Origin*, p. 47).

Darwin's thesis is that this "doubt and conjecture" is a consequence of speciation being an ongoing process. If new species are continuing to emerge and if this emergence is a slow, gradual process, then an unambiguous classification of all populations of all organisms into individual species will often not be possible. A naturalist who tries to pigeonhole every local form into either a variety, a subspecies, or a species will be confronted with a continuity of variation between forms that defies such rationality.

Doubtful Species

Darwin's evidence for the indistinctness of species includes the disputes between naturalists concerning how to classify them: "Compare the several floras of Great Britain, of France or of the United States, drawn up by different botanists, and see what a surprising number of forms have been ranked by one botanist as good species, and by another as mere varieties" (*Origin*, p. 48). He cites a Mr. H. C. Watson, who drew up a list of 182 British plants that had been classified by some experts as species but by others as varieties. He compares classifications of the same genera by a Mr. Babington, who divided them into 251 species, and a Mr. Bentham, who divided the same genera into only 112 species. This means that 139 of Mr. Babington's "species" were relegated to the status of varieties by Mr. Bentham. Darwin cites other examples of ambiguity. For example, many types of plants and animals found in Europe can also be found in North America. The populations on each continent are often different in some way, but he wonders how one can decide whether or not these differences are sufficient to consider them to be different species. Likewise, plants and animals found on different islands often differ, but how much difference is enough to call them separate species? Because such populations are geographically isolated from one another, it is never possible to find populations that are intermediate in character between the two. The presence or absence of such intermediates between two forms found on the same landmass is often cited as evidence for deciding whether they are the same species. Darwin relates his experiences working with systematists on the classification of the birds of the Galapagos Islands and those from South America: "I was much struck how entirely vague and arbitrary is the distinction between species and varieties" (*Origin*, p. 48). "A wide distance between the homes of two doubtful forms leads many naturalists to rank both as distinct species; but what distance, it has been well asked, will suffice?" (*Origin*, p. 49).

One might think that this debate is a consequence of systematists having inadequate information about the species under study. If more is learned about them, then a clear classification will emerge. Darwin points out that the opposite is true: "it is in the best-known countries that we find the greatest number of forms of doubtful value" (*Origin*, p. 50). There is "no clear line of demarcation" (true at the time he was writing) between varieties, subspecies, and species. "These differences blend into each other in an insensible series; and a series impresses the mind with the idea of actual passage" (*Origin*, p. 51). By "passage," Darwin is implying that this "insensible series" represents the continuous transition from differences between populations to differences between species.

Darwin argues that individual differences should be thought of in a new way: not as annoying departures from some imagined ideal form, but as the raw material for selection to act upon. Varieties, which are often locally adapted populations, are the products of the process of natural selection having caused some individual differences to become characters shared by all members of the population. Varieties will become progressively more distinct as such individual differences are accumulated through the process of natural selection, in the same way that humans practicing artificial selection can accumulate individual differences to create distinct breeds of plants and animals. It is this same progressive accumulation of differences via natural selection on natural populations that causes varieties to become distinct subspecies, then distinct species. "I attribute the passage of a variety, from a state in which it differs very slightly from its parent to one in which it differs more, to the action of natural selection in accumulating . . . differences in structure in certain definite directions. Hence I believe that a well-marked variety may be justly called an incipient species" (*Origin*, p. 52).

I formally separated natural selection from speciation in this reader's guide because the two topics are so distinct in the modern study of evolution. Evolution by natural selection leads to adaptation. However, such evolution does not necessarily lead to speciation. Speciation is only sometimes a consequence of evolution by natural selection. Darwin makes this same point when he states: "It need not be supposed that all varieties or incipient species necessarily attain the rank of species" (*Origin*, p. 52). He imagines that very few attain this status. Varieties may go extinct, may endure for the long term as just varieties, may expand and exterminate the parent species, or may attain independent species status and coexist with the parent species. Nevertheless, he sees the passage from individual variation to speciation as a continuous one.

Darwin concludes his argument with his own concept of species: "From these remarks it will be seen that I look at the term species, as one arbitrarily given for the sake of convenience to a set of individuals closely resembling each other, and that it does not essentially differ from the term variety, which is given to less distinct . . . forms. The term variety, again, in comparison with mere individual differences, is also applied arbitrarily, and for mere convenience sake" (*Origin*, p. 52).

I cannot argue with Mayr's assessment (1942; see chap. 8) that Darwin's *Origin of Species* does not define species and lacks our modern concept of species. However, Darwin is concerned here with the process of speciation, rather than with the definition of a species. Every bit of his argument is built on the premise that the distribution of variation in nature reflects this ongoing process of natural selection, causing diversification among populations and hence the formation of new species. The continuity of this differentiation of populations made it difficult to define species boundaries. This continuity means that species cannot be the products of acts of special creation, since such acts would have resulted in species that were discrete and easily defined entities.

Darwin then seeks evidence for this continuity of change by looking for patterns that might link a species' distribution, its abundance, and the degree of variation that occurred within it: "Guided by theoretical considerations, I thought that some interesting results might be obtained in regard to the nature and relations of species which vary most by tabulating all the varieties in the several well-worked floras" (*Origin*, p. 53).

A "flora" is a summary, usually in book form, of all of the plants found in a given area. It generally consists of a description of each species, including its varieties and subspecies, and reports when each variety/species is found. The modifier "well-worked" implies that Darwin concentrated on floras developed for parts of the world that had been well studied by a number of investigators and hence were well known.

If speciation really is an ongoing process and species are just arbitrary points in a continuum, then a systematic tabulation of species by their properties should give some indication of this process. Some species should, because of some unknown property, be in the process of expansion and diversification, while others may be headed down the path toward extinction. Darwin felt that his tabulations might reveal something about such patterns of change. He classified all the plant species by the size of their range and the number of described varieties. He then considered their genus-level classification. A genus (plural genera) is the next step up the taxonomic hierarchy from species and comprises species that are all more closely related to one another than they are to species in different genera. For example, lions and tigers are different species in the genus *Panthera*. We customarily refer to a species by its Latin binomial, which is its genus plus its species designation, so we call the lion *Panthera leo* and the tiger *Panthera tigris*.

Darwin characterized each genus first by the number of species it contained, then by the range of each species, and finally by the number of varieties per species. A species that consists of many varieties is one that, by Darwin's interpretation, is actively diversifying and is likely to spawn new species. A genus that contains many species is, likewise, one that has recently experienced such diversification. Darwin does not actually present the tables; as the *Origin* is just an "abstract" of his longer work in progress (but in the end never published), he simply summarizes his findings. I present them below under the subheadings he used in his second chapter.

Dominant Species Vary Most

"Dominant" refers to species that are the most abundant and/or widespread, and hence the most frequently encountered and recognized. Darwin found that widespread species were more likely to have named varieties, and he posits that this occurs because they encounter a greater diversity of environments and biotic interactions that in turn select for locally adapted populations. Species that are the most abundant are also the ones most likely to have multiple recognized varieties. Darwin had concluded earlier (chap. 2) that large populations facilitate the development of new breeds through artificial selection and new varieties through natural selection because they contain more individual variations for natural selection to act upon. Being geographically widespread and abundant are thus two properties that promote the formation of distinct varieties.

Species of Large Genera Are Variable

Darwin subdivided all genera into two categories—large and small, meaning that they contained either many or few species. Such a dichotomization is a crude form of classification because there is such a large range of genus sizes within the "large" and "small" subdivisions, but the dichotomy was sufficient to reveal a pattern. He found that large genera were more likely to contain species that were abundant, had multiple varieties, and spanned large geographic ranges. In this section he argues that "where many large trees grow, we expect to find saplings," meaning that if speciation is more likely in a given genus, then the individual species in that genus will inherit whatever properties promoted diversification and hence also be more likely to diversify into distinct varieties, which are in turn the precursors of new species. "On the other hand, if we look at each species as a special act of creation, there is no apparent reason why more varieties should occur in a group having many species than in one having few" (*Origin*, p. 55).

Darwin performed additional analyses on plants and similar analyses on beetles and arrived at the same results: "These facts are of plain signification on the view that species are only strongly marked and permanent varieties; for wherever many species of the same genus have been formed, or where, if we may use the expression, the manufactory of species has been active, we ought generally to find the manufactory still in action" (*Origin*, p. 56).

These were general trends, rather than strict rules. Darwin still found large genera that consisted of species with few varieties and small genera that consisted of species with many varieties. The geological record tells us that such exceptions should be found, since the history of life reveals that fortunes of large, expanding lineages sometimes change in ways that lead to decline, then extinction. Likewise, small lineages may sometimes expand and diversify. Darwin summarizes by saying: "All that we want to show is, that where many species of a genus have been formed, on average many are still forming, and this holds good" (*Origin*, p. 56).

Species of Large Genera Resemble Varieties

The amount of difference between varieties plays a role in determining whether or not they should be classified as species. Varieties that are not too different from one another may be considered to be the same species even if we do not find evidence of intergradations between them. Darwin found that large genera were more likely to contain species that had the property of varieties, meaning they were less well separated from one another than were the species of small genera. Furthermore, the species of large genera were not all equally distinct. They were often instead divided "into sub-genera, or sections, or lesser groups" (*Origin*, p. 57). In this regard, species seemed to cluster as satellites around other species in the same way that varieties clustered as satellites around a parent species, although the different species within a subgenus were more distinct from one another than were the different varieties found within a species in that same genus. Furthermore, Darwin found that species within large genera tended to have geographic ranges quite similar in size to those of named varieties, which suggested that they had recently made the transition from variety to species. The implication is that such species groups within genera are the products of what had once been dominant species with many varieties.

Darwin sees the potential for a common mechanism underlying the way varieties cluster around a species and the way the species often separate out into clusters that are recognized by subgeneric or other types of designations. He proposes that the mechanism that causes this clustering is the "divergence of character" (introduced in chap. 4). He argues that the most severe competition for any organism will come from its closest relatives because they will be most similar in their requirements. Because of this competition between close relatives, be they varieties of the same species or closely related species, natural selection will favor the evolution of increasing differences between them. The clustering of varieties within a species, and of closely related species within a genus, are seen as the products of such divergence of character. This selection for closely related forms that utilize their environment in different ways in turn causes them to become morphologically distinct from one another. The different beaks of the Galapagos finches (see chap. 1 above) are an example of such a divergence of character. In the case of the finches, different beak shapes are associated with different diets. We will revisit this concept in the next chapter.

Darwin's conclusion is: "In all of these several respects the species of large genera present a strong analogy with varieties; and we can clearly understand these analogies, if species have once existed as varieties, and have thus originated: whereas, these analogies are utterly inexplicable if each species has been independently created" (*Origin*, p. 59).

Evolution Today: Plethodon Salamanders of the Eastern United States

Darwin leaves us with the impression that species often cannot be unambiguously classified, but this is a judgment he passed at a time when there was not even a concept of what a species was nor of what defined the boundaries between species. This is no longer the case. We now have the biological species concept, know the importance of reproductive isolation as a criterion for defining species, and know how to characterize reproductive isolation. We also have the genetic tools for identifying hybrids in nature and making inferences about the extent to which the gene pools of different species are separated from one another.

One example from modern systematics shows how the confusion and apparent indistinctness of species can be addressed with molecular methods. Richard Highton has devoted his career to the classification of salamanders in the genus *Plethodon*. This genus is large and widespread in North America, but most species are found in the woodlands of the eastern United States. Most amphibians lay their eggs in water and have a larval stage that metamorphoses into an adult that lives on land. Frogs generally begin life as tadpoles that later absorb their tails, sprout legs, then venture onto the land. The change in morphology of salamanders is less dramatic, but they are similar in starting life in the water, often with featherlike external gills, then metamorphose into terrestrial adults. Salamanders in the genus *Plethodon* are an exception: they have direct development, which means that the eggs are laid on land and hatch out tiny salamanders that are just like miniature adults. This freedom from water means that they are not tied to areas that are within reach of appropriate ponds and streams for reproduction and can have much larger ranges. They can also be incredibly abundant, perhaps because their abundance is not capped by the availability of larval habitat. In fact, they are often the most abundant vertebrate, in both numbers and biomass, with local population densities of up to thousands per hectare. They are invisible to most people because they spend most of their time underground, but they have often been a source of happiness to aspiring young herpetologists, as I can say from experience, because they can persist in small woodlots in the middle of dense suburbs and may be the only amphibian wildlife to be found in the neighborhood.

The history of classification of this genus is like that of the competing classifications of British plants by Mr. Bentham and Mr. Babbington cited by Darwin. A few species are polymorphic: one morph has a reddish stripe down the middle of the back while the other is solid black. This difference was perceived by some as sufficient to name them as different species. Sometimes different species have been classified as one because, while differences in color pattern and morphology could be seen between populations, these differences often seemed too small to define them as different species. In fact, it is often the case with these salamanders that species are hard to define because there are so few morphological differences between them.

Highton looked beyond morphology to define species. He extracted enzymes from their tissues, then separated them with starch gel electrophoresis, and in that way assessed one form of genetic variation. (Extracts of tissues were placed in small wells in a slab of gelatinized starch, then subjected to an electric field. A subset of the alleles found at any given locus coded for proteins that differed in their electrical charge, which influenced how fast they would migrate in an electrical field.) Highton could often differentiate between species that were very similar in appearance by examining the allelic differences at enzyme loci, which could be seen as bands of enzyme activity at different locations on starch gels. If the ranges of two similar species overlapped with one another, then it was possible to discriminate between them if they had different alleles at some of these enzyme loci. It was also possible to see if two "species" hybridized with one another. To look for evidence of hybridization, Highton identified 140 geographic regions, called zones of contact, where the ranges of closely related species overlapped with one another. In most of these areas of overlap, he collected salamanders along a path that crossed from the range of one species, through the area of overlap, and into the range of the second species. If each species was homozygous for a different allele of some enzyme, then their hybrids could be recognized as heterozygotes, which have one allele from each parent.

Sometimes Highton found species that were all but identical in appearance living in the same place, with no evidence that they ever interbred. That is, he found individuals that were homozygous for either one or the other allele that distinguished the two "species," but did not find heterozygotes that had one allele each from a parent of the different species. Because heterozygotes were not found, reproductive isolation appeared to be complete. For example, he found that the species *Plethodon websteri* and *P. dorsalis* had ranges that overlapped by around 1 kilometer along Highway 31 in Jefferson County, Alabama. In each collection made in this narrow region of overlap, he found members of both species present, but never a hybrid between them. In other contact zones between other species, he could sometimes find hybrids, but only along a narrow zone of overlap between the two species. Here, premating isolation was not complete, but there was physical evidence that interbreeding was limited and the gene pools of the two species remained distinct, perhaps because of postmating isolation mechanisms. Conversely, he discovered one hybrid zone where the two forms were genetically less distinct from one another compared to the species in other contact zones he had studied. He found that the zone of hybridization between them was broad and that there was evidence of a complete mixing of the gene pools in the hybrid zone. In this case, it appeared that he was looking at the reunion of two formerly isolated populations of the same species.

The composite of all the studies by Highton and his colleagues (Highton 1995) revealed a portrait of the history of this group. Such measures of genetic differences between species can yield estimates of overall "genetic distance," and these results can be synthesized with geological history to yield estimates of how long it has been since they shared a common ancestor. Highton and his colleagues applied these principles to the genus *Plethodon* and found that the eastern populations divided into four "species groups," three of which were descended from a common ancestor that lived during the Miocene epoch (28.8–5.3 million years before the present). The fourth species group was derived from a common ancestor that lived during the Oligocene, the epoch that came before the Miocene. It was represented by at least two descendant species by the time of the Miocene. The four species groups now contain from two to twenty-eight species, for a total of forty-five known species alive today in the eastern woodlands. There may well be more to be discovered, since so many were discovered in the past two decades.

Much of this diversity seems to be the result of a burst of speciation during the Pliocene epoch. The Pliocene was a time of gradual cooling and drying when grasslands replaced lowland forests. These salamanders are confined to woodlands, so their habitats contracted into "sky islands" on isolated mountaintops that were separated by expanses of dry savanna. The favored hypothesis for the cause of the burst of speciation of *Plethodon* during the Pliocene is that populations of the four ancestral species were isolated on these sky islands and that speciation occurred during this interval of enforced allopatry, or geographic isolation. There may well have been more species of *Plethodon* alive at the beginning of this epoch, but all except four disappeared, leaving no living descendants. These are small, delicate animals that rarely appear in the fossil record, so we have no way of knowing what their diversity was like in the past. The skeletal differences between the living species are so small that even if we had a fossil record, it would not have proved very informative about species diversity. The most recent retreat of the glaciers and the subsequent expansion of moist woodlands have allowed many of these salamanders to expand their ranges. As many as five species can be found together in some places.

Highton's work shows that one can now apply such molecular methods alongside traditional assessments of morphology to often resolve the apparent continuum between variety and species. Having a biological species concept, in which species are defined by reproductive barriers, has played an important role in our ability to recognize species boundaries. Whereas Darwin perceived species as often being indistinct, we can now often define them unambiguously and show that species are not just subjective or an arbitrary point on a continuum, as was perceived by Darwin.

The common ground between then and now is that we still have an abundance of "forms" that defy such unambiguous classification and that serve well as a record of what Darwin described as a passage across species boundaries. Highton's salamander work shows how the added benefit of genetic data can help characterize the diversity of possible relationships between genetically distinct populations that represent these passages. In some cases, forms that can be seen as distinct when found in geographic isolation from one another can also be shown to freely interbreed when they come into contact with one another. In other cases, when distinct populations come into contact they hybridize, but the zone of hybridization is narrow and there is no evidence of continued hybridization between the two entities, perhaps because of postmating isolation. In yet other cases, populations coexist, and yet premating isolation is so complete that hybrids are never seen.

References

Highton, R. 1985. The width of the contact zone between *Plethodon dorsalis* and *Plethodon websteri* in Jefferson County, Alabama. *Journal of Herpetology* 19:544–46.

Highton, R. 1995. Speciation in eastern North American salamanders of the genus *Plethodon*. *Annual Reviews of Ecology and Systematics* 26:579–600.

Chapter 10

Natural Selection II

As with "Variation under Nature" (*Origin*, chap. 2), the fourth chapter of the *Origin* cuts across the topics of natural selection and speciation. I reviewed those components that dealt with natural selection in my chapter 5. Here I deal with those parts that pertain to speciation, using Darwin's original subheadings.

Circumstances Favourable to Natural Selection

Because Darwin constantly emphasizes the continuity between individual differences and the formation of locally adapted races, then varieties, then species, he consistently treats the subjects of natural selection and speciation together. In my chapter 5, I summarized Darwin's thoughts about the conditions that would best promote the formation of a locally adapted population, which is the first step to the formation of a distinct variety. He thought that varieties would arise more readily in large populations, since they would harbor more variation for natural selection to act upon. He also argued that isolation would facilitate the formation of a locally adapted population, since a constant influx of individuals from other areas, where the natural selection might favor the evolution of different traits, would slow down local adaptation.

This logic leads directly to Darwin's most prominent model for circumstances that promote speciation. Darwin's model links his emphasis on large populations and isolation with a Lyellian perspective of geological history. Charles Lyell argued that the surface of the earth oscillates, with some portions of the earth experiencing gradual increases in elevation while others were subsiding. Such oscillations caused the distribution of dry land and ocean to be in a state of flux. When continents were elevated, they would be uninterrupted expanses of dry land. As they subsided, they would become archipelagoes as rising oceans filled lower-lying land.

Darwin saw clear evidence for such changes in elevation at the *Beagle's* first landfall, on the island of St. Jago (see the introduction above). This early experience convinced Darwin of the reality of Lyell's ideas about geological processes and primed him to see the world though the lens of Lyell's theory for the remainder of the voyage.

Later, Darwin witnessed dramatic examples of the turbulent and fluid nature of the earth and changes in the elevation of land. On January 19, 1835, the *Beagle* was anchored in the Bay of San Carlos in the island of Chiloe, off the coast of Chile. During the night, the volcano Osorno erupted and presented the "magnificent spectacle" of plumes of cinders exploding from the top of the mountain and masses of molten rock being cast into the air and tumbling down the mountainside. Darwin was surprised to learn that the volcano Aconcagua, 480 miles north on the mainland of Chile, erupted on the same night "and still more surprised to hear, that the great eruption of Coseguina (2700 miles north of Acongua), accompanied by an earthquake felt over 1000 miles, also occurred within six hours of this same event" (Darwin [1860] 1962, pp. 293–94).

On February 20, 1835, Darwin was hiking in the forest near the Chilean town of Valdivia. He was lying on the ground resting when an earthquake struck and shook the ground for two minutes, "but the time appeared much longer." He observed: "A bad earthquake at once destroys our oldest associations: the earth, the very emblem of solidity, has moved beneath our feet like a thin crust over fluid (Darwin [1860] 1962, p. 303).

Most of the houses in the town of Valdivia were made of wood and weathered the quake without falling. However, when the *Beagle* entered the harbor of Concepcion, north of Valdivia, on March 4, Darwin and his crewmates found that it and neighboring towns had been destroyed by the quake and the three tsunamis that followed. They also found that the harbor had been elevated, so that a rocky shoal that had once been fully submerged was now exposed. The elevation was most remarkable on the nearby island of Santa Maria, where Darwin saw rocks covered with fresh mussel shells now well above the high-water mark. Darwin found other seashells on the island at elevations up to 1,000 feet and concluded that many such earthquakes over a long interval of time had pushed the now-dry land out of the ocean.

Darwin later learned that the earthquake of February 20 was coincident with earthquakes and volcanic eruptions elsewhere. "The island of Juan Fernandez, 360 miles to the N.E., was . . . violently shaken, so that the trees beat against each other, and a volcano burst forth under water close to shore" (Darwin [1860] 1962, p. 313). He concluded that the coincidence of volcanic eruptions and earthquakes on January 19, then again on February 20, all along the same mountain chain, in association with the presence of "upraised recent shells along more than 2,000 miles of the western coast" of South America, suggested a cause-and-effect relationship between volcanoes, earthquakes, and changes in the elevation of the land: "From the intimate and complicated manner in which the elevatory and eruptive forces were shown to be connected during this train of phenomena, we may confidently come to the conclusion, that the forces which slowly and by little starts uplift continents, and those which at successive periods pour forth volcanic matter from open orifices, are identical" (Darwin [1860] 1962, p. 314).

A few days later the *Beagle* docked in Valparaiso, and Darwin embarked on a trek through the Andes while those on the ship took soundings in the harbor. On March 20 he reached the highest point of the road through the mountains and found fossilized shells: "In these upper beds shells are tolerably frequent; and they belong to about the period of the lower chalk of Europe. It is an old story, but not the less wonderful, to hear of shells which were once crawling on the bottom of the sea, now standing nearly 14,000 feet above its level" (Darwin [1860] 1962, pp. 321–22).

It had long been known that portions of the surface of the earth that were once under the ocean could now be found at high elevations in mountain ranges such as the Alps; their rocks contained the fossil remains of marine origin as documentation of their former subsidence. These were the sorts of observations that motivated Lyell to propose that the surface of the earth was in a constant state of flux. Darwin saw evidence for such change when he witnessed firsthand how volcanic eruptions and earthquakes could elevate land. He inferred that very large changes in the surface of the earth could be attributed to the cumulative influence of many such events; all that was required was the passage of an immense interval of time and repeated events. After the voyage, he published a monograph on the geology of South America in which he proposed that the Andes were a product of events like the earthquake that he had witnessed, repeated again and again over a time span of millions of years, and that portions of Patagonia were a former seabed that had been uplifted along with the mountains.

Elevation in one part of the world had to be matched by subsidence elsewhere if Lyell's theory was correct. It was this logic that led Darwin to propose a new hypothesis for the origin of coral atolls (see the introduction), which is that they had grown around the emergent tip of a volcano that later subsided into the ocean. As the volcano sank, the coral continued to grow to the surface. Darwin published this theory in a different monograph after the voyage. It, too, has stood the test of time.

Darwin later combined his geological observations with his emerging ideas about the transmutation of species. He envisioned such cycles of changing elevation as being ideal for species formation. During the continental phase, large populations, which harbor more variation, would begin the development of locally adapted populations. Subsidence and island formation would isolate these populations from immigration and give them the opportunity to perfect local adaptations. Subsequent elevation would reunite them with other such isolates and generate a round of competition for jointly used resources. Darwin saw this cyclical pattern of change—during which large, extensive populations are isolated into smaller, isolated populations that later reunite, intermix, and compete—as creating the ideal incubator for the formation of new, highly competitive species. Along with these repeated cycles of elevation and subsidence, we now have evidence for other processes that support Darwin's model of speciation. Recall that Highton's work on salamanders in the genus *Plethodon* (chap. 9) revealed that climate change created "sky islands" that had the same effect as Darwin's oscillation cycle on species distribution and speciation. In fact, islands can be thought of in a broader sense as isolated habitats, which come in a variety of configurations. They can be land separated by oceans; mountaintop "sky islands" that harbor moist, cool habitats separated by warm, dry lowlands; or ponds of freshwater separated by dry land. Darwin's hypothesis will apply whenever such "islands" are sufficiently large to sustain a viable population and when changing circumstances periodically reunite and divide them.

Darwin's proposal here, in chapter 4 of the *Origin*, that continents are the primary site for the formation of new species seems to contradict his personal experience with the Galapagos Islands, which were home to hundreds of endemic plants and animals whose closest relatives were found in South America (see chap. 1 above). This pattern of island endemism, where new, lifeless volcanic islands are colonized by immigrants that diversify into new species, is repeated on oceanic islands around the world. It fosters the impression that the more permanent isolation on oceanic islands creates a more effective stage for the formation of new species. Darwin argues, however, that the prevalence of endemic species on islands may be deceptive: "to ascertain whether a small isolated area, or a large open area like a continent, has been most favourable for the production of new organic forms, we ought to make the comparison with equal times; and this we are incapable of doing" (*Origin*, p. 105). His argument is that the strange endemics that characterize islands are products of their prolonged retention, facilitated by their isolation, rather than their more rapid formation. He thinks instead that "largeness of area is of more importance [than isolation], more especially in the production of species, which will prove capable of enduring for a long period, and of spreading widely" (*Origin*, p. 105).

He supports his argument with observations on the success of invading species. The smaller continent of Australia had already been repeatedly invaded by species from the much larger Eurasian landmass. Plants and animals from Eurasia were also invading (and continue to invade) isolated islands. Darwin argues that Australia and oceanic islands are smaller and hence are inhabited by species that tend to be less competitive than those spawned on larger landmasses, where populations are persistently larger and where the struggle for existence is more intense. The species of continental origin are successful at invading smaller landmasses and displacing the native species because they are products of this more intense struggle for existence.

Darwin also found that islands or island-like habitats were the more likely haunts of "living fossils"—relics of ancient forms of life that had once been more abundant and widespread. He argues here that such relics can persist in small, isolated pockets of habitat where they are shielded from competition with more recently evolved groups of organisms. His examples include denizens of freshwater basins, which tend to be distributed as small, isolated islands surrounded by land. Here is where we find the remainders of orders of freshwater fishes like gars, bowfins, sturgeons, and paddlefish, which are remnants of groups of fishes that were once much more widespread and diverse. These are also the habitats where we find lungfish (*Lepidosiren*), which are among the last remnants of the fish group from which amphibians were descended, and the duck-billed platypus (*Ornithorhynchus*), which is one of the few surviving egg-laying mammals.

Darwin maintains that continental landmasses will be the most potent incubators for the formation of new species in spite of the apparent origin of new species on islands. The large populations that continents sustain will be host to the most variation and will be most able to diversify and exploit the available resources. Subsidence and rising sea levels create periods of isolation, shielding populations from emigration and enabling the process of natural selection to perfect local adaptation. Periods of increased elevation and falling sea levels will reunite the islands and allow the newly formed varieties to expand their ranges and come into competition with the other former isolates. Such repeated cycles will spawn the highly competitive, dominant species and successful invaders that we associate with continents today.

Darwin entertained other possible modes of speciation. We can infer that he thought isolation alone was sufficient to set the stage for speciation, from his consideration of endemism on islands and from the role that Galapagos endemics played in causing him to think about the transmutation of species in the first place. He concluded that immigrants from South America had invaded the Galapagos and then evolved into new species once isolated in their new habitat. However, his emphasis on Lyell's oscillating continent model and the importance of large populations led him to conclude that speciation was more likely on continents.

In this chapter, Darwin also argues that speciation can occur without any enforced allopatry. For example, he envisions a population of wolves that range from lowlands to mountains. These differences in habitat would be associated with differences in the types of prey available, which would in turn result in the evolution of specialized varieties of wolves best suited to these different habitats. These varieties could be expected to "cross and blend where they met" (*Origin*, p. 91). However, the hybrids might be at a disadvantage because they would be less effective than the parents in utilizing either habitat. If this was true, they would suffer in the struggle for existence and hence would be selected against. In this fashion, the two specialized varieties could eventually rise to the status of being separate species.

Extinction

Darwin introduced the topic of extinction in his third chapter ("Struggle for Existence," discussed in my chap. 4), then expands on it here in his fourth chapter. Extinction was an accepted but disturbing reality at the time the *Origin* was published. If your worldview is that all species are products of divine acts of special creation, then extinction has an ominous and mysterious quality. Why would God create an organism and then allow it to die out? Georges

Cuvier is credited with proving the reality of extinction earlier in the nineteenth century with his descriptions of fossils of species that were related to but clearly different from those alive today. Likewise, Richard Owen described some of the gigantic reptiles that he united under the classification "dinosaur," or "terrible lizard." The weight of evidence in the form of fossils, often from giant and bizarre organisms that could no longer be found living anywhere on earth, established extinction as a reality. But why does extinction occur?

To Darwin, extinction is a by-product of evolution. If the struggle for existence, or interactions between living things, is the engine of evolutionary change, and if it can cause the origin of new species, then it can cause the extinction of others as well. Darwin's is a competitive universe in which every gain by some population is attained at the expense of some other population. The losers in such a struggle will become less abundant. "Rarity," Darwin says, "is the precursor to extinction" (*Origin*, p. 109). He argues that any living species that is rare will be more susceptible to extinction. As seasons and the fortunes of its enemies vary, so will its own abundance. A prolonged period of unfavorable conditions, which may be no more than conditions relatively more favorable to a competitor or predator, can tip the balance and push a rare species to extinction. Darwin argues, from the geological record, that the number of species is not progressively increasing, as would be the case if new species were formed and old species were not lost. Since the number of places in nature is finite, it follows that the origin of each new species will be matched by the extinction of some other species and that there will be an approximate equilibrium in diversity.

Darwin observes that it does not follow that any region of the earth is now fully stocked with plants and animals. He knew of the successful invasion of organisms into a wide variety of habitats, even in continental regions that already had a high diversity of species. For example, the Cape of Good Hope in Africa was known to have the most diverse plant communities on earth, yet it had been host to successful foreign invaders without the extinction of any resident species; the key was always to find a different way of utilizing available resources. Furthermore, Darwin's tabulations of species reported in the previous chapter showed that abundant, widespread species had more varieties and hence were more likely to spawn new, locally adapted populations than were species that were rare and/or had small ranges. "Hence, rare species will be less quickly modified or improved within any given period, and they will consequently be beaten in the race for life by the modified descendents of the commoner species" (*Origin*, p. 110).

His argument is thus that the formation of successful new varieties and species via the struggle for existence will be matched by the decline of others. The most severe competition will be between close relatives, since they will be the most similar in their resource needs: "each new variety or species, during the progress of its formation, will generally press hardest on its nearest kindred, and tend to exterminate them" (*Origin*, p. 110). To make his point, Darwin turns again to his analogy with artificial selection. New breeds of cattle or sheep, or new varieties of flowers or grain, often emerge to displace older, inferior domesticates. Darwin quotes an agricultural writer who describes how long-horned cattle "were swept away by the short-horns, as if by some murderous pestilence" (*Origin*, p. 111).

Darwin revisits the phenomenon of extinction when he deals with the fossil record; his views on the fossil record appear in part 3 of this book (chap. 18). There I offer a modern view of the phenomenon as well.

Divergence of Character

Here again we revisit a concept that Darwin introduced in his "Struggle for Existence" chapter. Darwin's "principle of divergence" represents what he thought to be one of the most potent components of the struggle for existence, which means that it would also be an engine behind the formation of new varieties and species, and an ultimate cause of extinction. He considered it to be as important as natural selection to the full development of his theory of evolution, yet it is rarely recognized today as a distinct feature of the *Origin*. It represents a late insight, dating to the mid-1850s.

Darwin observes that all organisms are constantly striving to increase in number, and do so by finding different ways of earning a living. A consequence of this striving is the formation of distinct varieties that utilize the habitat in a slightly different fashion. These varieties "are species in the process of formation, or are, as I have called them, incipient species" (*Origin*, p. 111). However, the differences between species are far more distinct than are those between varieties. How do the small differences among varieties become magnified into the larger differences among species?

Darwin begins his argument by drawing an analogy with artificial selection. He imagines an ancestral horse selected on by two breeders with different goals, one for speed, the other for strength. With the passage of time and continued selection, each breeder will develop breeds that are better suited to either specialization. At the same time, horses that fall between the two, being neither as fast nor as strong as the emerging breeds, will tend to disappear. In time, we will see the development of two distinct breeds with no evidence of continuity between them.

He then applies this logic to a hypothetical natural population of a carnivore. The Malthusian principle dictates that there will be an inexorable expansion of the population to occupy all available space and exploit all available food. Individuals may vary by seizing on different ways of sustaining themselves, such as by pursuing different types of prey, scavenging dead prey, occupying new habitats like trees or aquatic environments, or perhaps shifting to noncarnivorous diets. As their numbers and varieties multiply, they will invade new niches.

This same process of expansion and diversification applies to plants as well. Darwin observes that a given plot of ground will support a larger biomass of plants if it is sown with species from distinct genera. It is also known that more biomass of wheat can be grown if a plot is sown with multiple varieties of wheat, rather than one. This increase in biomass is presumably a consequence of different species exploiting the habitat in a different fashion. If we instead imagine a single species of grass invading a vacant patch of habitat, we can envision the events that follow as its population expands to fill the available space. Each individual plant will annually produce enormous numbers of seeds, far greater than required to replace itself in the next generation "and thus, as it may be said, is striving its utmost to increase in numbers" (*Origin*, p. 113). While the large majority of these seeds will perish, the few that succeed will tend to be those that are able to find a distinct place for themselves in their increasingly packed universe. What began as a single individual that was different in a way that allowed it to use the environment in a slightly different fashion grows into a subpopulation that shares this lucky attribute. This subpopulation may eventually emerge as a recognizable variety. In a similar fashion, many such varieties may be spawned by the original invader. The question is, what determines whether or not a given variety will persist? "I cannot doubt that in the course of many thousands of generations, the most distinct varieties of any one species of grass would always have the best chance of succeeding and of increasing in numbers, and thus of supplanting the less distinct varieties; and varieties, when rendered very distinct from each other, take the rank of species" (*Origin*, pp. 113–14).

The principle of divergence has had strong detractors. Ernst Mayr singled it out as a failed theory. His reasoning paralleled his argument that *On the*

Origin of Species is not about the origin of species. Darwin saw each species as an arbitrary point on a continuum of populations that are diverging from one another as a consequence of evolution by natural selection. For this reason, he saw divergence acting as readily between individuals within a population as between populations or species. He did not distinguish these levels of interaction. The critical contribution of Dobzhansky and Mayr was to recognize that speciation involves both divergence and the origin of discontinuity, or reproductive isolation (chap. 8). Mayr argued that individuals within a population cannot diverge from one another, because they are part of an interbreeding gene pool. If one accepts that the prospective reproductive isolation of full species is a prerequisite for divergence, then Darwin's principle of divergence must be modified. It can happen only after closely related populations have attained some measure of reproductive isolation from one another. This condition does not invalidate Darwin's principle; it merely delays its action to the point that the reproductively isolated descendants of a common ancestral lineage begin to interact competitively.

Darwin sees evidence for this principle of divergence in the diversity of organisms that coexist with one another in nature. He argues that plants and animals that live together in the same place tend to be more different from one another than one might expect if their coexistence were the product of chance. For example, he once identified all plants found on a 3-by-4-foot piece of turf and found 20 species in 18 genera and 8 orders (*Origin*, p. 114). Species are generally distinct in how they utilize the environment, but genera, as the next step up the taxonomic hierarchy, are presumably more different from one another in their requirements than are the species within a genus. Orders are two steps up the taxonomic hierarchy from genus, and, on average, different orders are much more distinct from one another than are genera within an order. The 20 species found on this single plot of land are derived from almost as many genera (18) and a large number of orders (8), so it seems that their diversity is disproportionately large and that closely related species are unlikely to be found together. If we just grouped species together at random, we would expect to see more species from the same genus, particularly from genera with many species, than appears to be the case.

Darwin also seeks evidence for his principle of divergence in enumerations of species that are successful invaders. Successful invaders tend to be distinct from the native fauna. Asa Gray's *Manual of the Flora of the Northern United States* reports 260 species of invaders. They come from 162 different genera, which implies that these species tend to be very different from one

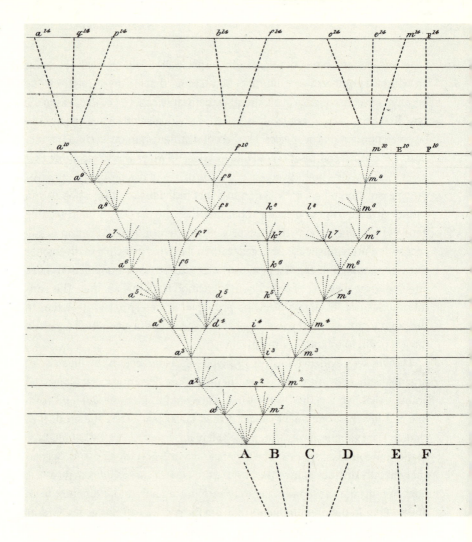

another since so few are from the same genus. One hundred of these genera are new to North America, which implies that they are also quite distinct from the native flora.

Darwin illustrates how this process of divergence, in combination with extinction, will play out over large expanses of time with the only figure in the *Origin*, duplicated here (fig. 12). You should study this figure as you read the text so that you can follow his argument. His figure begins with an imaginary genus that consists of species A–L arrayed along a horizontal axis that represents some imaginary form of environmental variation. Species that lie next to one another on this axis are presumably more closely related and are more similar to one another in habitat requirements than those that are farther apart. His imaginary genus has one of the properties of a real genus, which is

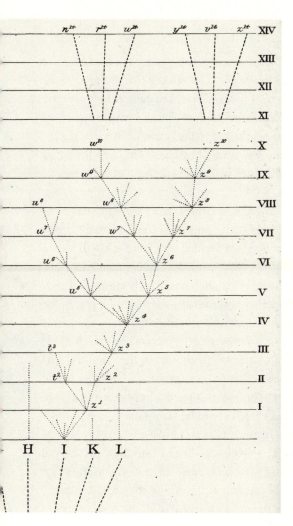

Figure 12
Darwin's "Tree of Life" Figure
This is the only figure to appear in the *Origin*. Darwin uses it in his fourth chapter to illustrate the long-term consequences of the combined action of evolution by natural selection, the principle of divergence, and extinction (see text for details). He also illustrates anagenesis, or change within a lineage over time (species F and its descendants) versus cladogenesis, or the splitting of lineages (species A and I and their descendants). He uses this figure in many other ways throughout the book, so it would pay to dog-ear this page to make it easy to find later.

that the species are not equally spaced; many instead fall into "species groups" that are more similar to one another than they are to other species in the genus. Species A–D represent one such group while species G–L represent a second. Darwin describes this genus as "a genus large in its own country," meaning that it consists of comparatively many species, some of which are abundant and widespread. His earlier tabulations (my chap. 9) suggested that such genera should include dominant species that are likely, by virtue of their dominance, to be the source of new species in the future.

Given his observation that species of such genera are also more likely to have diversified into varieties (chap. 9), he argues that some of the species in his imaginary genus will have this property. Two of these, A and I, begin to split into distinct varieties. These two "fecund" species are each a member of one of the two species groups. The existence of these groups in the first place implies that they were once varieties of the same species that diverged to the point of being more than one distinct species, so we see the continuity of a fecund parent species producing fecund descendants. Recall Darwin's argument that "where many large trees grow, we expect to find saplings" (*Origin*, p. 55).

Species A initially diverges into six incipient varieties. Although the figure shows them all originating at once, they would actually vary in the timing of their appearance, in their degree of distinctness, and in how long they persist. Here is where the principle of divergence comes into play. After one time interval on the vertical axis, which Darwin supposes could represent 1,000 or even 10,000 generations, species A has divided into two persistent forms, a^1 and m^1. Of all the incipient varieties that arose during this time interval, a^1 and m^1 are the farthest apart on the horizontal axis, which means that they are the most divergent in character and presumably the most different in the way they utilize the environment. "Moreover, these two varieties, being only slightly modified forms, will tend to inherit those advantages which made their common parent (A) more numerous than most of the other inhabitants . . . they will likewise partake of these more general advantages which made the genus to which the parent-species belonged, a large genus in its own country. And these circumstances we know to be favourable to the production of new varieties" (*Origin*, p. 118).

During the next time interval, a^1 begins to diverge into five varieties, but only a^2 remains at the end of the interval. Likewise, m^1 initially diverges into four varieties, and two, s^2 and m^2, remain at the end of the second interval. The varieties a^2 and m^2 are again the most divergent of all the descendants of A and lie farther apart than a^1 and m^1; s^2 lies between them. As the differences between the a and m lineages increase, enough ecological space opens up between them to admit the persistence of the new lineage s. The diagram depicts the continuation of this process for 10 time units, then in an abbreviated form for 4 more units. At each horizontal break in the series, Darwin summarizes those varieties that persist under a new superscript that represents their having changed to some degree during the previous time interval. Some lineages die out, some spawn new varieties, and some persist without branching. The overall structure of the figure shows that the incipient varieties that persist tend to be those that are most different from their neighboring lineages. The figure also shows that the genus as a whole fans out and occupies a progressively larger span of the horizontal axis, which implies that the descendants are ecologically more diverse than the ancestors. The ultimate cause of these patterns is that "natural selection will always act according to the nature of the places which are either unoccupied or not perfectly occupied by other beings; and this will depend on infinitely complex relations. But as a general rule, the more diversified in structure the descendants from any one species can be rendered, the more places they will be enabled to seize on and the more their modified progeny will increase" (*Origin*, p. 119).

Said differently, natural selection causes species to find vacancies and fill them. After 10 units of time, the original species A has diversified into three surviving lineages, a^{10}, f^{10}, and m^{10}. They may now be sufficiently distinct to be classified as subspecies. Darwin leaves open the question of when to consider them to be distinct species. To Darwin, species are, after all, just a point in this continuum of change. He imagines that each of these three lineages inherits the parent species' property of continuing to vary and diversify so that, after 14 time units, they have diversified into eight distinct species distributed into three distinct species groups.

Darwin also suspects that a large genus will contain more than one initial species that shares this tendency to diversify, as represented here by species I. This species diversifies into two distinct lineages, w^{10} and z^{10}, after 10 time units, then six species after 14 time units.

Darwin's imaginary genus thus begins with eleven species and contains two distinct species groups. After 14 units of time it consists of fourteen species in five species groups plus a singleton lineage (F^{14}), which is the descendant of the original species F that persisted but never diversified.

The pattern of multiplication of species, as shaped by the principle of divergence, reveals only part of the story. Extinction has also played a key role in shaping this process: "As in each fully stocked country natural selection necessarily acts by the selected form having some advantage in the struggle for life over other forms, there will be a constant tendency in the improved descendants of any one species to supplant and exterminate in each stage of descent their predecessors and original parent" (*Origin*, p. 121).

The significance of assigning a new number to a given lineage at each time unit is that it is a modified descendant and therefore a replacement of the lineage that was present in the previous time unit. Thus a^5 is not just a^4 one time unit later, but is rather a new and improved lineage that has displaced and driven a^4 to extinction. Recall that the most severe competition in the struggle for existence will be between those lineages that are most similar to each other.

Extinction has also shaped other features of the genus. Only three of the original eleven species (A, F, and I) are represented by descendants 14 time units later. As any one species diversifies and invades new habitats, it often does so at the expense of others. As the original species A and I diversify, those that are the most similar to their descendants go extinct. The descendants of I wipe out in turn K, L, H, and G, while the descendants of A diversify at the expense of B, C, D, and eventually E.

The diversity present after 14 units of time is not uniformly distributed. There are three clusters of species descended from the original species A, and two clusters descended from I. These might represent two genera, one consisting of eight species and the other of six. They may even be so different from one another that they can be classified as different families, one consisting of three genera and eight species and the other consisting of two genera and six species. (Family is the next step up the taxonomic hierarchy from genus. Families consist of clusters of closely related genera, while genera consist of clusters of closely related species.) "Thus, as I believe, species are multiplied and genera are formed" (*Origin*, p. 120); and, by extension, genera are multiplied and families are formed, and so on. I shall develop this line of reasoning further in part 3 of this book.

The F lineage in Darwin's diagram merits special consideration. It persists by virtue of being sufficiently different from those species descended from the A and I lineages and is represented in the end by F^{14}, which is perhaps little modified from its ancestor. "Having descended from a form which stood between the two parent species (A) and (I), now supposed to be extinct and unknown, it will be in some degree intermediate in character between the two groups descended from these species. But as these two groups have gone on diverging in character from the type of their parents, the new species (F^{14}) will not be directly intermediate between them . . . ; and every naturalist will be able to bring some such case before his mind" (*Origin*, p. 124).

Darwin is suggesting that such a species will have some affinity for the other organisms that are its closest relatives, but we should not expect it to simply be intermediate between the descendants of A and I in appearance. Lineages A and I will have changed through time and may have left descendants that became quite different in appearance from their ancestors as they adapted to new environments. There will have been a continuity of change within any lineage, but not between them. All intermediate steps, meaning those species that formed the bridge between the common ancestor and the species that survive today, have been lost. We will not be able to see any evidence of the changes that took place between the common ancestor and the organisms alive today, so the surviving lineages may not reveal strong evidence of their former close affinity. It is like being left with the tips of some tree branches without having any knowledge of the tree beneath them that once joined them together. Because of all this lost information and the way the descendants of (A) and (I) diverge over time, F^{14} may well appear as an isolated species with little apparent affinity to all others that are alive today. Darwin cites the "living fossils," such as the lungfish (*Lepidosiren*) or duck-billed platypus (*Ornithorhynchus*), as examples of such species. In addition, what was initially a collection of eleven species similar enough to be classified in the same genus will now be a much more diverse assemblage of species that may be separated by what appear to be unbridgeable gaps. The branches that join them are ones that extend back through time to the ancestral lineages, which separated from one another in the distant past and then continued to diversify. All these branching lineages are now extinct.

It is important to recognize that Darwin is defining two different types of speciation. One is "anagenesis," or the change of one species over time. The F lineage, defined by the succession of f^1–f^{14}, can be thought of as a succession of species. Each time a new species evolves, it displaces its ancestor. At any one point in time we see a only single, variable species, but each point in time is represented by a different species in this lineage. In any one time interval, the different varieties that are competing and replacing one another will just represent diversity within a species.

The second type of speciation is "cladogenesis," or lineage splitting. The initial species A ultimately diverges into eight descendant species (a^{14}, q^{14}, p^{14}, b^{14}, f^{14}, o^{14}, e^{14}, m^{14}) through a progressive splitting of lineages. It is the principle of divergence that causes cladogenesis, or the multiplication of an individual species into multiple, related species that become recognized as distinct genera or even higher levels in the taxonomic hierarchy. Lamarck's concept of species transmutation allowed only for anagenesis. A. R. Wallace, who independently discovered evolution by natural selection in 1858, appears to be the only other person to have proposed cladogenesis prior to the *Origin*, but not with Darwin's clarity.

Before Darwin, others had noticed that the hierarchy of animal and plant classification could be represented with a tree diagram. Orders could be shown to branch into multiple families, families into multiple genera, and genera into multiple species. Darwin's tree diagram differs from all that came before it because he postulates a biological cause of the taxonomic hierarchy. The hierarchy is a result of the combined action of divergence and extinction. Divergence causes cladogenesis and generates multiple species from a single ancestor. Extinction removes many of the intermediate links between them and causes some of the descendants to appear as closely related clusters of species that are well separated from other such clusters. It is extinction that creates the impression of the unbridgeable gaps that separate species and higher levels in the taxonomic hierarchy.

Darwin argues that the analogy between a tree growing in nature and his "tree of life" diagram is much more organic than a simple similarity of appearances. The rules that govern tree growth and shape the adult tree are like the principles of divergence and extinction his diagram illustrates. Young trees produce a profusion of buds as they grow. Some buds produce just a leaf for one season; the only evidence of their existence is a scar on their resident branch that is visible for a subsequent season or two. Some buds become twigs, but most twigs die and are shed. The few that are retained grow into branches that become a new source of buds and twigs. As the tree gains height, some branches overshadow others. This competition for light causes most lower branches to be shed, but some thicken to support the expanding growth of the branches and twigs that they have spawned. The adult tree, with one trunk that divides into a few major limbs that in turn divide into progressively finer branches, then twigs, then buds, is thus a product of the dynamics of the growth and branching of some limbs and the death of others. The trunk and main limbs represent the few of the many thousands of initial buds that were retained as the tree grew. The thousands of buds that form each spring are all candidates for new growth, but most will last only a season, and only very few will eventually mature into thick branches.

In the same fashion, Darwin's tree of life represents the synthesis of birth and death processes. The most fundamental birth process is the production of so many more offspring than are required for a parent to replace itself in the next generation. Most offspring will die without reproducing. Most parents will produce no surviving offspring, but some will produce many. Beyond the fate of individuals, the tree will be shaped and pruned by the combined action of divergence and extinction. Among the profusion of offspring, some will differ from others, and some of these differences will enable their possessors to use the environment in a new and different way. Some small subset of these individuals will be successful in "probing" the habitat for vacancies, just as the branches of a growing tree seek light. Those that find the necessary resources will produce young that inherit the parents' abilities to use the environment in a different way. A smaller subset of these lineages will emerge as distinct varieties that become recognizable subsets of the species because they share traits that adapt them to a particular environment. The "environment" is defined primarily by their interactions with other organisms, including the ones that they eat, the ones that eat them, or those that they compete with for food. For reasons that we do not entirely understand, some branches will proliferate more readily than others into new varieties and species. As some proliferate, they do so at the expense of neighboring branches, which die out and are shed from the tree of life. The varieties that persist will tend to be those that are most different from others in how they utilize the habitat, but will also be those that are better at utilizing resources exploited by other species.

Darwin envisions most extinction as occurring through parental lineages being displaced by their "new and improved" offspring, but extinction can have many other causes. It can be the fate of species that compete less effectively for resources, of prey species that are less skilled at evading predators, of predators who do not keep up with their prey in the race of life, or can happen as a consequence of all the other ways in which organisms depend on or exploit others.

Darwin's tree is a synthesis of logic and observation. He sensed from his observations, such as finding endemic species on the Galapagos Islands, that the transmutation of species was possible. He first discovered natural selection as a mechanism that could cause change over time. He then envisioned the struggle for existence between organisms as the most important cause of natural selection. Only later, when he had developed a hypothesis for the cause of extinction and his principle of divergence, was he able to fully describe mechanisms that could bridge the process of natural selection to cause the formation of new species. At all stages he sought evidence in nature for his ideas, such as with his tabulations of the number of species within a genus or varieties within a species. His figure combines all these ideas and observations to project how they will influence the history of life on a larger scale. What emerges are patterns that correspond well with the existing scheme for classifying plants and animals, which is a hierarchical clustering of varieties within species, species within genera, genera within families, and so on.

The synthesis that emerges in Darwin's figure and associated discussion, however hypothetical it may seem, has proven to be a remarkably accurate depiction of the history of life as we understand it today. It has held up to the test of time and has been well supported by the subsequent flowering of evolution as an independent scientific discipline. It represents a huge leap in our understanding of the history of life and the rules that govern it because it defines the connection between processes that happen within populations, beginning with the origin of varieties, then species, and continuing on to higher levels within the taxonomic hierarchy. In the same fashion that earthquakes, volcanoes, erosion, and other processes that we can see and characterize on a day-to-day basis can, over time, shape the surface of the earth, natural selection can shape the history of life.

Evolution Today: *Plethodon* Salamanders Revisited

Darwin's illustration is an abstraction, but the evolutionary history of any well-studied group of organisms reveals similar patterns. The salamanders in the genus *Plethodon* serve well as an example. The family tree of *Plethodon* from eastern North America (there are additional species in western North America), based on molecular evidence, reveals four species groups, each of which is descended from a single ancestral species. These four lineages are represented today by 2, 5, 10, and 28 described species each for a total of 45 species (figs. 13 and 14). The largest species group is in turn divided into four

Figure 13
Plethodon glutinosis **and** *Plethodon cinereus*
These are typical-size adults of each species. *P. glutinosis* (above) represents the largest and *P. cinereus* (below) the second-largest, in terms of the number of species, of the four species groups in the eastern members of the genus *Plethodon*. *P. glutinosis* adults range from 11 to 21 cm total length. *P. cinereus* adults range from 6 to 12 cm total length.

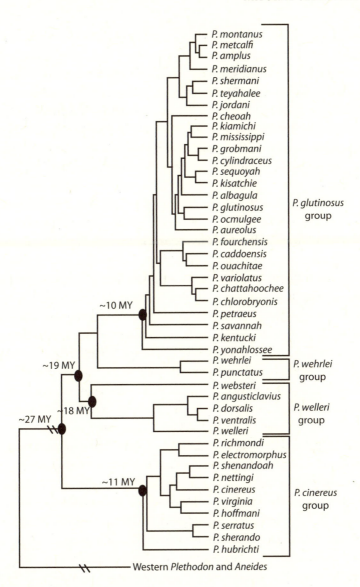

Figure 14

Phylogeny of the Eastern *Plethodon*

This figure presents a family tree of *Plethodon* salamanders found in eastern North America. Their classification into four species groups is based on data from allozymes, or different versions of enzymes that can be detected using starch gel electrophoresis, and from albumen immunology, which is another method for evaluating genetic relationships between species. The albumen data provide an estimate of the time when clusters of species shared a common ancestor. For example, the 10 MY label on the stem of the *P. glutinosis* group means that the 28 species in this group are descended from a single common ancestor that lived approximately 10 million years ago.

"species complexes," each of which contains species that are more closely related to one another than to others in the species group. The largest of these complexes contains 16 species. These patterns show that there were big differences between the original four ancestors in the rate at which new species accumulated. The most "fecund" of the four ancestors diversified into four descendant species, each of which diversified into a distinct species cluster. These four descendants also varied in their tendency to diversify. The most successful among them diversified into 28 of the 45 extant species. The uneven pattern of diversification is thus quite similar to that in Darwin's figure, with one lineage showing a continued tendency to multiply and diversify into clusters of related species. In fact, if you deleted all the extinct lineages from Darwin's figure, it would look similar to this tree. Furthermore, Highton's range maps show that the largest species groups also contain the individual species with the largest ranges. They include those species that have been shown to be the most abundant vertebrates in their habitat both in numbers of individuals and in biomass. There is also clear evidence of character divergence between the four species groups and within the largest group. Body size is a simple character for judging such divergence. Size is well correlated with many other features of an organism's ecology, such as the range of prey that it can feed on or the types of habitat that it can exploit. Two of the groups are consistently small bodied while two are considerably larger and stockier. The largest group has in turn diversified in size as it invaded new habitats, with those species found on the coastal plains tending to be smaller than those from the mountains.

The genus *Plethodon* is part of the salamander family Plethodontidae. The diversity of this genus is reflected in the diversity of the family, since Plethodontidae is by far the largest family of salamanders. It includes over half of the named species of all salamanders, and new species are being named every year. This family includes other very large genera, but also includes lineages that have the properties of Darwin's F lineage, which is the "living fossil" that persisted for a long time without diversifying to form new species. Early in his career, Highton discovered a new species that was found only in the Red Hills of Alabama on the slopes of steep, forested ravines between the Alabama and Conecuh rivers. It was so distinct from all others in the family that it could not be placed within any existing genus. It was instead classified as a new genus containing only this single species. The family Plethodontidae shares this property as well; it is distantly related to all other salamanders, so much so that it is difficult to define its closest relative among the living salamanders.

Speciation on Islands versus Continents

Earlier in this chapter, Darwin proposed that speciation is more likely on continents, or large landmasses that support large populations, than on islands. He did so in spite of the observation that oceanic islands often are inhabited by endemic species, or species found there and nowhere else. Such endemism seems to suggest that these islands are hot spots for the formation of new species. Darwin argued, however, that the presence of odd endemics on islands could as well be because they are relicts protected from the constant struggle for existence that so often confronts species on continents. He also argued that, if we really want to know where speciation is more likely to occur, then "we ought to make the comparison with equal times; and this we are incapable of doing" (*Origin*, p. 105).

What was not possible in Darwin's time is possible today. While we do not yet have a definitive answer to Darwin's proposal, I can offer a brief progress report.

What Darwin is implying in his argument is that species found living are actually the product of two processes—the gain caused by new speciation events and the loss caused by extinction. Seeing distinct species that occur only on isolated islands can mean either that evolution and speciation happens more quickly there or that extinction is less likely. What Darwin is calling for in the suggested "comparison with equal time" is a measure of species turnover rate. One way to get such a measure is to estimate how old species are. If islands have endemic species because speciation is faster, then the island species should be younger than those on continents. If islands have endemics because extinction is less likely, as proposed by Darwin, then island species should be older than those on continents. In the accompanying box, I offer a brief explanation of how we infer species ages and rates of speciation from a synthesis of molecular and paleontological data.

Cadena and coauthors (2005) evaluated the ages of species of birds found on the Lesser Antilles archipelago in the southern Caribbean. Species that were endemic to one island proved to be much older, on average, than species that were more widespread. This pattern suggests that the endemics really are relicts rather than being products of recent, rapid species formation. This observation alone addresses only part of Darwin's argument, since it does not include a comparison with species from continents.

Dolph Schluter (2000) summarizes data that are more directly relevant to Darwin's argument in his book *The Ecology of Adaptive Radiation*. He combed through the literature and found information on the numbers of species in "sister taxa," one of which was found on an oceanic island and the other on a continent. ("Sister taxa" in this case might be the two most closely related genera in the same family.) Following the logic of Darwin's figure, these genera are derived from some distant common ancestor. Since they are the product of such a branching process, they should be equal in age. The question is whether the ancestral species that colonized and proliferated into a genus on a continent spawned more or fewer surviving species than did the ancestral species that colonized an oceanic island. Schluter compared the finches on the Galapagos and Hawaiian islands with closely related finches from the mainland, as well as Hawaiian fruit flies with closely related continental lineages.

Schluter's hypothesis was that islands present colonizers with a "blank slate," or an empty environment, in which they can expand into many unoccupied niches and rapidly diversify into many new species. His prediction was thus the opposite of Darwin's. He found that the island lineages had a tendency to contain more rather than fewer species, which supports his hypothesis and seems to contradict Darwin's. One catch in such a comparison is that it looks at the net rate of accumulation of new species, which is actually the difference between the addition of new species and the loss of species due to extinction. The number of species present today is a function of both of these processes, so that number alone does not tell us which group is undergoing more rapid speciation.

Schluter also found that the island lineages tended to be more ecologically diverse than were their most closely related counterparts on the mainland. Again, this pattern suggests that these islands presented opportunities to the early colonists. Fewer species were present to begin with, and the struggle for existence was less intense, so the lucky few that had managed to reach the new island found more opportunities to adapt to unoccupied ecological niches. Island colonists often evolve into life-forms that are quite different from their mainland relatives. The finches that colonized the Galapagos or Hawaiian islands confronted a habitat in which there was a diverse menu of seeds to choose from and few other organisms that consumed them. They diversified into an array of species that utilize every available size of seed. Their mainland relatives instead fit into a complex community where they have to duke it out not only with very different species of birds, such as pigeons, quail, and parrots, but also very different types of organisms, such as granivorous rodents and ants, for the largest and smallest seeds. As a result, they utilize a much smaller range of seed sizes and are much less diverse in their morphology.

Darwin's ideas about what causes speciation and speciation rates are thus partly upheld. The struggle for existence on continents appears to be more intense because of their greater biological diversity. Individual species groups may be more constrained in how they utilize their habitats as a consequence of this competition. Islands appear to have reduced extinction rates and to be host to relict endemic species; but islands, or any new, sparsely inhabited environment, also present the kinds of opportunities that Darwin predicted would promote speciation. Darwin's logic is that speciation is a consequence of the constant probing of the environment for vacancies, then filling them. Vacancies are more abundant in new, sparsely colonized habitats, so they appear to be host to the most rapid speciation. There is not strong evidence to support Darwin's counterargument that speciation rates will instead be persistently higher on continents because of the more intense struggle for existence. Such evidence is lacking mostly because it has not been sought; the scientists of today have ideas of their own and are trying to address other questions.

DNA Sequences and the Molecular Clock

The possibility of knowing the relative rates of speciation on islands versus continents was inconceivable to Darwin, but we can now provide such estimates with a synthesis of information from the fossil record and from estimates of genetic differences between species using modern molecular methods. The most common first step is to obtain the nucleotide base-pair sequences (see below) for the same genes in different species. The sequence for the same gene in different species will change over time. Some of these changes can be attributed to the effects of natural selection on the evolution of the proteins that the gene codes for, but many changes are "neutral" in the sense that they have little or no effect on the organism.

DNA is a string of four different types of paired molecules, or building blocks, called nucleotides. This string of "base pairs" functions like a code that is translated into proteins, which are made of amino acid building blocks strung together in a linear series. Three DNA nucleotides code for each amino acid in a protein. Since there are four types of nucleotides and three of them code for each amino acid, there are 64 ($4 \times 4 \times 4$) possible codes.

(continued)

There are only 20 different amino acids commonly found in the proteins that comprise all living organisms, so this means that there is often more than one triplet code, called a codon, for the same amino acid. The duplicate codes almost invariably have the same first two nucleotides but differ in the third nucleotide.

The differences between species revealed by DNA sequences are caused by changes in the sequence of nucleotides. Some changes cause a change in the amino acids that are coded for, which means a change in the protein that is produced. Some changes, particularly in the third nucleotide of a three-letter codon, cause no changes in the amino acid that is coded for, so it appears that they are "neutral": they have no effect on the protein coded for or on the organism itself.

The first thing that we can do with DNA sequences is to reconstruct the most likely "family tree" that describes how organisms are related to one another. If two species are derived from a recent common ancestor, such that they are part of the same species group within a genus, then their DNA sequences will tend to be more similar to one another than if they are from different genera or different families. Comparisons of DNA sequences from multiple species can be used to construct a treelike figure that summarizes the genetic similarities of species. Closely related species are joined by short branches, while those that are more distantly related are joined by longer branches. These trees also depict the most likely evolutionary relationships between different groups of species.

This information on the degree of genetic differences between organisms can be translated into estimates of when in the past they shared a common ancestor. If the rate of substitutions of one nucleotide base pair for another was constant, then the change in a DNA sequence would be constant, like the ticking of a clock. The number of differences between two lineages could provide an estimate of how long ago they were derived from a common ancestor. The difficulty with this interpretation is that the rate of nucleotide substitution is not constant, but can be subject to natural selection. Thus the rate of evolution of different genes in the same organism is not the same, the rate of evolution of the same gene in different organisms can also vary, and the rate of evolution of the same gene in the same organism is not constant over time. A given gene may evolve rapidly at a time when it is a target of an episode of selection, but much more slowly at other times. There can also be differences in the rate of substitution of nucleotides that do not cause a change in the amino acid coded for, versus substitutions that do cause such a change—we refer to these as synonymous versus nonsynonymous substitutions. All these factors mean that DNA change does not simply tick like a clock. They also mean that the application of the molecular clock hypothesis has aroused a great deal of controversy and debate.

In spite of these limitations to the potential accuracy of the molecular clock, we can claim some significant successes from our efforts to infer time intervals based on the degree of similarity or difference between the DNA sequences of different organisms. These successes show that, if the concept is carefully applied, some information about time can be obtained from such molecular data. For example, molecular biologists discovered the remarkable similarity in amino acid and DNA sequences of humans and chimps and inferred that they shared a common ancestor approximately 5–6 million years ago. At the time, anthropologists thought that the morphological differences between chimps and humans were so great that their common ancestor must date to a time more than twice as ancient as inferred by molecular biologists. Subsequent results from the paleontological record and far more extensive molecular analyses have shown that the molecular biologists were far closer to the correct answer than were the anthropologists.

To apply the molecular clock, it must be "calibrated," such as by matching dated fossils or geological events with particular branches of the family tree. It is through such a synthesis of molecular, paleontological, and geological data that we get the best information on the ages of lineages.

References

Cadena, C. D., R. E. Ricklefs, I. Jimenez, and E. Bermingham. 2005. Is speciation driven by species diversity? *Nature* 483:E1–E2.

Darwin, C. [1860] 1962. *The voyage of the* Beagle. Annotated and with an introduction by Leonard Engel. Garden City, NY: Doubleday and Company.

Highton, R. 1995. Speciation in eastern North American salamanders of the genus *Plethodon. Annual Reviews of Ecology and Systematics* 26:579–600.

Mayr, E. 1994. Reasons for the failure of theories. *Philosophy of Science* 61:529–33.

Nei, M., and S. Kumar. 2000. *Molecular evolution and phylogenetics.* Oxford: Oxford University Press.

Rudwick, M.J.S. 2005. *Bursting the limits of time: The reconstruction of geohistory in the Age of Revolution.* Chicago: University of Chicago Press. (Pp. 364–88, Cuvier's discovery of extinction.)

Schluter, D. 2000. *The ecology of adaptive radiation.* Oxford: Oxford University Press.

Chapter 11

Hybridism

Here I skip to chapter 8 of the *Origin*. I do so because this chapter, titled "Hybridism," is a continuation of Darwin's argument for the transmutation of species and against the idea that species are products of acts of special creation. This chapter is actually part of a three-chapter sequence in the *Origin* (chaps. 6–8) that is a preemptive strike against anticipated challenges to his theory. I review this chapter here because it deals with an integral part of the modern definition of species. "Hybridism" refers to the consequences of making crosses between different species and the quality of the offspring that are produced. The subject matter might cause you to think that Darwin is pursuing in this chapter the same goal as Dobzhansky and Mayr eighty years later, which is to define species as reproductively isolated units. His goal is the opposite. He instead reinforces his thesis that species are just an arbitrary point on a continuum of variation, from individual variation within a population to variety to species. Darwin argues that the continuum from variety to species and beyond is evident in the ability of some species to interbreed with one another to form hybrids.

Darwin's contemporaries thought that each species was the product of an act of special creation and that all species were absolutely reproductively isolated from one another. Varieties within species were products of mysterious "secondary processes"; the "primary process" was special creation. The view of Darwin's contemporaries was that species were endowed with barriers to interbreeding "in order to prevent their crossing and blending together in utter confusion" (*Origin*, p. 255). If they were correct, then we should always find that two species could never successfully interbreed.

Darwin argues that because speciation is an ongoing process, such absolute barriers do not separate species. He cites results that show that varieties within species usually interbreed and produce viable, fertile offspring, but not always. Different species usually cannot be successfully hybridized, but sometimes they can. It is even sometimes possible, although very rarely, to cross species from different genera. Darwin also cites evidence that there is variation among individuals within a species in their ability to breed with other species; some individuals can be easily crossbred while others cannot. The ability to interbreed can thus be described as a trait that varies among individuals along a continuum of more to less fertile; so the ability to reproduce is like any other feature of an organism, rather than being endowed with some special property of being totally possible within species, absolutely impossible between species, as one would expect if species were perfectly distinct and noncontinuous–that is, specially created. He argues that when hybrids from crosses between varieties within a species or between species are sterile, it is not a direct consequence of natural selection. It is instead incidental to all the other differences that arise between the varieties or species as they follow their separate evolutionary paths.

Darwin's treatment of "hybridism" often has an alien quality to the modern reader for three reasons. The first is the nature of the audience he is addressing. His audience was a scientific establishment and general public that advocated special creation, and he was trying to convince them that, contrary to the prevailing belief, species are not defined by absolute reproductive barriers. The second is that he defines "species" differently from the way we do today. We define species as reproductively isolated units, but he defines them as simply points in a continuum. The third is that he uses a more restricted definition of reproductive isolation than we use today. We now distinguish between pre- and postmating isolation. It is possible for species to be reproductively isolated because of premating factors, such as where or when they breed or behavioral interactions that precede breeding. Species that are well isolated by premating factors may still be able to produce viable, fertile offspring if we somehow bypass these barriers. Postmating barriers include failure of sperm to fertilize an egg, or if fertilization occurs, disruption of the normal development of the zygote, or if the zygote develops normally and gives rise to viable, healthy offspring, their sterility or incapability of breeding for some other reason (see chap. 8). Here, Darwin focuses exclusively on postmating isolation. He deals with experiments in which the pairing of species was accomplished with artificial means that preempted all or most forms of premating isolation. For example, the plant studies reviewed by Darwin involved the artificial pollination of one plant with pollen from another. These circumstances circumvented most aspects of premating isolation and instead illuminated only the viability and fertility of offspring after the gametes of different species were brought together.

In chapter 8 I reported that many of the organisms we classify as different species are genetically compatible but are kept separate from one another by premating reproductive isolation. If a premating barrier is what normally keeps two species apart and we thwart it though some artificial means, we may well find that they are capable of producing healthy, fertile offspring. This means that there is no contradiction between our understanding of what a species is today and Darwin's argument that different species are not separated by absolute reproductive barriers. I summarize the specifics of Darwin's arguments with the use of the same subheadings that appear in his chapter 8.

Sterility

Darwin first argues that one must distinguish between two types of sterility. One type occurs when two mated individuals fail to produce offspring. The second type occurs when they produce hybrid offspring that are sterile. The original parents have "their organs of reproduction in perfect condition" (*Origin*, p. 246), so a failure to produce healthy offspring must mean that their gametes are in some way incompatible. If they succeed in producing viable hybrid offspring, then these offspring may "have their reproductive organs functionally impotent" (*Origin*, p. 246).

Darwin is distinguishing between different sources of reproductive incompatibility, but he cuts the deck differently from the way we do today. The sterility of a pairing of individuals of different species includes what Coyne and Orr called "postmating, prezygotic" isolation, plus any "postzygotic" isolation that results in the death of an embryo. The sterility of hybrid offspring includes all other postzygotic mechanisms that do not affect the viability of the embryo but instead impair the ability of the hybrid offspring to reproduce. Crosses between a horse and donkey are an example of the second type of sterility. These crosses produce offspring, mules or hinnies, that are healthy but sterile. We know now that the sterility of the hybrids is caused by horses and donkeys having different numbers of chromosomes; this difference prevents their offspring from producing viable gametes.

The bottom line of sterility involving the parents' inability to produce viable offspring, versus sterility involving the inability of hybrid offspring to reproduce, is the same, which is that the species are reproductively isolated; however, the distinction is important because it tells us that mating incompatibilities can be caused by a diversity of mechanisms.

Darwin cites the work of Joseph Kolreuter and Karl von Gartner, German scientists who performed extensive hybridization experiments between species of plants. Kolreuter published his results in a series of papers between 1761 and 1766; Gartner published a single volume in 1849. A common method was to emasculate flowers by removing the anthers, or pollen-producing organs, to prevent self-fertilization, then to fertilize them by brushing the surface of the stigma, which is joined to the seed-producing organs, with the pollen of a different species. Darwin observes: "It is impossible to study the several memoirs and works of these two conscientious and admirable observers . . . without being deeply impressed with the high generality of some degree of sterility" between species (*Origin*, p. 246). Darwin is always a gentleman in the *Origin* and begins all his critiques with such a compliment. He then points out that there are many wrinkles in their work. Kolreuter found ten cases in which his crosses between what were allegedly different species could produce viable and fertile offspring. But he was an advocate of special creation and assumed at the outset that all species were defined by the inability to interbreed, so he simply reclassified all these allegedly different species as being varieties within a species.

Gartner was equally dedicated to the doctrine of special creation, but he disputed some of the pairings that Kolreuter alleged were interfertile. One difference between the two of them is in how they defined the ability to interbreed. Kolreuter apparently just scored whether offspring were produced. Gartner instead counted the number of seeds produced in each cross between parents of the same or different species, then scored the number of seeds produced by their hybrid offspring. He found that the hybrids often produced fewer seeds than the parents and concluded from this decline in fertility that the parents were indeed different species. To make this argument, he implicitly defined sterility in crosses between species as a relative rather than an absolute term: sterility meant a decline in fertility rather than a total lack of reproduction; some fertility in crosses between species was possible after all, but it would amount to less than the fertility in crosses within a species.

Darwin capitalizes on this disagreement over which organisms are or are not the same species, which reproductive outcome is or is not sterility, by pointing out that what these researchers have really shown is that sterility can be measured in degrees—it is a continuous trait like any other feature of an organism, rather than being a trait endowed with some special, absolute quality. Darwin concludes: "It can thus be shown that neither sterility nor fertility affords any clear distinction between species and varieties; but that the evidence from this source graduates away, and is doubtful in the same degree as is the evidence derived from other constitutional and structural differences" (*Origin*, p. 248).

Darwin has already argued that the morphological differences between varieties and species are a matter of degree, rather than being absolute. Varieties within a species are more similar to each other than they are to other species. The same is true for the ability of different "forms" to interbreed. Interfertility is a variable trait like any other.

Darwin also challenges Kolreuter's and Gartner's techniques. For example, they began by mating two individuals from different species. If the cross produced viable offspring, then they self-fertilized the offspring generation after generation to see whether or not their fertility declined. If it did, they interpreted this declining fertility as evidence that the original parents were indeed different species. We have already reviewed Darwin's arguments that sexual reproduction rather than self-fertilization is important for the long-term fitness of organisms and that breeding close relatives can result in poorer-quality offspring (chap. 5 above). Any organism can show such a decline in fertility when inbred, so seeing this decline in self-fertilized hybrids tells us nothing about the compatibility of the parents.

Darwin cites the work of the "most experienced hybridizer . . . Hon. and Rev. W. Herbert" to counter the results of Gartner. Herbert worked with some of the same species as Gartner, but with the benefit of hothouses and perhaps superior horticultural skill. He showed that some species that Gartner found to be absolutely sterile when crossed were actually perfectly interfertile. Darwin also cites the common experiences of horticulturists. Many of the cultivars that they were developing, and that we find in our gardens today, were hybrids of recognized species that proved to be vigorous and fertile generation after generation when reared in gardens. Horticulturists often kept plants in natural gardens and let natural pollinators take care of fertilization. A pollinator like a bee would distribute pollen among many different individuals of different species. Any hybrids that arose in such conditions would be protected from the inbreeding depression that would result from artificial self-fertilization. The gardeners of Darwin's day produced many hybrids that were perfectly fertile for many generations, rather than displaying the declining fertility that Kolreuter and Gartner observed when they selfed hybrids for successive generations.

Darwin was not able to find the same sort of experiments done on animals. He makes only general reference to some studies and states that they suffer from the same weaknesses as the plant studies done by Kolreuter and Gartner, such as the failure to avoid close inbreeding in the hybrids. Darwin does, however, cite examples of the less formal crossing of different species of animals, some in circumstances equivalent to those of the horticulturists who developed domestic cultivars of plants derived from the hybrids of different species. One example he cites is crosses of *Cervulus* (now *Acervulus*) *vaginalis* and *C. reevesii*, which are two species of muntjac antelope. We now have a number of examples of successful hybridizations between different species of deer and antelope. One occurred on its own when two closely related species were introduced into New Zealand. Others are products of deliberate pairings that produced hybrids with desirable properties for domestication (see Hartwell n.d., www.messybeast. com/genetics/hybrid-deer-antelope.htm). Darwin cites similar examples of crosses between geese that were considered to be so distantly related as to be classified into different genera. He is likely referring to what are now classified as the Chinese (*Anser cygnoides*) and common or greylag (*A. anser*) geese: "I am assured by two eminently capable judges, namely Mr. Blyth and Capt. Hutton, that whole flocks of these crossed geese are kept in various parts of the country [India]; and as they are kept for profit, where neither pure parent-species exists, they must certainly be highly fertile" (*Origin*, p. 253).

Darwin's goal up to this point has not been to say that hybridization between species is common. (In fact, it is much more common that different species cannot interbreed.) His goal is instead to take on those who advocate belief in special creation and in the related notion that absolute reproductive barriers must separate species. If it can be shown that different species can sometimes interbreed and can sometimes produce hybrids that are as fertile as the parents, then these exceptions will argue that species cannot be defined as discrete units separated by absolute reproductive barriers. "Finally, looking to all the ascertained facts on the intercrossing of plants and animals, it may be concluded that some degree of sterility, both in first crosses and in hybrids, is an extremely general result; but it cannot, under our present state of knowledge, be considered as absolutely universal" (*Origin*, p. 254).

Laws Governing the Sterility of First Crosses and of Hybrids

Darwin now considers the rules that govern the sterility of crosses between species. "Our chief object will be to see whether or not the rules indicate that species have been specially endowed with this quality, in order to prevent their crossing and blending together in utter confusion" (*Origin*, p. 255).

We will begin with the punch line. Darwin faithfully summarizes and embellishes the rules proposed by Gartner, based on his "admirable work on the hybridization of plants," but in the end Darwin concludes that there really are no rules. The alleged "rules" really argue that the supposed barriers to interbreeding between species are not absolute and are just by-products of other, unknown differences between them. Gartner intended his rules to characterize the distinctness of species, as revealed through their inability to interbreed, but Darwin converts them into his own "antirules" to argue that no such distinct boundaries exist.

Darwin's first antirule is that sterility comes in shades of gray, rather than being black or white, as assumed by his predecessors. "When the pollen of one family is placed on the stigma of a plant of a distinct family, it exerts no more influence than so much inorganic dust" (*Origin*, p. 255). But when different species in the same genus are crossed, the outcome can vary from absolute sterility to complete fertility. In some strange cases, the hybrids produced even more seeds than either parent.

Darwin's second antirule is that the ability of parent species to cross and produce viable hybrids is a separate trait from the fertility of those hybrids. Some species can be crossed only with great effort to produce just a few hybrid offspring, yet the offspring are highly fertile. Conversely, some species cross very readily to produce viable hybrids, but many of those offspring are sterile.

The third antirule is that the ability of two species to cross and produce viable, fertile offspring is "innately variable," meaning that there are differences between individuals within a species in their ability to cross with other species. "It is not always the same when the same two species are crossed under the same circumstances, but depends in part upon the constitution of the individuals which happen to have been chosen for the experiment" (*Origin*, p. 256). This means that if one makes the same cross between two species ten times, each time with different individuals, then sometimes the cross will succeed and sometimes it will not. Those crosses that succeed might do so to varying degrees. This pattern means that the ability for two species to cross and produce viable, fertile offspring varies between individuals, rather than being an absolute property of the species.

The fourth antirule is that more closely related species are more successful than more distantly related species in producing viable, fertile offspring when crossed, except when they are not. Species in the plant genus *Dianthus* readily hybridize with one another, while species in the plant genus *Silene* can rarely be hybridized. Many species within the plant genus *Nicotiana* will readily hybridize, but *Nicotiana acuminata* could not be hybridized to eight other species from the genus.

The fifth antirule is that there is no apparent association between any character of a plant species and its ability to hybridize with another species: "plants most widely different in habit and general appearance, and having strongly marked differences in every part of the flower . . . can be crossed. Annual and perennial plants, deciduous and evergreen trees, plants inhabiting different stations and fitted for extremely different climates, can often be crossed with ease" (*Origin*, p. 257).

The sixth antirule is that reciprocal crosses do not always yield the same result. Crossing a male from species A with a female from species B may produce viable, fertile offspring, but crossing a female A with a male B may fail entirely. Kolreuter easily fertilized *Miarbilis jalappa* with the pollen of *M. longiflora*, then tried over two hundred times for the following eight years to fertilize *M. longiflora* with pollen from *M. jalappa* and failed every time. Darwin concludes that "these cases clearly show that the capacity for crossing is connected with constitutional differences imperceptible to us, and confined to the reproductive system" (*Origin*, p. 258).

Darwin notes that there are more antirules to be gleaned from Gartner's writings, but the ones he has already extracted are sufficient to make his point:

> Now, do these complex and singular rules indicate that species have been endowed with sterility simply to prevent their becoming confounded in nature? I think not. For why should the sterility be so extremely different in degree, when various species are crossed, all of which we must suppose it would be equally important to keep from blending together? Why should the degree of sterility be innately variable in the individuals of the same species? Why should some species cross with facility, and yet produce very sterile hybrids; and other species cross with extreme difficulty, and yet produce fairly fertile hybrids? Why should there often be so great a difference in the result of reciprocal crosses between the same two species? Why, it may even be asked, has the production of hybrids been permitted? (*Origin*, p. 260)

Darwin's reply to all these questions is that the sterility of crosses between species or of the hybrids that such crosses produce is simply incidental to or dependent on some other "unknown differences" between species. The sterility has no meaning of its own.

Causes of the Sterility of First Crosses and of Hybrids

Here Darwin enumerates reasons why crosses between members of different species are sometimes sterile. His reasons are the same as the ones we recognize today. First, something might keep sperm (pollen) and eggs (ovules) apart. Pollen that is deposited on a stigma, for example, must "tunnel" through the pistil to the ovary, but these pollen tubes cannot form properly if there is a mismatch in the genotypes of the flower and pollen. Second, sometimes the sperm (pollen) can reach the ova but cannot initiate development. Third, sometimes the sperm (pollen) can initiate development, but the embryo that is produced cannot develop normally. While all these reasons why two species might fail to produce viable hybrid offspring were known to Darwin, the factors governing whether viable hybrids would be fertile and capable of producing offspring of their own were not known.

Darwin's remaining arguments in this section are based on accurate observations but an inaccurate interpretation of their meaning because the laws of inheritance were not yet known. Because of the archaic nature of the arguments, I will not summarize them here.

Fertility of Varieties When Crossed and of Their Mongrel Offspring

First, an update on Darwin's lexicon of variation. The offspring produced by a cross between species are called "hybrids." The offspring produced by crosses between varieties within a species are "mongrels." If species really were the products of acts of special creation and were defined by absolute reproductive barriers, then it would follow that varieties within a species should always be perfectly fertile when crossed with one another. Darwin concedes that this is most often true, but not always. Yet the exceptions, where crosses between varieties within a species have reduced fertility, contribute to the disproof of the doctrine of special creation.

When Gartner crossed some of what were "considered by many of our best botanists as varieties" (*Origin*, p. 268), he found that the crosses sometimes had reduced fertility. Gartner's response was to simply reclassify them those varieties as separate species. Darwin argues that the underlying logic is a product of circular reasoning. If many experts had ranked these forms as varieties, then there must have been a reason for it. The usual reason was that the varieties had overlapping ranges and that individuals with intermediate morphology could be found in those areas, so that in nature they were behaving like varieties within a species, rather than as different species.

Darwin supports his argument with examples from domesticated plants and with observations made by Gartner and Kolreuter, whom Darwin describes as "hostile witnesses." Domestic varieties are generally of recent origin and represent variation within species (except in those cases where the domesticate is a hybrid product of different species). Domestic varieties are sometimes shown to vary in how readily they can be crossed. For example, Gartner grew a dwarf corn variety with yellow seeds and a tall variety with red seeds right next to each other. When he artificially fertilized flowers from one variety with pollen from the other, twelve of thirteen crosses produced no seeds. The thirteenth produced only five seeds. All five grew into plants with normal fertility, so Gartner considered the yellow- and red-seed corn to be the same species. Kolreuter hybridized five varieties of common tobacco with one another and found that all had normal fertility. He then hybridized all five with a different species in the same genus and found that four of the five crosses produced sterile offspring while the fifth resulted in offspring that were not sterile. Darwin concludes that varieties within a species can be very much like different species in their ability to cross with one another and that any incompatibility between varieties "is not a special endowment, but is incidental on slowly acquired modifications" (*Origin*, p. 272) between two varieties.

Hybrids and Mongrels Compared, Independently of Their Fertility

Gartner also cataloged rules that discriminated between the properties of mongrels and of hybrids. Darwin again argues that the only true rule is that there are no rules. Crosses between varieties are much more likely to be successful than those between species, but otherwise show the same trends. One such "rule" is that mongrels tend to be more variable than hybrids. Another is that mongrels are more likely than hybrids to revert to one or the other parent form. In both cases, Darwin argues that if such differences between mongrels and hybrids are real, they are only differences of degree rather than being absolute rules discriminating one from the other.

Darwin's general conclusion is that there are no clear distinctions to be made between the properties of mongrels versus hybrids: "If we look at species as having been specially created, and at varieties as having been produced by secondary laws, this similarity would be an astonishing fact. But it harmonizes perfectly with the view that there is no essential distinction between species and varieties" (*Origin*, pp. 275-6).

His bottom line is again that species are just points in a continuum and are only well-marked varieties, so the only distinctions between varieties and species, regardless of the character being considered, are ones of degree. Furthermore, there is nothing special about reproductive isolation, at least not the particular aspects of postmating reproductive isolation considered here. It is a character that varies continuously, like any other. Reproductive isolation is not selected for; it arises instead as a by-product of the independent evolution of varieties or species as they adapt to their own environments.

Darwin's arguments about reproductive isolation are fully consistent with our modern perspective in all regards save one. Whereas Darwin emphasized the continuity between the formation of varieties and the formation of species, we now follow Dobzhansky and Mayr's argument that speciation has the distinct property of causing the origin of discontinuity in nature. Darwin saw no necessary role of reproductive isolation in defining species boundaries, whereas today we define species as groups of individuals that are reproductively isolated from one another. Furthermore, his claim that species were not strictly reproductively isolated from one another was a keystone of his argument for species transmutation and against special creation. Darwin was trying to smash the idea of special creation and discontinuity at all levels, including the role of reproductive isolation in defining species. Since the doctrine of special creation included the idea that species were immutable and absolutely reproductively isolated from one another, Darwin did his best to show that they were not. Recall that Darwin defined reproductive isolation differently than we do today because he excluded many forms of premating isolation. Because of this difference in definition, his conclusions are not in any way incompatible with our current definition of species.

Evolution Today: Reconciling Darwin's and Modern Concepts of Speciation

This chapter causes us to confront again the differences between Darwin's concepts of species and speciation and ours today. Biologists often present Darwin's theory of the origin of species and our modern theory of speciation alongside each other without realizing that our modern models for speciation may include significant departures from what Darwin proposed.

Whereas Darwin saw natural selection as the engine of all divergence and the cause of speciation, today we think that the importance of natural selection depends on how speciation happens. Natural selection plays little or no role in causing some types of speciation, but is the primary driver in other types of speciation. Consider, first, the most widely accepted mechanism for species formation, which is allopatric speciation. Strict allopatric speciation, as originally defined by Mayr, means that a species is subdivided into two or more populations that are geographically isolated from one another and that each then follows its independent evolutionary pathway. Over time, these populations become progressively more different from one another, perhaps because they are adapting to different environments, or because they adapt to the same environment in different ways, or for a variety of other reasons. For example, they may become different because of sexual selection causing differences in male displays and female preferences. As a result, males of one population may be unable to mate with the females of the other population. If any such changes occur during a period of isolation, then these populations will not interbreed if they are reunited, and will be recognizable as different species. When the cord was cut and one species split into two, the split was caused by chance events that occurred while the populations were geographically isolated from one another. Natural selection never played a direct role in causing these populations to become reproductively isolated from one another.

Darwin and Mayr agree on the causes of reproductive isolation; both see reproductive isolation arising as an incidental by-product of the genetic differences that arise between isolated populations rather than being caused by the direct action of natural selection. They disagree about the role of natural selection in causing the divergence between populations that results in speciation, with Darwin seeing natural selection as the sole cause of divergence while Mayr sees no necessary role for it, at least in this strict form of allopatric speciation.

A second version of allopatric speciation occurs when two populations diverge while they are isolated from one another, but can still interbreed when they come back into contact. If this occurs, then the process of reinforcement can finalize the evolution of reproductive isolation (see chap. 8). If the reunited populations are adapted to use the environment in a different way or have different attributes that make males attractive to females, then a hybrid between the two may be "neither fish nor fowl" in the sense that it will not be as good as either parent in earning a living or finding a mate. If this occurs, then natural selection can act against those individuals from either population that make the mistake of mating with a member of the other population, since the offspring they produce will be inferior to those produced from matings within each population. Natural selection against such mistakes cuts the cord that joins the two incipient species. Reinforcement thus represents a departure from Darwin's view on the evolution of reproductive isolation, since in this case it is natural selection that causes reproductive isolation to evolve, but at the same time reinforcement restores natural selection as playing an important role in speciation.

Speciation can also happen without geographic isolation. One version of "nonallopatry" is sympatric speciation, or the formation of reproductively isolated populations within what had once been a single species. We think that sympatric speciation will most readily occur when there is a genetic association between habitat choice, fitness in alternative habitats, and mating preference. Imagine an organism that has two habitats to choose from. It has genetic variation for habitat choice, genetic variation for traits that influence how successful it will be in either habitat, and genetic variation for the mate that it will choose. When an individual chooses habitat A, its success at earning a living in that habitat will be a function partly of whether it has the right genotype to perform well there. It will produce more successful offspring if it also has a genotype that causes it to prefer to mate with an individual that shares these attributes. When all these conditions are satisfied, then it is possible for a single population of organisms to split into two, each specialized on one of the alternative habitats and each preferring to choose a mate from its own population without the two incipient species ever having been geographically isolated from one another. However, the number of conditions that must be satisfied for sympatric speciation is many more than the number required for allopatric speciation. Allopatry requires nothing beyond populations being geographically isolated. Newly developed theory argues that sympatric speciation can occur. Laboratory experiments have documented the evolution partial reproductive isolation in sympatry, and there are some undisputed cases of sympatric speciation in nature. None of this evidence tells us how often it occurs relative to other forms of speciation.

Two examples of sympatric speciation that are very strongly supported are the clusters of cichlid fishes found in two tiny lakes, Bermin and Barombi Mbo, in Cameroon. Both lakes fill volcanic craters. Both lakes were formed when a single river penetrated the wall of the volcano and flooded the crater. The lakes are shaped like a steep-sided, inverted cone with no obvious topography that could subdivide them into geographically isolated habitats, especially for mobile fish. Each lake contains a cluster of closely related species of cichlids that are found only in that crater. Genetic analyses show that the nine endemic species in Lake Bermin are recently derived from a single common ancestor that colonized the lake. Similar analyses divide the eleven species in Lake Barombi Mbo into two clusters, implying that they were derived from two different colonizing species. There is no evidence of hybridization between the species within each lake, so they are behaving like different species. The association of the lakes with a volcano means that they can be accurately dated using radiometric techniques, at less than 2.5 million years for Lake Bermin and 1 million years for Lake Barombi Mbo, making it possible to estimate the rate of formation of these new species.

The species in each lake differ mostly in terms of their diet and mode of feeding. While their present state does not reveal how they became what they are today, their circumstances strongly suggest that the species within each lake evolved in those lakes without ever being geographically isolated from one another. It is easy to envision their origin as being the product of Darwin's principle of divergence. In the same way that Darwin imagined plants on a single patch of ground expanding in population size, then probing the environment for different ways of earning a living, then specializing on different facets of a continuous habitat, we can also imagine that the one species of cichlid that invaded Lake Bermin and two that invaded Lake Barombi Mbo established populations that grew and then diversified to exploit all available resources. This form of speciation is thus almost identical to what Darwin proposed, except that it includes an even greater role for natural selection than envisioned by Darwin, since natural selection also caused the evolution of the cichlids' reproductive isolation.

References

Coyne, J. A., and H. A. Orr. 2004. *Speciation*. Sunderland, MA: Sinauer Associates.

Hartwell, Sarah. N.d. Hybrid deer and antelope. www.messybeast.com. Referenced on 3/14/09.

Chapter 12

Evolution Today: The Mosquitoes
of the London Underground

Public use of the London Underground began on January 10, 1863. That date, or perhaps some earlier date when the tunnels were being readied for traffic, marks the beginning of the path toward the formation of a new species of mosquito. We often wonder how long it takes to form a new species; Darwin speculated timescales on the order of tens of thousands to hundreds of thousands of generations. The mosquitoes of the London Underground show that if conditions are right, the process can be much faster. I chose this example because it happened on Darwin's home turf and postdates the publication of the *Origin*. I will not pass judgment on whether these mosquitoes should now be defined as a distinct species, but will instead show that they have moved far down the path toward forming a reproductively isolated network of populations, which is the currently accepted definition of a species. My emphasis here is the same as Darwin's in the *Origin*: we are concerned with the process of speciation rather than naming of species.

Culex pipiens is the most widespread mosquito in the world. It is found on all continents except Antarctica. It is a vector of diseases, including West Nile virus and St. Louis encephalitis. It breeds wherever it can find stagnant pools of water, such as in untended birdbaths, forgotten buckets in the backyard, discarded automobile tires, clogged rain gutters, or wherever else fetid, stagnant water accumulates. It does very well in polluted urban settings. It lays rafts of eggs that float on the water, then hatch into wriggling larvae that feed on microbes. They emerge as adults around ten days later.

Females seek victims from whom they obtain a blood meal, then use the nutrients to form new batches of eggs. Anywhere they find a good supply of stagnant water and hosts for blood meals, a few of these mosquitoes can quickly multiply into swarms.

In London, *C. pipiens* is found as a surface form (*C. pipiens pipiens*) and as a subterranean form (*C. pipiens molestus*). The adult surface mosquitoes get their blood primarily from birds. They mate in large swarms in open areas. They live in a seasonal environment, which means they have a seasonal diapause, a time when they stop reproducing, store fat, and hide away for the winter in sheltered areas that stay warm enough to keep them from freezing.

The mosquitoes that ventured into the London Underground found excellent, unoccupied pools of stagnant water to lay their eggs in, but encountered a very different habitat for the adults. There were no birds to feed on, so they instead began feeding on mammals, primarily rats and humans. Their mating occurred in closed areas. Their environment had no seasons and was always fairly warm, so they lost their seasonal diapause and remained active all year long. They also evolved the ability to produce a clutch of eggs without first getting a blood meal, perhaps because their larval environments were rich but their prospects of finding a blood meal as adults were poor. While their presence was generally only an issue for rats and the few who worked in the tunnels, these subterranean mosquitoes added to the misery of the Londoners who sought refuge in the Underground during the nighttime bombings of World War II.

In the 1990s, British geneticists Katharine Byrne and Richard Nichols became interested in the origin of the Underground mosquitoes and in their relationship to those found on the surface (Byrne and Nichols 1999). Subterranean (*molestus*) mosquitoes can be found in similar sheltered habitats, such as caves and sewers, throughout western Europe and in other parts of the world. These populations share some of the attributes of the London Underground mosquitoes, so it is possible that some of them made their way to the London Underground. On the other hand, it may be that the resident surface mosquitoes invaded the London Underground on their own and independently adapted to the subterranean environment.

Byrne and Nichols used electrophoresis, the same method used by Highton to study *Plethodon* salamanders, to compare the Underground and surface populations. They sampled and quantified genetic variation at twenty loci in mosquitoes from seven sites dispersed throughout the more than 110 miles of tunnels, ranging from Shepherd's Bush in the west to Finsbury Park in the northeast to Elephant and Castle in the southeast. They also sampled from twelve surface populations found in gardens and ponds close to the location of the Underground sites. Comparing results, they found that all the alleles in the Underground populations could also be found in the surface populations. If mosquitoes from elsewhere had colonized the Underground, then we would expect those long-distant colonists to have brought some unique alleles into the Underground gene pool that were not seen in the surface population from London. The absence of such foreign alleles suggests that the original Underground mosquitoes were colonists from the surrounding countryside and not long-distance migrants from the mainland. The thought of a mosquito flying across the English Channel may seem silly, but people can inadvertently transport mosquitoes. For example, the paradisial qualities of Hawaii once included having no mosquitoes, but people accidentally transported them there as unwanted baggage.

Byrne and Nichols also found that there was much less genetic variation in the Underground mosquitoes than in those on the surface. All the underground populations were genetically more similar to one another than they were to the surface populations. This is the pattern we would expect if a small number of surface mosquitoes moved into the Underground to establish the new populations. There may have been only a single successful colonization event. A small number of colonists would carry only a small amount of genetic variation into their new environment. Byrne and Nichols's conclusion was thus that the founders of the Underground population were most likely to have been a few colonists from the surface that moved into the new habitat provided by the tunnels and spread throughout the Underground as the tunnel network expanded.

Byrne and Nichols collected egg rafts from the breeding sites of the Underground and surface populations, raised the young to maturity, then performed crosses between different Underground populations and between Underground and surface mosquitoes. The different populations of Underground mosquitoes readily bred with one another and produced viable eggs. The hatchlings grew up and proved to be as fertile as their parents. They all thus behaved as if they were different populations of the same species.

Every pairing of a female from the Underground with a group of males from the surface failed to produce eggs. Each time a pairing failed, the female was then mated to males from the Underground and, each time, produced viable offspring; the genotypes of the babies confirmed that an

Underground male was the father. This means that all the Underground females were fertile, so their failure to mate with males from the surface indicated some form of premating reproductive isolation.

Darwin envisioned organisms in nature as always producing a surplus of offspring that are in turn always probing the environment for new opportunities. Byrne and Nichols found evidence for this perennial struggle for existence in the form of surface and Underground populations that were each probing the other's habitat for new opportunities. The Oval Station in the Underground was a site where surface mosquitoes appeared to be establishing a new beachhead for invading the Underground. They were found in a flooded service tunnel at the bottom of a shaft that opened to the surface. This was an Underground site, but the mosquitoes in that shaft were genetically like those on the surface. They required a blood meal to produce eggs. They were also reproductively isolated from the other Underground populations, since all but one pairing between them and mosquitoes from other Underground populations failed to produce offspring. Males from the Oval Station did succeed in mating with one of the Underground females they were paired with. Her raft of eggs produced viable offspring, but the offspring did not produce offspring of their own. Byrne and Nichols also found a surface population of mosquitoes in Beckton, southeast London, that was invading houses and biting people. The genotypes of these mosquitoes grouped them with the Underground populations, which suggests that they had recently emerged from the Underground and established a surface population. They did not produce eggs in captivity, so no mating trials were done on them.

Byrne and Nichols were careful to say that their goal was not to determine whether the Underground and surface populations represented different species, and they did not offer conclusions about how these populations should be classified. I agree with their caution, since it is the process that is important, not what we call the two forms. The important results are that the data in hand argue that the Underground population was derived from a small number of colonists from the surface, and that these colonists spread throughout the tunnel system. They have clearly moved far down the path toward becoming reproductively isolated from surface mosquitoes, or toward the discontinuity that defines a species. Since the Underground represents a newly formed habitat of known age, we can also conclude that all this happened between the publication of the first edition of the *Origin* in 1859 and the mid-1990s, when Byrne and Nichols collected their samples. This is far less time to form a species, by orders of magnitude, than imagined by Darwin.

Why did the Underground mosquitoes evolve so quickly toward reproductive isolation from the surface mosquitoes? One obvious mechanism is "disruptive selection," or selection for different phenotypes in the two environments. If each environment demands very different adaptations, then any time a surface and Underground mosquito met and mated, they would produce offspring that were not well suited to either environment. The list of specific adaptations to the Underground is long. The two forms feed on different hosts. Mosquitoes use chemical cues to track down hosts, so there may be differences in the cues that each type of mosquito uses to locate the source of a blood meal. The Underground mosquitoes live in an environment that is warm enough for them to be active year-round, so they no longer have a seasonal diapause; the surface mosquitoes have an obligatory diapause, which is necessary to survive the winter. Byrne and Nichols suggest that Underground mosquitoes that colonize the surface, as they appear to have done in Beckton, will likely be wiped out every winter because they lack the adaptations for diapause. Underground mosquitoes can produce eggs without a blood meal, while surface populations cannot. This adaptation evolved perhaps because the mosquitoes are less certain to encounter hosts in the Underground than on the surface. I imagine that the root of this adaptation is in the larval life stage, since the larvae would first have to pupate, then emerge as adults with sufficient reserves to support the development of eggs. This adaptation would give the Underground mosquitoes a big advantage over any new surface mosquitoes that invaded the Underground and required blood meals to reproduce, if blood meals were hard to obtain.

A second possible mechanism for the rapid evolution of reproductive isolation is that it is a by-product of the evolution of reproductive behavior. Adaptation to the Underground could have altered mating behavior or mechanisms of mate choice in a way that made it less likely for the Underground mosquitoes to mate with surface mosquitoes. Byrne and Nichols specify that surface and Underground mosquitoes form mating swarms in different types of settings. The surface mosquitoes swarm in open areas, while the Underground mosquitoes breed in confined spaces. This difference in behavior by itself could help to reduce the gene flow between the two populations. However, the way the two forms were mated in Byrne and Nichols's experiments would have circumvented such differences in mating sites, so there must be something else that contributes to their premating isolation than just where they choose to mate.

Byrne and Nichols reviewed other studies of the *pipiens* and *molestus* forms from throughout Europe, the Middle East, and North Africa and found that *pipiens* and *molestus* from northern latitudes tend to be more distinct from one another than those from southern latitudes. The northernmost *molestus* and *pipiens* populations show genetic differences between neighboring populations similar to what we see in London. At midlatitudes, in the northern Mediterranean, the differences are less extreme, and genetic studies yield evidence of some gene flow between the two forms. At more southern latitudes, in the Middle East and North Africa, there are few differences between the two forms. One key change in the southernmost populations is that the surface mosquitoes are more like the underground mosquitoes in that neither group has a diapause, probably because of the warmer winters and reduced differences in winter and summer day length. Byrne and Nichols propose that the increasing severity of winters in the northern parts of the range and the split between diapausing surface populations and nondiapausing subterranean populations is a key factor that separates the two forms in northern latitudes. Subterranean mosquitoes breed year-round, while surface mosquitoes have a short, summertime breeding season, so there is only a short period of overlap when they could crossbreed. Furthermore, any subterranean mosquitoes that invaded the surface habitat too late for summer mating would not likely survive until the next mating season. Lacking the ability to diapause, they would be exterminated come winter.

I feel the burden of Coyne and Orr's critical gaze as I consider the implications of these results. What might the alternative explanations be? The main alternative is that the mosquitoes that moved into the London Underground were migrant *molestus* mosquitoes from somewhere else and arrived with a full suite of adaptations to the Underground. The best available counterargument is the genetic data. Unique alleles among the twenty sampled loci have been found in foreign populations, but none of them appear in the London Underground. All alleles in the underground are a subset of what is seen on the surface. Further, their genetic similarity supports the theory that the Underground population could have been derived from only a few and perhaps just a single successful colonization event from one of the surface *pipiens* populations found near the Underground. Likewise, other *molestus* mosquitoes from other habitats throughout their multicontinental range may also represent the independent invasion of a subterranean environment and the independent evolution of adaptation to that environment. We have seen this pattern before. In chapter 6, I reported on work by Culver and associates on the small amphipod crustaceans inhabiting different cave systems in the southeastern United States and by Verovnik and associates on isopod crustaceans from different cave systems in Albania and Italy. The cave forms were all well adapted to their new environment, with reduced eyes but enhanced antennae for navigating in an environment without light. Both sets of authors also found that each cave population was genetically most similar to the nearby surface population, rather than to other cave populations, so each represented an independent colonization of and adaptation to the cave environment. (Although the genetic data on foreign mosquitoes seem to rule out foreign *molestus* migration to the London Underground, one other source of migrants seems possible. One could imagine that the Underground mosquitoes are descendants of some other local *molestus* population that got a head start in adapting to a subterranean habitat before the Underground—perhaps, say, the dungeons under the Tower of London.)

A second alternative explanation is that even if the mosquitoes of the London Underground represent an independent colonization event and a very young, fast drive toward speciation, some of the genetic building blocks that caused their adaptation to the Underground may not be unique. Byrne (1996, in Byrne and Nichols 1999) discovered what might be a "preadaptation" to life in the Underground. She identified an allele of one enzyme that is rare on the surface but occurred at a frequency above 90% in the Underground. Chevillon and colleagues performed an analysis of the distribution of the same allele in *molestus* and *pipiens* populations from the French Alps and found evidence of selection in favor of this allele each time a population adapted to the underground environment, but selection against it in the surface populations. They suggest that this substitution of one allele for another at this genetic locus could contribute to the evolution of the ability to reproduce without obtaining a blood meal. In this way, a population of invading mosquitoes could establish a beachhead in a subterranean environment. The allele may simply be a rarity found in surface (*pipiens*) populations, but it may also be more common in the *pipiens* populations than it used to be, because of occasional matings between *molestus* and *pipiens* forms wherever they come into contact with one another.

Even though such genetic variation that facilitates adaptation to subterranean habitats may exist in surface mosquitoes, this does not change the fact that the London Underground mosquitoes still represent the rapid evolution of reproductive isolation. The idea of a preexiting surface variation just clarifies how one particular mechanism might facilitate such rapid evolution. It is a special case of our more general understanding of how such adaptation to a new environment occurs. The most general explanation for how organisms adapt is the same as the one proposed by Darwin, which is that all natural populations of organisms possess differences between individuals in all traits. The suggestion that a population may be using genetic variation that resulted from adaptations made by past populations to environments similar to those confronting the current population is really just a new variation on this old theme. The adaptation of mosquitoes to subterranean habitats may represent independent events in that each case is an invasion of the new habitat by the local population, but the actual adaptation may involve, in part, a reassembly of traits that evolved under similar circumstances but in a different place and time.

The mosquitoes of the London Underground are not the only example we have of organisms invading a new habitat with the aid of genetic building blocks derived from other populations that adapted to similar environments long ago. I present one additional example here because it emphasizes the potential rapidity with which reproductive isolation can evolve, plus offers a more general message about what we have learned over the past few decades regarding the genetic diversity of species in nature. This example also highlights the ways in which we can now integrate modern molecular genetic methods with the study of evolution.

Three-spine sticklebacks (*Gasterosteus aculeatus*) are small, primarily marine fish that migrate into freshwater environments to breed. When the glaciers that capped the northern latitudes of the earth retreated around ten thousand years ago, they left behind a diversity of vacant freshwater habitats, including rivers, lakes, and marshes that were similar, in principle, to the vacancies created by the construction of the London Underground. The sticklebacks invaded thousands of these new environments and often evolved into permanent freshwater populations. Many of them also evolved morphologies that were distinct from those of their marine ancestors (fig. 15). The most obvious changes were a reduction of armor. The marine forms are well armored as a defense against fish predators. They have bony plates on their sides, spines along their backs (hence the name "stickleback"), and a pelvic girdle arrayed with stiff pelvic spines. If you look at them head-on, the spines form a triangle: the dorsal spines point up, while the pelvic spines point down and to each side. The bony plates on the sides bridge the distance between the dorsal and pelvic spines so that the whole structure is a stiff ring studded with spikes, a configuration that causes these fish to stick in the throat of would-be predators and sometimes be spit out again. However, the typical marine predators are often absent in freshwater environments, and all or part of the armor has disappeared in many of the freshwater sticklebacks. That said, there is considerable diversity among freshwater populations in the degree to which the armoring has been lost.

Colosimo and colleagues (2005) found that an allele at a single gene strongly influenced the freshwater sticklebacks' loss of the lateral bony plates that run along the sides of their marine ancestors. This allele was found in freshwater populations of sticklebacks throughout their range and was also found in the marine populations. They estimated that the allele originated around 2 million years ago. The pattern of its continuing occurrence in different places means that there may have been a single, ancient event in which sticklebacks adapted to a freshwater environment and that some of the alleles associated with this adaptation became enriched in the marine populations because of some continued interbreeding between ma-

Figure 15
Marine vs. Freshwater Sticklebacks (*Gasterosteus aculeatus*)
The upper fish is typical of the marine members of this species, with distinct dorsal spines and a spine on the pelvic girdle (indicated with arrow). The other bony armor has been stained so that it shows up as dark gray to black in this photograph. The flanks of the body of the marine form are almost entirely plated with armor. Freshwater sticklebacks vary in their amount of armoring. The one pictured here has reduced dorsal spines, no pelvic spine or even a pelvic girdle, and few lateral plates. The fact that marine and freshwater sticklebacks readily interbreed in captivity has greatly facilitated research that characterizes the genes that control the expression of these traits.

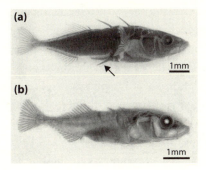

rine and freshwater fish. The next advance of the glaciers would have wiped out these ancient freshwater populations, but the genetic tool kit that had enabled them to adapt to freshwater environments could have remained as rare alleles in the marine populations. There have been several advances and retreats of glaciers over the past 2 million years. Each retreat would have opened up new freshwater environments that could be invaded by sticklebacks; each advance would have wiped out those freshwater populations. During each retreat phase, fish that retained the some of the genetic building blocks their ancestors had used to adapt to freshwater habitats of the past could invade some of the new freshwater habitats. By this hypothesis, each freshwater invader begins with a marine phenotype and independently evolves a freshwater phenotype as it adapts to its new environment, but at least some of the genes that contribute to these local adaptations do so with alleles derived from long-extinct freshwater populations that persisted in the gene pool of marine sticklebacks.

We now have many other examples of organisms that have moved into new habitats, experienced strong disruptive selection as they adapted to their new environment, and are now at least partly reproductively isolated from their ancestral population. Apple maggot flies (*Rhagoletis pomonella*) normally lay their eggs on the fruits of hawthorns (genus *Crategus*). In the 1860s, in the Hudson Valley of New York, it was discovered that some of them had shifted to laying eggs on apples that had been introduced from Europe. There is now a distinct "apple" race of these flies that is reproductively isolated from its hawthorn-preferring ancestors. Copepods (*Eurytemora affinis*) have moved from their marine environment into freshwater canals and reservoirs, Yucca moths (*Prodoxus quinquepunctellus*) adapted to an introduced species of yucca, and pea aphids (*Acyrthosiphon pisium*) adapted to different monocultures of crop plants, to name a few more. The impressive (and still growing) body of research on all these organisms and others supports the conclusion that local adaptation can be rapid and that it can in turn lead to the evolution of reproductive isolation, sometimes on timescales that are on the order of tens to hundreds of years. It also tell us that adaptation and speciation are going on now, right under our noses, rather than being something that happened in the past or happens so slowly that it is not detectable. All that is required for us to see speciation in progress is to look for it in the appropriate fashion.

One important message common to all these examples is that they illustrate the complexity of species in nature. Species are typically subdivided into many local populations that are adapted to local conditions. The extent of local adaptation will be a product of how strong selection is and the degrees to which local populations are isolated from each other. Recent studies that couple molecular genetic techniques with the study of local adaptation, as in Byrne and Nichols's study of the mosquitoes of the London Underground or Colosimo and colleagues' study of sticklebacks, have shown again and again that species in nature should be thought of as fluid mosaics of populations that are becoming locally adapted, sometimes with the aid of similar adaptations attained by long-extinct populations that adapted to similar environments. These locally adapted populations often go extinct, but sometimes, when the opportunity arises, establish a new beachhead in previously unoccupied territory. Some local beachhead populations are now verging on becoming reproductively isolated as they adapt to their new environment. Most such forays fail as the local population disappears or is pulled back into the fold of its species of origin by gene flow, but some cross the threshold of reproductive isolation to become new species. Once that threshold is crossed, an irreversible discontinuity has been formed. This modern view of a species as a fluid mosaic of differentiated populations is consistent with Darwin's view of the constant struggle for existence and the principle of divergence, which together cause the formation of distinct populations, varieties, and—on rare occasions—new species.

Most of these examples also contain a message about the influence of human activities on nature. Many people are now preoccupied with human-caused changes to the environment and the march of vast numbers of species toward extinction. Sadly, this is the dominant theme of our human legacy. However, as we are changing the environment we are also creating new environments that can become opportunities for some species. Most of the examples presented above are associated with recent changes caused by humans: the digging of the London Underground and creation of a new subterranean habitat; the introduction of new host plants (apples, Yucca) into North America; the digging of canals and reservoirs that provide new opportunities for invasion by marine organisms; and the cultivation of crops in monocultures, creating a discrete patchwork of alternative environments. At the same time that we are causing extinction, we are also acting as a potent agent of natural selection by changing the environment so rapidly and in so many ways. Extinction results when change is too rapid and too large for adaptation to be an option, but sometimes organisms can keep pace with human impact on the environment. Sometimes we create new opportunities and habitats that serve as blank slates for invasion and local adaptation and, in rare cases, the formation of new species.

References

Byrne, K., and R. A. Nichols. 1999. *Culex pipiens* in London Underground tunnels: Differentiation between surface and subterranean populations. *Heredity* 82:7–15.

Chevillon, C., Y. Rivet, M. Raymond, F. Rousset, P. E. Smouse, and N. Pasteur. 1998. Migration/selection balance and ecotypic differentiation in the mosquito *Culex pipiens*. *Molecular Ecology* 7:197–208.

Colosimo, P. F., K. E. Hosemann, S. Balabhadra, G. Villarreal, M. Dickson, J. Grimwood, J. Schmutz, R. M. Myers, D. Schluter, and D. M. Kingsley. 2005. Widespread parallel evolution in sticklebacks by repeated fixation of ectodysplasin alleles. *Science* 307:1928–33.

Feder, J. L., S. H. Berlocher, J. B. Roethele, H. Dambroski, J. J. Smith, W. L. Perry, V. Gavrilovic, K. E. Filchak, J. Rull, and M. Aluja. 2003. Allopatric genetic origins for sympatric host-plant shifts and race formation in *Rhagoletis*. *Proceedings of the National Academy of Sciences* 100:10314–19.

Groman, J. D., and O. Pellmyr. 2000. Rapid evolution and specialization following host colonization in a yucca moth. *Journal of Evolutionary Biology* 13:223–36.

Lee, C. E. 1999. Rapid and repeated invasions of fresh water by the copepod *Eurytemora affinis*. *Evolution* 53:1423–34.

Palumbi, S. 2001. *The evolution explosion: How humans cause rapid evolutionary change*. New York: W. W. Norton & Co.

Thompson, J. N. 2005. *The geographic mosaic of coevolution*. Chicago: University of Chicago Press.

Via, S., A. C. Bouck, and S. Skillman. 2000. Reproductive isolation between divergent races of pea aphids on two hosts. II. Selection against migrants and hybrids in the parental environments. *Evolution* 54:1626–37.

Wikipedia. N.d. London Underground. http://en.wikipedia.org/wiki/London_Underground. Referenced on 7/20/2008.

Part Three

Theory

Chapter 13

Preamble: What Is a Theory?

On June 19, 1987, the Supreme Court of the United States handed down a decision on *Edwards v. Aguillard*. This case pertained to a law titled the Balanced Treatment for Creation-Science and Evolution-Science Act (Balanced Treatment Act) that had been passed by the Louisiana state legislature. The U.S. Court of Appeals for the Fifth Circuit ruled that the Balanced Treatment Act was unconstitutional. Seven of the nine Supreme Court justices voted to uphold the circuit court decision. The purpose of the act was to forbid a science instructor in a public school classroom from teaching aspects of either "creation-science" or "evolution-science" without offering instruction in the other discipline, hence the word "balanced." The argument that the majority of the Court accepted was that "creation science" is not science but is, rather, thinly veiled religion. Hence, the act was implicitly promoting religion in science classrooms and violated the establishment clause of the First Amendment.

This case is but one episode in a long-standing debate over the teaching of evolution in public school classrooms in the United States. My goal is not to address this ongoing debate but rather to consider some details of the dissenting opinion that was submitted by Justice Antonin Scalia. Scalia's opinion encapsulates the popular misunderstanding of the theory of evolution and, more generally, of the meaning of the word "theory" when it is used with regard to science. His decision thus serves well as a foil for me to define "theory" as the word is used in science, then illustrate how the bulk of the *Origin* was devoted to articulating why evolution is a scientific theory that is central to the biological sciences.

Scalia cites the Balanced Treatment Act as declaring that evolution and creationism are the two competing theories for the origin of life. Furthermore, the act "requires that, whenever the subject of origins is covered, . . . evolution be taught as a theory, rather than as proven scientific fact" (p. 14). These statements, and his entire argument, contain two key misunderstandings.

First, Darwin's theory of evolution and the origin of species is not a theory about the origin of life. Darwin studiously avoided this subject. Everything about the way Darwin presented evolution in the *Origin* describes it as a process that requires an organism that can replicate itself and can transmit some form of blueprint for development from one generation to the next; so evolution cannot happen until *after* some form of life has arisen. He never considered how life began, nor do evolutionary biologists today. Research on the origin of life lies within the province of chemistry and biochemistry.

Second, there is a popular misunderstanding, even among those as well educated as Supreme Court justices, of the meaning of the word "theory." This word is like many others in having multiple definitions. The appropriate definition of the word depends on the context in which it is used. Unabridged dictionaries sometimes present over ten definitions for "theory," but I will illustrate my point with the three definitions supplied by the *American Heritage Dictionary of the English Language* (1970). "Theory" is most often used in a colloquial fashion to mean: "2. Abstract reasoning; speculation. 3. Broadly, hypothesis or supposition." This is the sense in which Scalia and the proponents of creation science use the word. To Scalia and others, evolution always carries the epithet of "just a theory," meaning that it is mere speculation rather than fact. I am reminded of the way Homer always referred to Greek gods, heroes, and peoples with an epithet, such as the gray-eyed Athena, the horse-taming Trojans, or Apollo the far-darter.

The first definition provided by this dictionary is the one that applies to the use of the word "theory" in science: "1. *a.* Systematically organized knowledge applicable in a relatively wide variety of circumstances; especially, a system of assumptions, accepted principles, and rules of procedure devised to analyze, predict, or otherwise explain the nature or behavior of a specified set of phenomena. *b.* Such knowledge or such a system distinguished from experiment or practice" (p. 1335).

Scientific theories are not mere speculation. They are also not the opposite of "facts." Scientific theories are based on a body of well-established facts and are designed to elucidate some central, organizing principle of a discipline. In doing so, they should also make predictions, either explicitly or implicitly, that make it possible to test the theory. In the context of this definition, a theory can also be a fact. Evolution is a fact in the sense that we know that it happened; there is no reasonable scientific debate about its reality. The scientific debate is instead over the details of how it happened. An example of a theory from a different discipline might help make these points.

Virtually all chemistry classrooms display the periodic table of the elements at the front of the room, over the top of the blackboard, sometimes to be referred to in lecture, but always displayed as an icon of truth. No one disputes the factual nature of the rows and columns of this table or the information it incorporates about the structure of matter. At the same time, this table is the embodiment of a scientific theory. There are many parallels between the discovery and subsequent development of the periodic table (which is a way of envisioning the relationships between the elements) and Darwin's discovery of evolution and the subsequent development of evolutionary biology. Yet there are also differences. One significant difference is that chemistry was a large, well-established discipline before the discovery of the periodic table of the elements, so the periodic table is a significant signpost on the road to our understanding the structure of matter. The *Origin* had some precedents, but it was not embedded in a continuous, well-developed chain of progress in the same fashion as the periodic table. It was a new beginning, and Darwin can be credited with giving birth to a new discipline.

The facts that preceded the discovery of the periodic table were the discovery of sixty-three of the elements, the estimation of their atomic weights, and the finding that there were patterns in how the elements reacted with one another and that they did so in regular proportions; for example, two volumes of hydrogen reacted with one of oxygen. Sometimes elements that were very different in atomic weight, such as carbon and silicon, were very similar in the way they reacted with other elements. These patterns suggested some underlying regularity in the structure of matter.

Dimitri Mendeleev was a chemistry instructor and was writing a textbook on chemistry when he too tried to find some organizing principles for these facts. He created a card for each element that included its name and atomic weight, then arranged and rearranged them, as in a game of solitaire. In the end product, there was a grid of columns and rows. Each column included elements that were similar in chemical properties but with a progressive increase in atomic weight from the top to the bottom of the column. Each row included elements that were different in their chemistry, with an incremental increase in atomic weight from left to right. Organizing the elements in this fashion—a table with columns and rows—was a step in identifying the underlying regularity in the structure of elements, a regularity that dictated each element's chemical properties. When he published this result in 1869, Mendeleev inferred that there were some elements yet to be discovered; he left gaps in his table that were to be filled by missing elements, with predicted atomic weights and chemical properties. He also inferred that the estimated atomic weights of some elements were incorrect. More often than not, his predictions were correct. The most important aspect of the periodic table is that it served as a central model for subsequent growth of knowledge about the structure of the elements, the discovery of new elements, the discovery of new phenomena (such as radioactivity, subatomic particles, fusion, and fission), and the ultimate integration of all this information into our modern understanding of the structure of matter.

One important parallel between Mendeleev's discovery of the periodic table and Darwin's discovery of evolution by natural selection is that both are logical constructs that were sufficiently supported by scientific evidence to stimulate further research. Both discoveries had some precedents in proposals with similar properties, even though they differed in the extent to which they were a departure from earlier work. Six years before Mendeleev, John Newland proposed his "law of octaves," which represented his discovery of groups of eight elements that were different in atomic weight but had similar chemical properties. In the same year as Mendeleev, Lothar Meyer defined a very similar periodic table, but he delayed a year in publishing and presented a less well-developed theory. Newland and Meyer, like Alfred Russel Wallace, have tended to be forgotten. Both the periodic table and evolution were ideas that were clearly in the air. Both represent such fundamentally important discoveries that they spawned the development of new disciplines.

Another point of similarity between the periodic table and the theory of evolution is that they were both predictive but not perfectly so, because they were shaped by an incomplete knowledge of their subject. Mendeleev did not know about the existence of subatomic particles and mistakenly arranged the first periodic table strictly by atomic weight, which implied that something about atomic weight dictated the chemical properties of the elements. Decades passed before chemists realized that it was the number of electrons in the outermost orbital of the atom that really determined the chemical properties of the elements, and that it was atomic number (the number of protons in the nucleus), rather than atomic weight (the number of protons plus neutrons), that was the appropriate quantity for organizing the elements into rows and columns. Likewise, Darwin did not know the laws of inheritance and placed undue emphasis on the role of the environment in causing the random generation of heritable variation. Nevertheless, both theories were based on accurate observations and premises that were correct enough to withstand the test of time and to be built on by those who followed.

The key property that the theory of evolution shares with Mendeleev's theory of regularities in the chemical properties of the elements, as embodied in the periodic table, is that it is based on observed facts. Darwin organized these facts to propose a process, natural selection, that could then explain how species change over time. It is not mere speculation about how species form. The facts behind natural selection are: (1) All organisms produce many more offspring than are required to replace themselves in the next generation. This "overproduction" of offspring means that all populations have a capacity for rapidly increasing in number, that they will soon pack their environment to capacity, and that there will be an ensuing struggle for existence to obtain the resources required for survival and reproduction. (2) Within every population of organisms one can find differences between individuals in characters that affect their ability to utilize their environment. (3) Some of these individual variations influence the ability of individuals to survive and reproduce, and may enable some individuals to penetrate new environments. (4) These variations are heritable, meaning that they are faithfully transmitted from parent to offspring.

Darwin observed that all organisms share these properties. When these conditions are satisfied, evolution by natural selection is inevitable.

Darwin then explored the consequences of this process. Natural selection provided an explanation for why organisms appear to be so beautifully adapted to their environment and hence provided an alternative to the prevailing idea of the time, as presented by Archdeacon William Paley in his *Natural Theology*, first published in 1802. Paley's famous analogy was to compare a watch with an eye. Their shared intricate complexity implies a common mechanism of creation. When we see such design, then there must be a designer. Paley's principles were given new momentum at the time of the voyage of the *Beagle* with the publication of the *Bridgewater Treatises*, which were "commissioned in the will of the eighth earl of Bridgewater, aimed at demonstrating the 'Power, Wisdom, and Goodness of God, as manifested in the Creation'" (Ruse 1999, p. 71). William Whewell, the Cambridge mentor who so influenced Darwin with his *Philosophy of the Inductive Sciences* (1840), updated and reaffirmed Paley's arguments in his contribution to the treatises.

A critical difference between Darwin's and Paley's proposals, and a way to distinguish between them as explanations for adaptations, is that creation by design leads to the expectation of perfection, whereas natural selection leads to the expectation only of a descendant that is in some way better than its ancestor in a given set of circumstances. Evolution by natural selection is a process that is laden with history, since it involves the progressive modification of preexisting structures to adapt them to new circumstances. Recall Jacob's distinction between tinkering and engineering (chap. 5). Evolution is tinkering, or making do with what is available, meaning that it begins with a preexisting organism and modifies it with whatever variation is available at the time. Darwin's explanation makes a prediction, which is that the adaptations we see in organisms today reveal this historical process and the underlying relationships between organisms. Adaptations often have quirky properties that can be understood only if they are seen as modifications of preexisting structures. Paley's argument for design and for the doctrine of species being the products of individual acts of special creation leads to no such expectation of a relationship between organisms but, rather, to the expectation that all organisms will be perfectly suited to their environment.

Darwin expanded on his theory with the introduction of his "principle of divergence" to explain how this process of natural selection could cause the transmutation of species (speciation) in two different ways. A single species could be transformed into a new single species over time (anagenesis, or linear replacement), or a single species could spawn multiple descendant species (cladogenesis, or branching into new lineages). Darwin also presented a few different models for how evolution by natural selection could lead to the formation of new species. With these conceptual building blocks, Darwin then explored the implications of the theory and the reach of its explanatory powers. One application was his new proposal for the cause of extinction. In his view extinction was a by-product of evolution by natural selection, the principle of divergence, and the formation of new species. As one lineage expanded, it often did so at the expense of another. His theory also provided a biological explanation for the already well-established system of nomenclature that we still use for classifying organisms.

Coming this far already qualifies Darwin's theory of evolution by natural selection as being one that has made a major and enduring contribution to our understanding of biology, but Darwin went far beyond this point. The bulk of the *Origin*, in terms of the number of pages, builds on his theory by showing how it explains many other phenomena that were well known in Darwin's time but had never been fully understood. Darwin showed that his theory could unite them all under a single explanatory framework. Some of the new phenomena that he thought his theory could explain included instinctive behaviors, homology of structure and comparative anatomy, the fossil record, the biogeography of living organisms, the embryonic development of living organisms, and more. He also addressed what he imagined to be the greatest challenges to his theory and proposed how it might accommodate them. At each step, he matched his theory against the alternative of special creation. If all organisms were products of the whims of the Creator, then we would expect no pattern of relationships between them, be they viewed through history in the fossil record, or as they are distributed on the surface of the earth today, or in their anatomy, or as they are classified, or as they develop. If, instead, organisms came into being through the process of evolution, then there would always be an imprint of such relationships that would be revealed in a diversity of ways.

The remaining chapters in part 3 of this book address Darwin's efforts to demonstrate that his theory satisfies William Whewell's concept of what constitutes the best science. Whewell was part of the scientific establishment that accepted special creation, but he was also a leading philosopher of science of the day. He argued that the best scientific theories were ones that were based on empirical evidence in combination with the inference of the general laws that explained them. He carried the argument a step further by postulating a "consilience of inductions"—a theory that could unite diverse, seemingly unrelated phenomena under a single explanatory framework. The more a theory could explain, the better it was. Gravity could explain the motion of planets around the sun, but it could also explain the rotation of the moon around the earth and the tides on the oceans of the earth. Darwin's goal was to define a theory that was on par with Newton's. The remaining chapters of the *Origin* present the full development of Darwin's "consilience of inductions."

References

Edwards v. Aguillard. 1987. 482 U.S. Supreme Court 578, 610–40. Dissent
     in *Edwards v. Aguillard* (1987) (Justice Scalia, joined by Chief Justice
     Rehnquist).

Ruse, M. 1999. *The Darwinian revolution*, 2nd ed. Chicago: University of Chicago
     Press.

Spronsen, J. W. van. 1969. *The periodic system of chemical elements: A history of the
     first hundred years*. Amsterdam and New York: Elsevier.

Chapter 14

Difficulties on Theory

In chapter 6 of the *Origin*, Darwin pauses to address possible problems before continuing with the development of his theory: "Long before having arrived at this part of my work, a crowd of difficulties will have occurred to the reader. Some of them are so grave that to this day I can never reflect on them without being staggered; but, to the best of my judgment, the greater number are only apparent and those that are real are not, I think, fatal to my theory" (p. 171).

Darwin's approach in this chapter adheres to our modern dictum that the best defense is a good offense. This chapter raises problems and poses solutions, so it serves well as a defense against anticipated criticism. At the same time it extends the explanatory reach of Darwin's theory to new areas of biology.

The common theme of all the difficulties raised here is the appearance of what seem to be unbridgeable gaps in the natural world. One such gap is the discontinuity that often separates two closely related species. A second is the appearance of complex adaptations, such as the eye of a vertebrate or the placenta of a placental mammal, with no apparent transition between organisms that lack this complexity and those that have it. If evolution works as claimed in the *Origin*, then one might think we should see transitional forms everywhere, but we rarely do. We instead see well-defined species and complex features with no apparent evidence for how they arose. The appearance of such apparently unbridgeable gaps represented a substantial challenge because they were consistent with the idea of species being the products of special creation without transitions ever having existed. Indeed, the presence of such gaps in nature has continued to be the root of many past and current debates within the community of evolutionary biologists, and between scientists and the religious community, about the role of natural selection in evolution. The gaps that separate species are what motivated DeVries (chap. 3), Goldschmidt (chap. 8), and others to propose that new species arise in a single step as a consequence of single mutations that have large effects. The dilemma of organs such as the eye is also a main source of fuel for the modern intelligent design movement, which focuses on Archdeacon Paley's argument that when we see something that appears to be irreducibly complex and hence the product of design, then there must be a designer.

Darwin's general answer to all such gaps is to maintain, throughout the *Origin*, that evolution is indeed a gradual process, that big changes happen through a series of small, intermediate steps, and that the discontinuity that exists today is a product of the extinction of these intermediate forms. If you revisit Darwin's figure (fig. 12 in chap. 10) and think about the transition from variety a^1 to a^{14}, you can imagine this as an example for any transition from one type of organism to another. It can represent the formation of new species, but it can also represent the evolution of a complex organ. If you place the fate of this lineage in the bigger "tree of life" picture illustrated in the figure, then you can envision why, under Darwin's theory, such progressive evolution over a long interval of time causes surviving lineages to become more and more divergent from one another. Divergence of character drives lineages apart, while extinction erases evidence of the transition. It is this erasure of the transition through the extinction of intermediate forms that creates the appearance of discontinuity. When we look at organisms alive today, we see only the tips of the twigs on each branch of the tree of life. The missing parts of the tree are where the gradual transitions occurred. Those gradual changes occurred in a past that is lost, or at best is imperfectly preserved, in the fossil record. We will see that this figure, however abstract it may seem, represents Darwin's general explanation for how all forms of discontinuity arise in nature.

In the discussion that follows, I use Darwin's original subject headings, then conclude with an example of my own research on the evolution of a complex organ.

On the Absence or Rarity of Transitional Varieties

Darwin's persistent emphasis on the gradual, continuous formation of new species and his description of species as arbitrary points in a continuum is contradicted by the observation that species are most often distinct, with there being little evidence of a transition between them. This contradiction can be reconciled with a more complete picture of how Darwin envisioned the process of evolution. While Darwin thinks that evolution, when it occurs, will be a gradual process, he also thinks that organisms will not always be evolving. Only some species will be evolving at any one time because the kind of variation that sustains evolution will only sometimes be available. When some population evolves to the point of being a distinct variety, it will displace the variety it was derived from and drive it to extinction. Victory in this struggle for existence will erase evidence of the transition from one variety to its successor. The evidence for the transition in living, transitional forms will thus always be short-lived relative to the intervening periods of stability. A more contemporary view of the sporadic nature of evolution is that always being able to evolve, because variation is always present, does not mean that organisms are always evolving. A second contemporary perspective, well illustrated by the Galapagos finch research, is that evolution is not necessarily always in the same direction. Year to year changes in direction can cancel out.

If we look in detail at the distribution of related species, such as along a north-to-south gradient or from the bottom to the top of a mountain, we tend to see one closely related species replacing another with only a small region of overlap, rather than a continuity of transitional forms between two species. (Recall Richard Highton's studies of *Plethodon* salamanders [chap. 9], in which he found just what Darwin described: where the geographic ranges of closely related species met, one species often replaced the other with only a narrow zone of overlap and little or no evidence of interbreeding.) If two neighboring species were derived from a common ancestor, as Darwin proposed, then why do we see no evidence of the transition between them?

One explanation that Darwin suggests for abrupt transitions between species is that the distribution of organisms might not stay the same throughout history. In his fourth chapter (chap. 5), Darwin embraced Lyell's proposal that the surface of the earth oscillates, so continents alternate between being continuous landmasses when elevated, then archipelagoes when the land subsides. Some populations may have evolved into distinct species while isolated on different islands. The narrow zones of overlap would be found where those now-distinct species met when their separate islands were reunited by falling sea levels. Here in chapter 6, Darwin revisits that scenario. "But I will pass over this way of escaping from the difficulty; for I believe that many perfectly defined species have been formed on strictly continuous areas" (*Origin*, p. 174).

Darwin then proposes a model for how species might form and become distinct without prior geographic isolation. He offers a hypothetical example to help us picture his argument. As you move up a mountainside from low to high elevations, you might encounter meadows in the valleys between mountains, then forested foothills covered with deciduous trees, then evergreen forests on the slopes at higher elevations, then grassland above the tree line, near the top of the mountain. Darwin's hypothetical species range from the meadows to above the tree line, then diverge into varieties that specialize on each type of habitat. Each type of habitat blends into the next across a narrow transition zone. The deciduous forests on the foothills might be separated from the evergreen forests on the higher slopes by a narrow region of mixed deciduous/evergreen forest. Distinct varieties can evolve that specialize on these transition zones. Because these transitional zones occupy small areas relative to the habitats above and below them, the varieties that occupy transition zones will always be represented by small populations wedged between the much larger populations of varieties living on either side.

Darwin gives two reasons for why varieties in transition zones will be prone to extinction. First, the inexorable struggle for existence causes the population size of each variety to fluctuate in response to any changes that occur in the species that the variety interacts with. If an organism is less numerous and has a narrow geographic range, then local fluctuations are more likely to drive it to extinction than if it is more abundant and widespread. Because transitional habitats occupy a smaller area than the pure habitat types that flank them, the varieties that are adapted to transition zones will be more prone to extinction.

The second argument pertains to Darwin's earlier consideration of the relationship between population size and the potential rate of evolution. Earlier, Darwin argued that adaptive evolution would proceed more quickly in larger than in smaller populations because larger populations would harbor more variation for selection to act upon. By this logic, the varieties adapted to deciduous and evergreen forests should be capable of more rapid evolution than a variety adapted to the transition zone between them. "Hence, the more common forms, in the race for life, will tend to beat and supplant the less common forms, for these will be more slowly modified and improved" (*Origin*, p. 177). The two flanking varieties will expand their range into the transitional habitat as they drive the transitional form extinct, leaving us with nothing more to observe than two distinct species in abutting ranges, and no evidence of transition between them.

There is a modern model of speciation without geographic isolation, called "parapatric speciation," that bears some similarity to Darwin's proposal. Under this model, speciation occurs in populations that border on one another and have some limited gene flow between them. We now entertain a few different variations on the theme of parapatry, but in one we envision a species adapted to an environment that varies continuously, such as along an elevation gradient. One hypothesis is that a species can become subdivided into varieties that are adapted to different parts of the gradient, such as deciduous versus evergreen forest. If this occurs, then reinforcement, or selection against interbreeding between the two varieties where their ranges overlap, can cause the evolution of reproductive isolation (see chap. 8 above). This process, for which there is theoretical and empirical support, is different from what Darwin proposes because there is never an intermediate variety in the transitional habitat that is driven to extinction. The distinctiveness of species is instead a consequence of natural selection favoring those who do not make the mistake of mating with the other variety. Coyne and Orr suspect that this form of speciation may be relatively common.

Darwin thus offers three explanations for how natural selection can cause the observed discontinuity between species. First, the formation of a new species is slow and demands a combination of special circumstances that will rarely occur. At any one point in time, only a small number of species in any community will be in the process of forming new species, and only these few might reveal some evidence of transition. Second, discontinuity can be a product the reunification of species that evolved during a period of geographic isolation. Third, discontinuity can be a product of the extinction of transitional varieties.

On the Origin and Transitions of Organic Beings with Peculiar Habitats and Structure

"It has been asked by the opponents of such views as I hold, how, for instance, a land carnivorous animal could have been converted into one with aquatic habits; for how could the animal in its transitional state have subsisted?" (*Origin*, p. 179).

Carnivorous land mammals, such as wolves or lions, have limbs suited for life on land, while specialized aquatic carnivores, such seals or sea lions, have limbs modified into flippers. Neither kind can survive for long in the other's habitat. While it may be difficult to imagine how a life-form of one kind could have evolved into the other, we have living carnivores that split their time between terrestrial and aquatic habitats and serve well as conceptual models for such a transition. Darwin cites the North American mink (*Mustela vison*) as having such a split identity. It has fur and webbed feet like an otter. It spends its summers diving in water and preying on fish and other aquatic organisms. During the winter it lives on land and feeds on terrestrial prey.

Some organisms present greater difficulties because they offer little in the way of living examples that provide a record of transition or even a conceptual model for how the transition might occur. For example, how did bats evolve from a terrestrial ancestor? All bats have well-developed wings and flapping flight. There are no close relatives that show how the transition from life on land or in trees to a life of sustained flight arose. Darwin ventures into other branches of the animal kingdom in search of examples to fill this gap. Squirrels display part of such a transition. Some live entirely on the ground, but there are also partly or wholly arboreal species. Among the arboreal squirrels, some have tails that are flattened while others have widened bodies with flaps of skin on the flank, all of which can serve as gliding surfaces when they jump from branch to branch. At the extreme, the flying squirrels have elastic skin that extends from wrist to ankle, then to the base of the tail. When they jump from a tree and spread their limbs, the skin serves as a parachute that enables them to glide "astonishing distances" (p. 180). This enhanced form of transport between trees enables them to escape predators, move more efficiently in search of food, and break accidental falls.

If circumstances change in some way, such as a climate change that alters the structure of the forest, or the arrival of a new species that competes for food, or the appearance of a new predator, then the seemingly well-adapted flying squirrels may decline in numbers and go extinct. Alternatively, they may adapt to their changing circumstances by utilizing the environment in some different way. Under the appropriate conditions, we might see "the continued preservation of individuals with fuller and fuller flank-membranes, each modification being useful, each being propagated, until by the accumulated effects of this process of natural selection, a perfect so-called flying squirrel was produced" (*Origin*, p. 181).

Other parts of the tree of life offer glimpses of what the next stage in this transition might look like. The flying lemurs have an elastic membrane that stretches all the way from the jaw to the base of the tail. They also have webbing between their fingers and toes. They are able to glide hundreds of feet from tree to tree with little loss of altitude. Darwin observes that no other lemurs have any adaptations to gliding, so there is no evidence of a transition between them and the flying lemurs. (In fact, we now know that the flying lemurs are not lemurs at all. There are only two species of flying lemur, and they are now classified in an order of their own, the Dermoptera, which means that they have no close relatives among living organisms.) Nevertheless, Darwin continues, we can imagine what the transition to a flying lemur might have looked like from the variation in squirrels. We can also look forward to the possibility that continued selection for enhanced aerial maneuvers might favor the evolution of elongated forelimbs and fingers, which could then represent the transition to the evolution of a batlike wing.

As a second example, the diversity of uses of wings by birds serves well "to show what diversified means of transition are possible" (*Origin*, p. 182). Imagine that all birds except penguins went extinct. We would look at the penguins' forelimbs as being well adapted for life in water and would not be able to imagine that they were modified from forelimbs that were once specialized for flight. Conversely, if all penguins and other types of birds that use their forelimbs for something other than flight went extinct, then we would be left only with birds with wings that are so specialized for flight that we would have difficulty imagining that they could ever evolve to serve a different function.

Darwin argues that transitional varieties, such as those that link birds with their wingless ancestors, are less likely to be seen alive or in the fossil record because they were few in number and were quickly replaced by the "new and improved" varieties. They were few in number because the transition began in a single population. Population sizes in the subsequent evolutionary sequence of varieties were likely to have remained small until the trait, such as the wing, was perfected. It is only then that we should expect to see the lineage expand and exploit its new adaptation by diversifying and adopting many different ways of utilizing the environment, such as the birds or bats that now subsist on a diversity of foods and occupy a diversity of habitats. "Hence the chance of discovering species with transitional grades of structure in a fossil condition will always be less, from their having existed in lesser numbers, than in the case of species with fully developed structures" (*Origin*, p. 183).

What would the first stage in a transition look like? Darwin argues that the transition to a new lifestyle and associated morphology might begin with something as small as "changed habits in the individuals of the same species" (*Origin*, p. 183). Such a change in habit may then favor the evolution of a change in structure that improves the possessor's ability to utilize a new environment. Alternatively, some initial variation in structure may cause a change in habit: "it is difficult for us to tell, and immaterial for us, whether habits generally change first and structure afterwards; or whether slight modifications of structure lead to changed habits; both probably often change simultaneously" (*Origin*, p. 183).

This first step could begin the transition to the evolution of a new variety, then species: "As we sometimes see individuals of a species following habits widely different from those both of their own species and of the other species of the same genus, we might expect, on my theory, that such individuals would occasionally have given rise to new species, having anomalous habits, and with their structure either slightly or considerably modified from that of their proper type. And such instances do occur in nature" (*Origin*, p. 184).

Darwin observes that many British insects have shifted their food preferences from native plants to exotic plants imported from elsewhere. We know now that such shifts have indeed initiated the evolution of specialized races that are now reproductively isolated from those that fed on native hosts, such as the apple maggot flies of North America that switched from feeding on hawthorn to feeding on apple trees imported from Europe (chap. 12). Darwin also reports that tyrant flycatchers (*Saurophagus sulphuratus*) in South America sometimes feed by hovering, like a kestrel, while at other times they stand stationary on the waterline, like a kingfisher, and then dash to catch fish. If either form of feeding were particularly profitable and if there were heritable differences between individuals in their ability to catch prey in one or the other fashion, then natural selection could act on this variation to initiate the evolution of a specialized variety. Finally, Darwin relates an observation by an individual named Hearne who saw black bears swimming for hours, catching insects with their mouths as a whale catches krill in the ocean. He concludes: "Even in so extreme a case as this, if the supply of insects were constant, and if better adapted competitors did not already exist in the country, I can see no difficulty in a race of bears being rendered, by natural selection, more and more aquatic in their structure and habits, with larger and larger mouths, till a creature was produced as monstrous as a whale" (*Origin*, p. 184).

The anatomist Richard Owen, who was once Darwin's mentor but later his adversary, singled out this example for special derision. In an anonymous review of the *Origin*, he said: "we look in vain for any instance of hypothetical transmutation in Lamarck so gross as the one above cited" (Browne 2002, p. 111). While Darwin is indeed imaginative here, Owen missed the point. We can see variation in structure and habits in living organisms, and this variation represents the raw material available for natural selection to act upon. Whether this will occur depends on how faithfully the trait is transmitted from parents to offspring, the consistent availability of the new resource, and the absence of competitors for that resource. For example, insects evolved flight before any predator existed that specialized in feeding on flying insects. The first animal that took to the air in pursuit of them gained access to a new and abundant resource.

Nature also offers examples of animals whose morphology carries a clear signature of their having made a transition from one type of lifestyle to another. The water ouzels, or dippers, live in northern latitudes and are circumpolar in their distribution, meaning that different species in this genus (*Cinclus*) can be found in North America, Europe, and Asia. They look like an ordinary thrush, but they feed underwater in rocky, fast-moving streams on insects that live on the submerged surfaces of rocks. I once found myself in Finland in early winter, just 10 kilometers south of the Arctic Circle and 10 kilometers west of the Russian border, watching some of these birds feeding in an icy stream. They disappeared under the surface of water as they foraged on rocks, using their wings as rudders in the flowing water, then popped out onto the surface, looking like an ordinary songbird.

The upland goose (*Chloepaega picta*), which Darwin saw in southern South America and the Falkland Islands, often lives far from water but has webbed feet like all other geese. Darwin recalls seeing a woodpecker that lived in the treeless plains of Patagonia, "which, in every essential part of its organization, even in its colouring, in the harsh tone of its voice, and undulatory flight, told me plainly of its close blood-relationship to our common species; yet it is a woodpecker which never climbs a tree!" (*Origin*, p. 184).

In each such case, the retention of adaptations to one environment that are used differently in another, like the wings of an ouzel used as rudders in the water, or are just an anomaly in the new home, like the webbed feet of the terrestrial upland goose, reflects the organism's historical transition from one lifestyle to another. "He who believes in separate and innumerable acts of creation will say, that in these cases it has pleased the Creator to cause a being of one type to take the place of one of another type; but this seems to me only restating the fact in dignified language" (*Origin*, pp. 185–86).

Darwin's explanation of such mismatches between habit and anatomy is that they are products of the struggle for existence and the principle of divergence. All species are constantly producing a surplus of offspring that are probing the environment for new ways to earn a living. They seize on whatever opportunities are available to fill vacancies in nature. In doing so, they retain features of their ancestors that had different lifestyles. The webbed feet of the upland goose are nothing more than evidence of their having an ancestor that, like all other geese, lived in water and had feet adapted for swimming. Evolution by natural selection means tinkering, or working with what is at hand. Descendants need only be better than their immediate ancestor in utilizing their environment, rather than being be perfectly suited to it.

Organs of Extreme Perfection and Complication

Here Darwin addresses a different kind of apparently unbridgeable gap: "To suppose that the eye, with all its inimitable contrivances for adjusting the focus to different distances, for admitting different amounts of light, and for the correction of spherical and chromatic aberration, could have been formed by natural selection, seems, I freely confess, absurd in the highest possible degree" (*Origin*, p. 186). The dilemma of the eye is that it is a composite of many individual adaptations, such as protective lids, tear ducts, optic nerves, photosensors in the retina—all of which must be present and all of which must be appropriately integrated to produce a functioning, durable eye. If any one of them were missing, the eye would function poorly if at all, and organisms with such a useless, complex, and costly appendage would not be favored in the struggle for existence. This seems to argue for the simultaneous evolution of every feature in the eye all at once—or, as seemed more likely to Darwin's contemporaries, for the special creation of such an exquisitely designed organ. How does Darwin explain the transition between having nothing more than a rudiment of some single ocular feature, such as a cluster of a few cells sensitive to light versus dark, to the perfection the eye?

His answer is that such complexity arises through a process that is like climbing a long, winding staircase one step at a time, rather than leaping a tall building in a single bound. The reason nature appears to have leapt is that all the intermediate steps have been lost through the paring process of extinction, leaving behind some elevated plateau, seemingly suspended in the sky, with no evidence for how evolution got us there. Darwin imagines that there was continuing variation among individuals in the structure of their proto-eyes, as there was (and is) for any other part of the body. Any such variation that conferred some advantage on its possessor would have been seized on by natural selection and transmitted to its offspring. All that was required, every step along the way, was for the selected variation on the theme of the eye to be better in some way than what preceded it.

The ideal way to study the gradual evolution of a trait would be to compare a living organism that had the trait with all its "lineal ancestors," thus revealing the evolution of the trait. This can never be done because such ancestors are now extinct. An alternative would be to compare close living relatives in which the trait is expressed to varying degrees. Variant forms distributed among closely related species can represent intermediate stages in the evolution of a trait. If we were to try to apply this approach to the evolution of the vertebrate eye, however, we would not get far. All living vertebrates have inherited an eye from a single common ancestor that lived more than 450 million years ago. All stages in the transition from eyelessness to well-developed eyes have been lost. While some vertebrates now have reduced eyes or have lost their eyes entirely, the process of taking something away is quite different from evolving it, so the examples of reduced complexity offer little help.

Darwin instead looks beyond vertebrates and across the whole animal kingdom, where a broad diversity of photosensitive organs makes it possible to imagine how an eye could evolve through many intermediate steps. At one end of the spectrum, some animals have "eyes" that are nothing more than a patch of photosensitive cells that can discriminate between light and dark (fig. 16). Others have cells that line a concave surface, which gives them some ability to perceive the direction to a light source because a beam of light may fall on only part of the curved surface and stimulate only some of the photosensitive cells. Yet others have a layer of transparent tissue over the curved surface through which light passes and which can function

Stages of eye complexity in mollusks

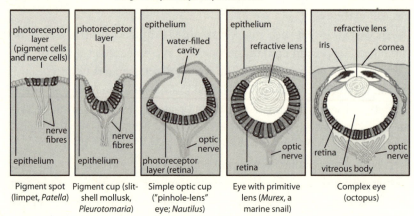

| Pigment spot (limpet, *Patella*) | Pigment cup (slit-shell mollusk, *Pleurotomaria*) | Simple optic cup ("pinhole-lens" eye; *Nautilus*) | Eye with primitive lens (*Murex*, a marine snail) | Complex eye (octopus) |

Figure 16

These drawings illustrate the sort of variation in the structure of photosensitive organs that was known to Darwin when he wrote the *Origin*. Darwin suggests that this diversity among living mollusk species captures some of the stages through which the evolution of complex eyes might have passed for any kind of organism, including vertebrates. An important implicit message is that all the less complex photosensitive organs that we see in nature serve the needs of the organisms that possess them, so there is no need to assume that an eye would have to attain a complex structure before it became functional. Limpets (far left) have just a flat patch of photosensitive cells, which can only detect light versus dark. The slit-shell mollusk (second from left) instead has photosensitive cells arrayed on a concave surface, which means that they have some limited ability to detect the direction from which the light is coming. The nautilus (third from left) has cells arrayed inside a chamber that admits light only through a pinhole, which greatly enhances the animal's ability to detect the direction that the light is coming from, since incoming light will stimulate only a few cells in a specific portion of the chamber. The murex (fourth from left) has a clear layer of tissue that entirely encloses the "eye," plus a ball of tissue, or lens, that lies between the opening and the light-sensitive cells. The lens can focus the light. The octopus has added complexities that produce a sense of vision that is as refined as ours.

like a lens. There is huge variation in the complexity of the eye structures that resolve images, and this corresponds with how well the images can be resolved. There is also great diversity in other features of eyes, such as the ability to see color. Eyes that are capable of very fine image and color resolution yet differ greatly in structure can be found in different animal groups, including vertebrates, insects, and some mollusks. This variation is a signature of acute vision having evolved independently in each lineage.

This enormous diversity in light-sensitive organs across the whole animal kingdom tells us two things. One is that an organism does not need a perfect eye to benefit from photosensitivity. What we might otherwise think of as highly imperfect eyes function perfectly well for many organisms; not all lifestyles require acute vision. The second is that the diversity of eyes that we see in nature provides some guideposts to how photosensitive organs could have evolved through intermediate steps, each of which conferred some advantage to its possessor.

Darwin now takes on Archbishop Paley and the assertion that seeing "design" in nature presupposes a designer. He argues by analogy to the telescope of his day, which had been improved over the years through the efforts of the "highest human intellects" (*Origin*, p. 188). Modern telescopes are not the product of a single invention but appeared first in crude form, then were perfected with many subsequent design modifications. Darwin argues that evolution by natural selection, given many millions of years, millions of generations of organisms, and endless variation to act upon, can also attain complexity and the appearance of design through such successive modifications of structure. If the process began with a cluster of light-sensitive cells, then natural selection would have favored each variant that in some way caused an organism's vision to be better than that of its immediate ancestor. If this process of incremental improvement continued for millions of years, and if the intermediate steps were lost through extinction and not preserved in the fossil record, then the end product could very well look like what others have interpreted as the act of a creator: "In living bodies, variation will cause the slight alterations, generation will multiply them almost infinitely, and natural selection will pick out with unerring skill each improvement. Let this process go on for millions on millions of years; and during each year on millions of individuals of many kinds; and may we not believe that a living optical instrument might thus be formed as superior to one of glass, as the works of the Creator are to those of man?" (*Origin*, p. 189).

Darwin is probably too conservative when he estimates that "millions on millions of years" are required to evolve an eye. Nilsson and Pelger (1994) presented calculations for the time required, based on theoretical considerations of how an eye works and what is required to make the transition from a patch of photosensitive cells to a complex eye. Their upper limit was a few hundred thousand years. Below, I present an example from my own work that documents the evolution of a complex organ on a similar timescale. Elsewhere in this book, other examples I have cited show that evolution by natural selection is a much faster process than Darwin dared to imagine.

Darwin then presents an alternative mechanism for the evolution of a novel, complex trait, which is to recycle old parts for new functions. The origin of many specialized organs can be traced to ancestral states in which the organ served a different function, then was co-opted to serve two functions, then became specialized for just the new function. Darwin offers many examples of such recycling, of which I will describe only one. The swim bladder of the modern bony fishes serves primarily to regulate buoyancy, so that fish can swim without either floating to the surface or sinking to the bottom. In some fishes, this same structure is an auxiliary breathing organ. If such a fish is in water that does not have enough oxygen for its gills to provide for its needs, it can gulp air from the surface to fill the swim bladder, then absorb oxygen through blood vessels on the surface of the bladder. In different lineages of fishes, there have been repeated reversals between the bladder serving as an organ of buoyancy versus an organ for respiration. In some lineages, it has even attained a third function, which is to augment hearing: it is joined to the inner ear by a series of bones. Vibrations in the water are transmitted to the swim bladder, then to the ear via the series of bones that bridge the two organs.

Nature provides us with many other examples of organs that have had multiple functions, only to become specialized for just one function. Sometimes, as in the case of fish that can respire with gills and lungs, multiple organs contribute to a single function, and as one becomes more specialized, the other is co-opted for a different function or is lost. Reptiles, birds, and mammals have no gills, but we can still see in their embryos' development how some of the former components of gills were recycled for other functions. The bones in our middle ears are recycled components of the gill arches of early, jawless vertebrates. Our facial muscles, which enhance our ability to express emotion, are recycled components of the pump that moved water over the surface of the gills.

A problem arises when we see specialized organs with the same function in distantly related organisms. For example, around a dozen lineages of fish have organs that generate electric fields. If these organs were all derived from one common ancestor, then we might expect all fish with such organs to be closely related, but they are not. "I am inclined to believe that in nearly the same way as two men have sometimes independently hit on the very same invention, so natural selection . . . has sometimes modified in very nearly the same manner two parts in two organic beings, which owe but little of their structure in common to inheritance from the same ancestor" (*Origin*, pp. 193–94). A close examination of the organs should reveal some "funda-mental difference" (p. 193) as a signature of their independent origin. In the sixth edition of the *Origin*, Darwin added details about these fundamental differences in the structure of organs that generate electric fields in different lineages of fish. We now refer to this phenomenon of similar but separately evolved traits as "convergence," which is defined as the independent evolu-tion in different lineages of some trait that ends up being similar in its func-tion and structure each time it evolves.

Darwin concludes with another attack on special creation: "Why, on the theory of Creation, should this [recycling] be so? Why should all the parts and organs of many independent beings, each supposed to have been sepa-rately created for its proper place in nature, be so invariably linked together by graduated steps? Why should not Nature have taken a leap from structure to structure? On the theory of natural selection, we can clearly understand why she should not; for natural selection can act only by taking advantage of slight successive variations; she can never take a leap, but must advance by the shortest and slowest steps" (*Origin*, p. 194).

Organs of Little Apparent Importance

Organs of extreme perfection are easy targets for a challenge to Darwin's theory, but he is as much concerned about organs that seem to have little or no function. How might natural selection account for these? The short tail of a giraffe constantly twitches and sweeps around its anus and thus seems to be specialized to serve as a flyswatter. How can natural selection explain the apparent perfection of something that seems so trivial? One explanation is that such features may be more important than we realize. Here Darwin re-minds us that cattle and horses were excluded from large portions of Patago-nia because of the attacks of insects (chap. 3). Short of such life-and-death encounters, the constant harassment of insects can reduce the strength of an individual, making it more subject to disease, predation, or starvation.

It also helps to consider where the tail came from and how it became modi-fied in different lineages. The tail originated as the chief engine of locomotion in our aquatic ancestors. Most fish propel themselves by sweeping their tail back and forth through the water. With the transition to land, the tail became transformed into many different things, ranging from flyswatter to auxiliary limb for climbing to an aid to balance when running. It has been entirely lost in the adults of some lineages, as in humans and the other great apes. It has even become a weapon. I was once hiking up a river in Trinidad and found what looked like a large, dead iguana lying among the rocks on the bottom. I picked it up and averted my gaze because I knew that it would stink as soon as it was out of the water. As I turned to look at it, it sparked to life and struck me on the side of the head with its tail, causing me to see stars. In a similar fashion, many organs that were of great importance when they first evolved have been recycled for various uses, some seemingly trivial, or have been reduced or lost entirely as organisms adapted to different environments.

A different sort of example is the way the skull develops in humans and other mammals. Our skull consists of several bones that knit together to form a single, bony box that protects the brain. These bones are not entirely joined until after birth. The separation of these bones gives the skull some needed flexibility as the infant passes through the mother's birth canal, so we might think that they evolved for this purpose. However, a braincase composed of separate, fused bones is characteristic of all vertebrates, most of which hatch from eggs, so it has an ancient origin that long predates our evolution. These preexisting skull sutures were co-opted for a new function.

It is also true that, while it may be difficult to infer the adaptive value of some traits, this difficulty may just be a function of our ignorance or of the trait being correlated with other traits. For example, the domestic animals that are kept by "savages in different countries" (p. 198) are subject to a measure of natural selection because they have to fend for themselves. A "good observer" told Darwin that the color of cattle was correlated with their susceptibility to attack by some insects or their ability to eat certain toxic plants. The prevalent coat color of cattle or horse herds kept in some regions might thus be more a function of such natural selection than of the tastes of their owners. Darwin concludes: "I have alluded to [these examples] only to show that, if we are unable to account for the characteristic differences of our domestic breeds, which nevertheless we generally admit to have arisen through ordinary generation, we ought not to lay too much stress on our ignorance of the precise causes of the slight analogous differences between species" (*Origin*, pp. 198–99).

Darwin cites sexual selection as a source of traits that are otherwise non-adaptive. All that is required for a trait to become exaggerated by this mechanism is for it to improve a male's ability to obtain mates, either through combat with other males or by appealing to the tastes of females. Darwin suggests that many of the marked differences between human races may be the product of such selection, "but without here entering on copious details my reasoning would appear frivolous" (p. 199). As I said before, Darwin knew that the *Origin* would be inflammatory enough even without reference to human evolution. He saved that subject for books that came later.

Finally, Darwin invokes "homology" as an alternative to adaptation to explain why some features of organisms look the way they do. Homologies are structures that different organisms possess because they inherited them from a common ancestor. They can be retained by descendants regardless of how the structure may have evolved to serve different functions. For example, all vertebrates with limbs (amphibians, reptiles, birds, and mammals) have a very similar arrangement of bones in their limbs (fig. 17). We see the same arrangement of bones in the arm of a human, wing of a bat, or flipper of a seal, so we can hardly interpret it as being an adaptation to a

Figure 17
Forelimb Homologies
Pictured here are the forelimb skeletons of three fish (a = *Eustenopteron*, b = *Panderichthys*, c = *Tiktaalik*) in the lineage from which amphibians evolved and of an early tetrapod (d = *Acanthostega*). All these animals lived during the upper Devonian period, approximately 385–360 million years ago, and had a similar arrangement of forelimb bones—which are homologous with those found in the forelimbs of all living amphibians, reptiles, birds, and mammals (H = humerus, which is our upper arm bone; R = radius and U = ulna, which are the two bones of our forearm). *Acanthostega* had eight digits, or fingers and toes; later amphibians and their descendants had only five. See figure 26 (p. 352) for an illustration of these homologies in an embryonic cow forelimb.

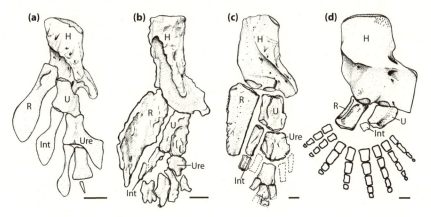

specific lifestyle. Darwin proposes that all limbed vertebrates inherited this configuration of bones from a single common ancestor. Their similarities are nothing more than a brand of relatedness.

Darwin argues that homologies enable us to distinguish the veracity of evolution versus special creation. Adaptations are the products of tinkering, or the modification of old structures for new functions, so they will carry evidence of their history. The forelimb bones of vertebrates are all the same, not because they are adapted for a specific lifestyle but rather because they are descended from a common ancestor. If organisms were the products of special creation, there would be no history. We would instead expect them to be perfectly suited to their environment, without the burden of quirks of inheritance.

This chapter of the *Origin* ends with a discussion of what natural selection cannot do. I covered this material in my chapter 6. The key element of Darwin's argument is that natural selection does not attain perfection. All it can do is favor the survival and reproduction of individuals that are in some way better suited to a particular environment than their immediate ancestor was.

Summary of Chapter

Although Darwin's stated goal is to deal with problems with his theory, this chapter presents important extensions of its explanatory power. The common theme to all these extensions is that the gaps we see in nature—those that separate species or those that are represented by the appearance of complex adaptations—can be explained by evolution by natural selection. He offers two categories of explanation for the origin of complex adaptations. One is that complex, novel traits can be built one step at a time, beginning with some simple precursor. The second is that natural selection is a process of tinkering that transforms one type of organ or organism into another.

Darwin adds something new in the last paragraph of this summary that represents yet another major extension of his theory. Prior to the *Origin*, it had been proposed that there were "two great laws—Unity of Type and the Conditions of Existence" (p. 206), which were competing explanations for the anatomy of organisms.

Georges Cuvier envisioned organisms as being like integrated machines, and he championed "conditions of existence" as an explanation for the features of organisms that made them so well suited to a given lifestyle. The requirements of flight explained the structural details of the wing of a bird, while the requirements of making tools explained the structure of the human hand. He did not consider how organisms came to be this way. He simply argued that their lifestyle dictated everything about how they were put together.

"Unity of type" refers to the similarities in anatomy that we see among large groups of organisms, such as the homology of bones in the limbs of vertebrates. Richard Owen argued that unity of type was a reflection of the Creator's plan and that this plan, rather than conditions of existence, was the primary law that dictated why organisms look the way they do. The similar arrangement of bones in the limbs of all limbed vertebrates reflected this ideal vertebrate design, or their unity of type. All species of vertebrates are variations on the theme of this ideal. He saw the conditions of existence as being secondary to this ideal design.

Darwin argues that both of these "great laws" can instead be explained by his theory of evolution by natural selection. The "unity of type" reflects traits inherited from a common ancestor. As the descendants of this common ancestor multiplied in number and invaded new environments, the tinkering process of evolution caused their common morphology to be modified for specialized functions. Darwin interprets the features Cuvier called "conditions of existence" as products of the way natural selection shaped organisms and adapted them for specific lifestyles. He argues that natural selection, which explains these conditions of existence, is the primary law. Owen's unity of type is just the echo of common ancestry.

I have referred before to Richard Owen and Darwin's other mentors as the scientific establishment that advocated special creation. We can imagine that each time any of them encountered such arguments against special creation and in favor of natural selection, there was an incremental increase in their blood pressure. For Owen, the worst was yet to come.

Evolution Today: The Evolution of the Placenta

With a combination of luck and modern molecular methods, it is sometimes possible to fill some of the gaps that troubled Darwin. Here I summarize one of my research programs designed to fill such a gap, by showing how and why a complex organ evolved. I am interested in the evolution of the placenta, but not the one seen in mammals. Mine is found in fish that are closely related to guppies.

One day I found a paper published in 1940 by C. L. Turner. Turner described differences between species in the guppy family, the Poeciliidae, in how they produced live young. He found that guppies and many other poeciliids fully provision an egg prior to fertilization. The mother simply retains the developing young until they are born alive, rather than depositing the eggs after they are fertilized. The embryos lose weight between fertilization and birth because they fuel their own metabolism and growth with reserves that were packed into the eggs before they were fertilized.

Turner found other species with eggs that were tiny when fertilized but developed into much larger young. In fact, the babies could be thirty or

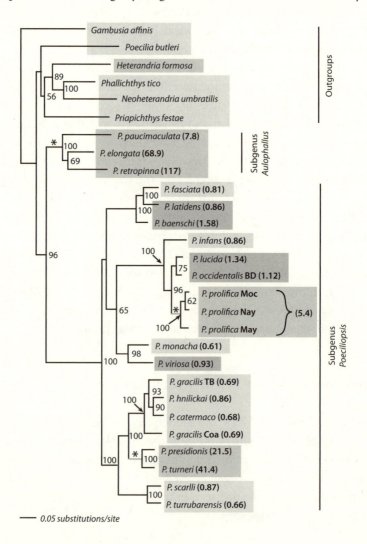

more times heavier at birth than the egg was at fertilization, so the mother had to continue to provision the young throughout development. These species have special structures, in both the mother and the embryo, that are not seen in species like guppies. The embryos develop inside an envelope of maternal tissue that has a blood supply and is lined with microvilli, like the inner surface of our intestines. The embryos of some species have externalized part of the membrane that surrounds the heart. It emerges from a seam along the middle of their belly and wraps around their body. It is well supplied with circulating blood and has microvilli that interlock with those on the inner surface of the maternal envelope. This combination of maternal and embryonic tissues enables mothers to transfer nutrients to the developing embryo, so it functions like the mammalian placenta. Very little work had been done on this discovery between 1940, when Turner published it, and 1987, when I discovered Turner's paper in the library.

We now have a DNA-based family tree for the Poeciliidae and have characterized maternal provisioning in over 150 species. We can use the combination of these two types of data to reconstruct the likely evolutionary history of placentas in these fish.

Here are results for just one genus in this family (fig. 18). Most of the species that Turner studied are in this genus. In fact, one of them carries the

Figure 18
Family Tree and Maternal Provisioning in the Fish Genus *Poeciliopsis*
This tree is based on DNA sequences from two genes. The 6 species at the top are in the family Poeciliidae but not the genus *Poeciliopsis*. The lower 19 species represent all but one of the species in the genus *Poeciliopsis* (some species are represented by more than one branch). The number in parentheses after each species is the ratio of the mass of individual offspring at birth divided by the mass of the egg at fertilization. Values less than one (e.g., *P. monacha* has a value of 0.61) show that the embryo loses weight between fertilization and birth. Values greater than one (e.g., *P. retropinna* has a value of 117) show that the embryo gains weight between fertilization and birth. The value for *P. retropinna* means that more than 99% of the materials in a baby at birth were transferred from the mother to the baby during development. Values greater than 5 were found in all three species in the subgenus *Aulophallus*, in *P. prolifica*, and in the sister species *P. presidionis* and *P. turneri*. These represent three different instances of the evolution of extensive postfertilization maternal provisioning. These species also have tissues in the mother and developing young not seen in the other species. The anatomy of these tissues suggests that they are adaptations that facilitate the transfer of nutrients from mother to embryo. The number at each branching point is a measure of the statistical confidence we can place on the conclusion that species downstream of that branching really share a single common ancestor, with 100 being the highest possible degree of statistical certainty.

specific designation of *turneri* in his honor. Our reconstruction of the evolution of placentas revealed that the common ancestor of this genus lacked a placenta and that the placenta evolved three times within it. We found two clusters of closely related species that either do or do not have a placenta. We have found similar patterns elsewhere in the family. We now know that the distant ancestor of the whole family was an egg layer, but that both egg retention and the process of bearing live young evolved in the common ancestor of the Poeciliidae. Placentas then evolved independently at least eight times throughout the family. We can even estimate the upper limit on the amount of time required for the placenta to evolve. In one case that upper limit is 750,000 years, but the real interval could have been much shorter than this.

The element of luck that facilitated our research was in finding this variation among close relatives. This means that we have caught the evolution of a complex trait while there are still closely related species that either lack the trait or have it in intermediate stages of development. In contrast, all placental mammals inherited their placenta from a single common ancestor that lived over 100 million years ago. The mammals alive at that time may well have had the sort of diversity we see in poeciliid fishes today, but if they did, then it has all been lost through extinction. We are left with just the placental mammals and two distantly related lineages of mammals: the marsupials, which have a much more primitive form of prenatal provisioning, and the monotremes, which lay eggs. There is considerable diversity among lineages of placental mammals in the structure of their placentas, but it represents evolution that postdates the evolution of the placenta, so it yields no evidence for how the placenta evolved in the first place. We can imagine that 100 million years from now the Poeciliidae might present the same picture. If any descendants of the Poeciliidae still exist, they might all share a much more elaborate placenta than we see in any fish alive today. There may be no clues to be found among living relatives about how such complexity arose.

What can we do with our discovery of such variation? Because the trait has evolved several times and because there are often close relatives that either lack the trait or express it in varying degrees, we can compare these close relatives to figure out what has changed. We can study them in nature to evaluate how the placenta might be an adaptation. We can do experiments that reveal the costs and consequences of having a placenta. We can make anatomical comparisons, or compare patterns of gene expression, to explore how the placenta evolved.

We could do the same by comparing a placental mammal with a marsupial mammal, but there would be some critical differences between mammals and our fish. Because placental mammals and marsupials last shared a common ancestor more than 100 million years ago, any differences we find between them may have played a role in the evolution of the placenta, but may also have arisen during the 100 million-plus years since these organisms shared a common ancestor. In our fish, placentas evolved much more recently, did so several times, and vary in their degree of development, even among close relatives. Each of our comparisons is of close relatives and represents a much more recent event than in mammals, so it is more likely to yield accurate clues about how and why the placenta evolved. We can also make the comparison again and again, and identify the common elements of each origination.

At this stage, we have not gone too far beyond establishing the pattern of evolution in the family and identifying those parts of the family tree that will hold the secret for how this transition was made. We have only begun the new phases of research that include addressing how and why this complexity evolved. Stay tuned.

References

Ayala, F. J. 2007. Darwin's greatest discovery: Design without designer. *Proceedings of the National Academy of Sciences* 104:8567–73.

Browne, J. 2002. *Charles Darwin: The power of place.* Princeton, NJ: Princeton University Press.

Coyne, J., and H. A. Orr. 2004. *Speciation.* Sunderland, MA: Sinauer Associates.

Nilsson, Dan-E., and S. Pelger. 1994. A pessimistic estimate of the time required for an eye to evolve. *Proceedings of the Royal Society of London Series B, Biological Sciences* 256:53–58.

Reznick, D. N., M. Mateos, and M. S. Springer. 2002. Independent origins and rapid evolution of the placenta in the fish genus *Poeciliopsis. Science* 298:1018–20.

Turner, C. L. 1940. Pseudoamnion, pseudochorion, and follicular pseudoplacenta in poeciliid fishes. *Journal of Morphology* 67:59–87.

Chapter 15

Instinct

Orb-weaving spiders spin intricate, circular webs with silken spirals overlying radial threads. The wagon-wheel-shaped webs, suspended from branches or rocks or blades of grass, are custom blends of secretions from different abdominal glands. Each blend is well suited to the particular demands on the specific parts of the web, be it to have high tensile strength to hold up the entire structure or stickiness to ensnare unsuspecting prey. Spiders, with no prior training, are able to expertly weave these complicated structures. Many other organisms display remarkable and complex instinctive behaviors. Caterpillars spin complex cocoons; ants, bees, and termites organize themselves into complex societies; many species of birds migrate long distances to places that they have never been to escape the winter, then back again to reproduce. All these things can happen without the benefit of prior experience.

Can evolution by natural selection cause the evolution of such instinctive behaviors as readily as other features of the organism? If not, then they represent a significant challenge to Darwin's theory. At the start of his seventh chapter, Darwin defines instinct as "an action, which we ourselves should require experience to enable us to perform, when performed by an animal, more especially a very young one, without any experience, and when performed by many individuals in the same way, without their knowing for what purpose it is performed" (*Origin*, p. 207). He also refines his argument as focusing on the role of natural selection in shaping the evolution of instinct, not the origin of instinct itself: "I must premise, that I have nothing to do with the origin of the primary mental powers, any more than I have with that of life itself. We are concerned only with the diversities of instinct and of the other mental qualities of animals within the same class (*Origin*, p. 207).

Darwin thus sets the stage for an argument that assumes that the basic trait is present, that it is variable, that this variation is heritable, and hence that it will evolve. Instinct is as important as morphology (e.g., the beak of a finch) in determining an individual's fitness, be it in terms of survival, acquisition of food, or success in reproduction. If instincts can be shown to vary like other characters, and if changes in the conditions of life favor some variant, then instinct will evolve in the same way as any other feature of the organism. Here Darwin extends the reach of his theory by arguing that instinct is an adaptation like any other.

Darwin envisions instincts as evolving via many small steps accumulated over a long span of time. Single populations of living organisms are not likely to contain individuals with intermediate types of behavior that show us how instinct evolved, "for these could be found only in the lineal ancestors of each species," which are now extinct. However, "we ought to find in the collateral lines of descent some evidence of such gradations; or we ought at least to be able to show that gradations of some kind are possible; and this we certainly can do" (*Origin*, p. 210). As with the evolution of the eye and the diversity of photosensitive organs that we see in different species (chap. 14 above), we can document differences between species in instinct that provide clues for how instincts evolved. As with complex organs, the first steps in the origin of a complex instinct might be in behavior that varies in different seasons, in different life stages, or in response to some change in environmental conditions. Selection can capitalize on such variation in behavior in the same way that it can capitalize on organs that have multiple uses, then refine them for one specialized use.

Darwin first argues that, by his theory, nothing in an organism can evolve solely for the benefit of other species. If such a case could ever be found, then it would annihilate his theory. Some instinctive behaviors seem to purely benefit others and hence qualify as spoilers of his theory. A prime example of what appears to be such altruistic, instinctive behavior is seen in interactions between aphids and ants. Aphids earn their living by sucking fluids out of plants and can be found in clusters arrayed along the stem of their food plant. Ants are often seen stroking the abdomens of aphids with their antennae; aphids secrete a sugary, nectarlike fluid in response. The ants consume the excretion, so it appears that aphids produce it for the ants' benefit.

Darwin performed an experiment to establish a cause-and-effect relationship between ants stroking aphids and aphids excreting a sugary substance to feed the ants. He isolated aphids on a potted plant; they no longer produced the excretion. He "tickled" them with a hair to imitate the way ants stroke their abdomens, but was not able to elicit the excretion. He then released ants on the plant. As soon as the ants found the aphids and stroked them with their antennae, the aphids excreted the nectar and the ants consumed it. So do the aphids produce nectar solely to benefit ants? Darwin speculates that the aphids were using ants as a waste-removal service: "But as the excretion is extremely viscid, it is probably a convenience to the aphids to have it removed" (*Origin*, p. 211). We now know that this excretion is actually like "protection money." In return for the sugary excretions, ants defend aphids from potential predators and parasites, such as wasps that inject their eggs into aphids. The wasp larvae keep the aphid in a state of suspended animation while they eat it from inside its body, then emerge as adults and leave the aphid to die. When ants are present, they tend the aphids like a herd of milk cows and ward off predators. When ants are absent, the predators prevail. What seems like altruistic behavior on the part of the aphid is part of a relationship that benefits aphids and ants. Darwin argues that there should be a similar explanation for any behavior exhibited by any organism solely for the apparent benefit of another species. Even a single exception to this rule would be fatal to his theory.

Darwin turns to variation under domestication for evidence of genetic variation in behavior: "it cannot be doubted that young pointers . . . will sometimes point and even back other dogs the very first time that they are taken out; retrieving is certainly in some degree inherited by retrievers; and a tendency to run round, instead of at, a flock of sheep by shepherd-dogs . . . the young pointer can no more know that he points to aid his master, than the white butterfly knows why she lays her eggs on the leaf of a cabbage" (*Origin*, p. 213). Darwin speculates that these specialized behaviors are products of humans having capitalized on the instinctive hunting behaviors of the wild ancestors of domestic dogs. Pointing may be a derivative of a tendency to freeze, then slowly approach unsuspecting prey, while herding may be a derivative of a "wolf rushing round, rather than at, a herd of deer, and driving them to a distant point." He also notes that crossing different breeds produces offspring that seem to inherit a mix of parental behaviors, such as the cross between a greyhound and a shepherd dog that gave "a whole family of shepherd-dogs a tendency to hunt hares" (*Origin*, p. 214).

Domestication can cause the loss of behaviors. Wolf packs have a fluid social hierarchy in which young wolves constantly vie with their elders for dominance. When humans try to keep wolves as pets, they can be as sweet as domestic dogs when young, but as they mature they will often treat their owner as they would the pack elders and vie for dominance, which makes them dangerous companions. Sometime early in the domestication process, humans successfully selected for individuals that lacked this instinct. Likewise, the wild progenitors of dogs readily attack domestic animals such as chickens and sheep. Domesticated dogs imported to England from places like Australia and Tierra del Fuego, where the natives do not keep domestic animals, also attacked domestic animals, but the domestic dogs of England and mainland Europe did not. Darwin speculates that any domestic dog with such an incurable tendency would have been destroyed, eliminating such hunting instincts from the population. At the same time, domestic chickens appear to have lost the instinctive fear of domestic dogs and cats, even though they retain an instinctive fear of other predators. Domestication can thus be associated with the elaboration of some behaviors and the loss of others. These changes can be associated with either deliberate or inadvertent selection by humans, and they all illustrate the same point, which is that behavior is like any trait in being variable among individuals, heritable, and hence modifiable by selection.

Darwin follows with three examples from nature of elaborate instinctive behaviors that can be understood as products of natural selection. Since the nature of his argument is the same each time, I will elaborate on only one example: that of the slave-making ants. Slave-making ants survive by stealing the worker-ant pupae from the nests of other species and bringing them to their own nest. The captive pupae complete development, then devote their lives to caring for their captors.

Darwin summarizes the observations of Pierre Huber, a Swiss naturalist, who discovered the first example of slavery in *Formica rufescens*. This species relies on slaves to execute all the necessary tasks required for colony survival, save for the production of reproductive males and females and the capture of new slaves. The slaves take care of the young, feed the adults, and maintain the nest. Ant colonies periodically move their nests to new sites. When this species does so, the slaves organize the migration, pick and excavate the new nest site, then carry their masters there.

Ant workers in non-slave-making species normally take care of the young, feed them, and feed each other, but the workers of *F. rufescens* do nothing other than raid the nests of other species to steal and enslave their young. Huber wanted to know if the workers of this species had the ability to perform the necessary nest maintenance but were too lazy to do so when slaves were around. He isolated thirty *F. rufescens* workers with food and larvae and pupae. In the absence of slaves, the *F. rufescens* workers did not feed themselves or their larvae. Many of them starved. Huber added a single slave, and it immediately fed the surviving workers and cared for the larvae. It thus appears that the reliance of *F. rufescens* on slaves is absolute. They cannot survive without them. The question is, how could such a complete reliance on another species have evolved? If there were only this one species of slave-making ant, it would be very difficult to develop any hypotheses to address this question.

Huber discovered a different slave-making species, *Formica sanguinea,* that was less reliant on its slaves. Since this ant could also be found in England, Darwin made observations of his own, plus took advantage of observations made in England by Mr. F. Smith of the British Museum. Among all their observations of colonies in different parts of England and Switzerland, there was variation in the degree to which the slave maker relied on its slaves.

Both Darwin and Mr. Smith observed that slaves were almost never seen outside the nests of the subterranean *F. sanguinea* colonies. When Darwin disturbed a nest, the slaves would join the slave makers in coming to the surface to defend the colony. If he dug more deeply and exposed the larvae and pupae inside a the nest, then the slaves would help the slave makers move the young to a safer place. The slaves thus seemed to feel at home in the colony of the slave makers, acting as house servants that stayed in the colony and cared for the young; they only occasionally set out on foraging trips with the slave maker. In Switzerland, Pierre Huber found that the slaves played a much larger role in caring for the colony. Both slaves and masters brought in materials for maintaining the colony, and both tended the herds of nectar-producing aphids, although the slaves played a much larger role. The nectar the slaves harvested from the aphids was fed to the whole colony. The British slave makers were thus less reliant on their slaves than their Swiss counterparts were; all the populations of *F. sanguinea* were in turn less reliant on their slaves than *F. rufescens* was.

Darwin interprets this range of reliance on slaves as being like the range of photosensitive organs that we see in different species of mollusks (fig. 16 in chap. 14). Among the ants, each population and species represents what might be stages in an evolutionary sequence. A first stage, shown by many species, is to collect the larvae of other species and carry them back to the raiders' nest to serve as food. Some of the stolen larvae may become mixed with the brood of the would-be predator and, instead of being eaten, may complete their development. The first step to enslavement will occur if the captive larvae-turned-adults then respond to the chemical environment of the host nest by taking up their normal nest-care behavior. Subsequently, the incipient slave maker can more routinely steal larvae from the other species, rear them to maturity, and then employ them in nest care. The British populations of *F. sanguinea*, then the Swiss populations of *F. sanguinea*, then *F. rufescens* represent different steps in a progressive increase in the reliance on slaves. By this scenario, each incremental increase in the responsibility assumed by slaves is matched by a decrease in self-care by the slave maker, until some slave-making ants reach the state of *F. rufescens*, which is totally reliant on slaves for its survival. In this fashion, variation among colonies in instinctive behavior provides Darwin at least a hypothetical evolutionary pathway from non–slave making to complete reliance on slaves, in the same way that variation in photosensitive organs among species of mollusks shows how increasingly complex eyes can evolve. E. O. Wilson (1971) reported that Darwin's remains the favored hypothesis for the evolution of slave making.

Darwin then takes on what he considers to be features of instinct that present the greatest challenges to his theory. In fact, when he first encountered them, he thought they might be fatal to his theory. These difficulties are all associated with the existence of complex ant societies. A typical colony of ants consists of wingless, nonreproducing workers, an egg-producing queen, and mature males and females that are winged and disperse to initiate new colonies. These different kinds of ants are referred to as "castes." The colony can be thought of as an individual whose fitness is determined by the number of males and females that it produces, since these individuals establish new colonies. Fitness increases with the size of the colony because larger colonies produce more dispersing males and females.

As ant colonies grow larger, they become increasingly attractive targets for predators because they are a concentrated source of food in the form of developing larvae and cocoons. The colony's success therefore demands a more effective defense. The solution in many ant societies has been the evolution of a second caste of workers, called soldiers, that specialize in colony defense. This increased level of specialization in the nonreproductive worker castes converts a baby factory into a fortress (Holdobler and Wilson 1994).

The existence of such colonies poses three types of difficulties to Darwin's theory. The first is the existence of sterile workers. If natural selection favors those individuals who are more successful in contributing offspring to the next generation, then how can it lead to the evolution of individuals that produce no offspring at all? Darwin postulates that selection can act on the colony as a whole. The family, rather than the individual, is the unit of selection. The specialized, nonreproducing worker caste can increase the ability of the colony to garner resources from the environment and convert those resources into mature males and females that will emigrate and establish new colonies.

Darwin draws an analogy between natural selection on colonies and humans' artificial selection on vegetables or cattle to produce food with desirable properties. It is not possible to cook and taste an individual vegetable or animal, decide that it is more delicious or nutritious than others, then use it as a parent of the next generation. However, we can set aside some seeds or some animals and breed those whose siblings proved to be desirable. We have done so repeatedly as we developed domestic cultivars of plants and animals. In this fashion, humans have often selected on families of plants or animals, rather than on individuals, when selecting for better sources of food.

Natural selection can do the same, Darwin argues. If some pairings of ants produce sterile workers that aid in the rearing of siblings, and if their doing so means that their family produces many more successful offspring than pairings that produce only fertile offspring that immediately strike out on their own, then the family with stay-at-home, nonreproducing offspring will produce more future families despite having sterile workers.

The second difficulty posed by the existence of ant colonies is that nonreproducing workers are anatomically distinct from either parent. The parents are winged, while the workers are wingless and smaller than the parents. The parents contribute nothing to the survival of the colony, while the workers devote their lives to maintaining and defending the nest, feeding the young, and foraging for food. How can natural selection modify the structure and behavior of a worker over successive generations of selection when workers never reproduce? Darwin proposes that if the nature of the sterile workers has some influence on the quantity and quality of sexual offspring that are produced by the colony, then selection at the level of the colony can alter the nature of workers in addition to favoring the evolution of nonreproducing workers. Colonies with a better workforce may produce more descendant colonies and hence prevail in the struggle for existence. "But we have not yet touched on the climax of the difficulty: namely, the fact that the neuters of several ants differ, not only from the fertile females and males, but from each other, sometimes to an almost incredible degree, and are thus divided into two or even three castes" (*Origin*, p. 238).

Darwin details the remarkable differences that can be seen between castes within a colony with an example provided by Mr. F. Smith, who sent him samples of workers collected from a colony of driver ants (genus *Anomma*) from West Africa. Driver ants, sometimes also known as army ants, lead an itinerant lifestyle of wandering through the forest in great columns. The moving front of a driver ant column can be a meter or more wide, and the length of the column can be tens of meters as it sweeps through the habitat like Sherman's army through Georgia, catching small animals in their path. The columns include giant soldier ants with scimitar-like jaws that patrol the perimeter of the column, plus smaller worker ants that carry the young and participate in foraging. The pupae and eggs are carried by the smaller workers, so the column is a moving household, complete with queen, off-spring, and provisions. (A colleague in Trinidad told me that he once had to temporarily abandon his accommodation in the forest when it was invaded by a column of army ants. The ants spread through the building, from floor to roof, cleared it of insects as effectively as an exterminator, then moved on.) Darwin documents differences between the soldiers and workers by comparing them with an imaginary group of workmen building a house, "of whom many were five feet four inches high, and many sixteen feet high; but we must suppose that the larger workmen had heads four instead of three times as big as those of the smaller men, and jaws nearly five times as big" (*Origin*, p. 241).

The distinctions between worker castes go far beyond the great differences in size (fig. 19). Members of the larger caste, often referred to as "major" workers or soldiers, often have jaw morphology different from that of the smaller, "minor" workers. The soldier jaws can be specialized for shearing and used for tearing off body parts or for piercing the armor of enemies. The top of their head may be shaped like a flat shield and used to plug entrance holes of the colony to keep out would-be intruders. Their spectrum of behaviors too may be very different from those of the minor workers, such that they only patrol or defend the colony but do not participate in caring for the young, maintaining the nest, or foraging.

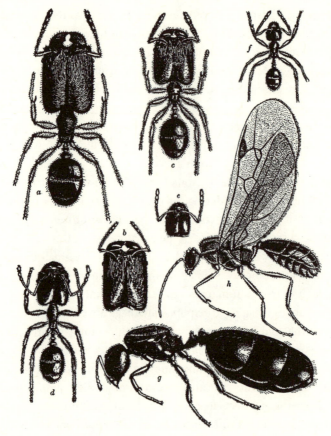

Figure 19
The Female Castes and the Male of the Myrmicine Ant *Pheidole tepicana*
Caption from Holdobler and Wilson, *The Ants* (p. 302): "In this classic drawing by
[William Morton] Wheeler, the worker caste is shown to be composed of subcastes
of successively diminishing size from the major workers [soldiers] (*a*) through media
workers (*b–e*) to the minor worker (*f*). The queen (*g*) and male (*h*) are also shown."
All this diversity among females (all except *h*) is induced by the environment that
they experience when developing.

How can natural selection explain not only the existence of a morphologically distinct worker caste but also multiple, nonreproducing worker
castes that differ in morphology, behavior, and the tasks that they contribute to colony survival? Darwin speculates that the evolution of distinct
castes of workers is preceded by variation within a colony. Such variation
was present in the examples of driver ants given to him by Mr. Smith. While

there were distinct large and small workers, there were also workers of intermediate sizes that filled the gap between them. There was also variation in the size and structure of the jaws such that, when the whole sample was considered, the castes "graduate insensibly into each other" (p. 241). Darwin envisions this as an intermediate state between having one worker caste versus two distinct, specialized castes. If the intermediate forms are less well suited to the alternative functions, causing colonies that have such variation to be less successful at producing males and queens than colonies with two discrete castes of workers, then continued selection will result in a species that consists only of colonies that have two discrete castes of workers.

Darwin's argument for the origin of castes is correct in the essential details, but the underlying mechanism is more bizarre than Darwin could have imagined. Our conventional way of thinking of evolution is that natural selection acts on genetic variation within a population, causing changes in the genetic composition of the population over time. If we envision the process in terms of the evolution of different species of Galapagos finches, then we can think of each species as having been shaped by natural selection so that it is specialized for a different lifestyle. Bill size and shape are key features of finch evolution because the bill is the main instrument for obtaining food. A key to this conventional version of evolution is that each species of finch is a product of natural selection acting on separate gene pools. The morphological differences between finch species are determined by differences in their genes.

In ant colonies, all the castes, including the fertile males, fertile females, and major and minor workers, emerge from eggs laid by the same queen, so they are all part of the same gene pool. The origin of male ants is easy to explain. Male ants are haploid, or have only one set of chromosomes, so they are produced when an egg is not fertilized. The fertile females and different castes of workers, however, are all females and have genotypes derived from the same gene pool. With only a single known exception, their caste is not influenced by their genotype.

The cause of the remarkable differences between castes is thus not natural selection acting on different gene pools to produce different products, but rather selection shaping a single gene pool with a single, flexible program of development that can produce individuals with very different phenotypes. A number of factors are now known to influence whether an egg will develop into a queen or into one of the different types of nonreproducing workers. One factor is the way the queen ant provisions each egg; queens sometimes provide some eggs with more nutrient reserves than others. Those with more reserves are more likely to develop into winged reproductive females that become the queens of new colonies. There are also a number of environmental factors that determine the outcome, including temperature, season, or how much and what the larvae are fed by the workers as they grow. Finally, there are hormonal interactions between the different members of the colony that can modify the ratio of different castes that are produced. The combination of all the factors that determine the fate of any one egg results in a colony that self-regulates the production of the different castes of ants, be they fertile males and females or nonreproducing workers; the self-regulating whole can adjust these ratios in response to changes in the environment or the season or the state of the colony.

The flexibility of this developmental program varies among species, as originally inferred by Darwin. In some, the variance among individuals in development results in continuous variation in the nature features of the workers rather than discrete castes of workers. In others, the pattern of development has discrete thresholds that channel the developing larvae in one or the other direction, such that the finished products fall into discrete castes with few or no intermediates between them. The variation among ant species thus ranges from those that have only a single type of worker, to those that have a worker population that is highly variable in body size and morphology, to those that have discrete castes of workers. This diversity provides the clues to how family-level selection on ants selected first for the evolution of a baby factory, then a defended fortress.

In this chapter, Darwin extends the reach of his theory to explain instinctive animal behaviors that were known to many but were probably viewed by Victorian society as just curiosities of nature. Darwin instead argues that such curious behaviors are one of the many complex adaptations, shaped by natural selection, that enable animals to survive and reproduce in a changing world.

Evolution Today: The Evolution of Instinct

I have already shown that evolution can proceed much more rapidly than Darwin anticipated. Lest you think that my examples are exceptional, I present another here, this time with respect to the evolution of an instinctive behavior. I will describe the rapid evolution of migratory behavior in a bird.

Blackcaps (*Sylvia atricapilla*) are small birds that breed in central Europe, then migrate to the southwest and overwinter either on the Iberian Peninsula (Spain and Portugal) or in North Africa. They were also known in other parts of Europe as summertime tourists, visiting after the breeding season was over. In the early 1960s, bird watchers began to notice that some individuals were overwintering in Britain and Ireland. The birds were attracted to the warmer environments around urban areas and the well-stocked feeding tables maintained by bird watchers. Seeing these birds was initially an exciting rarity, but a survey of backyard feeders during the winter of 2003–4 revealed that blackcaps appeared at a third of them. What began as opportunistic overwintering by a few individuals had grown into a sizable population that routinely overwintered in the British Isles. Some birds that had been caught and banded in Germany and Switzerland during the summer ended up in the British Isles over the winter, which established a connection between the summer breeding grounds and the new overwintering population.

It was easy to imagine that some of the summertime wanderers might decide to stay the winter in the British Isles if conditions were attractive, rather than risk the long flight back to Iberia or North Africa, but the growth of this population suggested that these were more than just opportunists. What had once been a single population that migrated between spring/summer breeding grounds in central Europe and overwintering grounds in Iberia and North Africa had split into two populations that spent the summer together in central Europe but migrated to different winter destinations. Since the direction in which birds migrate is known to be genetically determined (instinctive) in at least some birds, the presence of a new overwintering population in the British Isles raised the question of whether it was the product of a recent evolution of migratory instinct.

Berthold and coworkers (1992) asked if the birds that overwintered in Britain were indeed genetically different from those that overwintered in Iberia and Africa. They captured birds that were overwintering in England, then transported them to a site in Germany where this species normally spends the summer. The parents were housed in outdoor aviaries to breed and raise their young. In the autumn, when migration normally begins, Berthold and his colleagues quantified the preferred migratory orientation in the British parents, their offspring, and offspring from pairs of blackcaps that were naturally nesting in the neighborhood. The British birds and their young both oriented to the northwest, on a trajectory that would take them to England, while the young from nests in the neighborhood oriented to the southwest, on a trajectory that would take them to Iberia. The combination of the recovery in England of birds banded in central Europe and differences in migration tendency between the two groups of young reared in the same location and with no prior experience suggested that the overwintering British population was genetically distinct and a recent derivative from the Iberian/North African population.

In a later paper, Berthold and colleagues (Bearhop et al. 2005) found that it was possible to capture a bird from the summer breeding grounds, take a toenail clipping, then use details of the chemical composition of the clipping to determine whether that bird had spent the winter in England or Iberia/North Africa. They found that birds coming from England arrived at the breeding grounds around ten days earlier than those from Iberia and North Africa. Berthold and coworkers (1992) had predicted that this would occur because springtime day length increases more quickly in the more northerly latitudes of England and reaches the critical photoperiod that is known to trigger spring migration around ten days earlier than is the case in Iberia/North Africa. Since both females and males from England tended to arrive earlier, the British birds were much more likely to mate with each other than with the Iberian/North African birds, even though birds arriving from all overwintering areas were breeding in the same neighborhood. The British birds were also more successful in producing offspring. Their success might be because their earlier arrival gave them dibs on higher-quality breeding territories, or because their adaptation to the more severe winters of the British Isles gave them an advantage in the unreliable spring conditions of central Europe, or because the length of the migration from the British Isles to central Europe was shorter than that from Iberia/North Africa, so they arrived in better condition.

Whatever the reason, these new results show that what had once been a few opportunists overwintering in the changing environment of Britain has already grown into a successful subpopulation that instinctively migrates in a different direction. A consequence of this change in behavior is that they seem to be enjoying greater reproductive success and their population is expanding. While we have no idea where this split might lead, it is the sort of trend that could, in the long run, result in two distinct species, or perhaps in one population replacing the other. The results also show that behavior is a trait like any other. It is variable within populations, influences an individual's fitness, is subject to natural selection, and will evolve in response to a changing environment.

References

Bearhop, S., S. W. Fiedler, R. W. Furness, S. C. Votier, S. Waldron, J. Newton, G. J. Bowen, P. Berthold, and K. Farnsworth. 2005. Assortative mating as a mechanism for rapid evolution of a migratory divide. *Science* 310:502–4.

Berthold, P., A. J. Helbig, G. Mohr, and U. Querner. 1992. Rapid microevolution of migratory behaviour in a wild bird species. *Nature* 360:668–70.

Holdobler, B., and E. O. Wilson. 1994. *The ants*. Cambridge, MA: Harvard University Press.

Wilson, E. O. 1971. *Insect societies*. Cambridge, MA: Harvard University Press.

Chapter 16

Geology I: Background

In his ninth and tenth chapters of the *Origin*, Darwin explains the fossil record in light of his theory. This was a challenge because geology and paleontology were well-established disciplines, populated with strong-willed luminaries who had defined the key questions suggested by the fossil record and were debating the answers to those questions. Some of these key questions included how to explain the origin of species, why the species found in the fossil record were different from those alive today, and why there was an apparent succession of species in the record, with some suddenly disappearing and being replaced by others. Darwin believed that his theory answered all these questions and more. In his polite, Victorian manner, he took on the world with his reinterpretation of the past. He was well qualified to do so since he had a proven track record in geology, paleontology, and comparative anatomy. To appreciate the magnitude of his accomplishment, you need to have some background on the state of knowledge about the geological record in 1859.

By the seventeenth century, geologists recognized that the surface of the earth was covered by two categories of rock. One was "primary" rock, like granite. The other was secondary rock, which was composed of primary rock that had been weathered or otherwise broken down, then re-formed into other kinds of rock. It was inferred that secondary rock was the product of the erosion and redeposition of primary rock, which meant that it was younger than the primary rock from which it was formed. Thus sediments exposed on the side of a single eroded rock formation, like the walls of the Grand Canyon, were interpreted as a chronological sequence, with the oldest layer on the bottom and each layer above it progressively younger. Lay-

Figure 20

This photograph, taken in the Zanskar Valley of northern India, illustrates the kind of complexity seen in geological formations that suggested to geologists by the early eighteenth century that the earth is very old, even if they did not know how old. The lower rock unit exposed in the riverbank contains older strata that formed as the land subsided, which were then folded (note the undulations of the beds indicating folding) and uplifted as the Himalayan mountains grew. After these uplifted, folded strata were eroded, they subsided again and were buried by new strata deposited on top of them. This later deposition remains as horizontal layers resting on top of the older, folded formation. This entire sequence was then exposed by erosion caused by the river that now runs through the region. Given the extremely slow rate at which landscapes change during human lifetimes, early geologists correctly reasoned that sequences of events like these took millions of years to complete.

ers of secondary rock were spread over the landscape in incomplete sheets, often broken by the erosion of river valleys or by intrusions of primary rock, and they could also be twisted, distorted, and metamorphosed by events that postdated their formation.

Geologists devised models for the history of the earth to explain how a presumably even distribution of the original rock surface had become the chaotic jumble of primary and secondary rocks observable today (fig. 20). To some, the Bible suggested that the earth was only a few thousand years old. However, long before the *Origin*, the prevailing view was instead that the depth and complexity of the secondary rocks indicated that the earth was exceedingly old. Only the passage of vast intervals of time could account for the weathering of primary rock, its re-formation as secondary rock, and then the complex reworking and erosion of secondary rocks after their formation. The rocks alone documented the antiquity of the earth.

Fossils are embedded in secondary rocks. The kinds of organisms preserved as fossils changed over time. Older sediments contained organisms that were often quite different from anything found living today. Some of these strange organisms were of monstrous proportions. One example from North America was the Ohio animal, which was a large, elephant-like animal found on the banks of the Ohio River. Another was the Maastrich animal, found in a limestone quarry in the Netherlands, which had toothed, crocodile-like jaws measuring 1.2 meters in length. Others included giant cave bears, found in abundance in caves in Bavaria, and a rhinoceros-size *Megatherium*, or giant ground sloth, found in South America.

Jean-Andre de Luc (1779, in Rudwick 2005) proposed three hypotheses to account for the differences between fossil and living organisms. One hypothesis was that all the fossil organisms were still alive and were roaming about somewhere in the vast, unexplored regions of the planet. This idea was made more plausible by the occasional discovery of "living fossils," or living representatives of animals that were previously known only from fossils. Two such animals were the clamlike brachiopods and sea lilies, both of which were found alive in samples dredged from the ocean depths long after their discovery as fossils. Thomas Jefferson was aware of this hypothesis when he dispatched Lewis and Clark to explore the interior of North America. One of their missions was to search for living representatives of mastodons and mammoths, which Jefferson knew from fossils. However, this hypothesis faded as the catalog of strange fossil organisms grew and the extent of unexplored territory shrank without revealing living representatives.

A second hypothesis was that the strange organisms known only from fossils had gone extinct. A third hypothesis was that descendants of these organisms were still alive but had been "transmuted" over time into new species. These last two hypotheses remained viable and were being debated when the *Origin* was published.

Georges Cuvier (1769–1832) championed the second hypothesis and is credited with proving that extinctions had occurred. He spent his professional career as a member of the faculty at the National Museum of Natural History in Paris. His success was a by-product of Napoleon's conquests and efforts to make Paris the epicenter of science and culture. Napoleon endowed the museum with professorships and enriched it with collections plundered from museums in conquered territories. Cuvier had access to fossils and the best collection in the world of the skeletons of living animals. In addition he had "paper proxies," which were detailed drawings of fossils from other museums. Paper proxies proved to be a reasonable substitute for seeing the real specimens at a time when western Europe was at war and travel was difficult. Cuvier also had direct access to the rich fossil beds of the Paris Basin and a menagerie of live animals kept near the museum.

Cuvier "proved" extinction by first describing the range of anatomical variation present in living animals, then comparing them to fossils. In his earliest effort, he showed that the anatomy of the teeth and mandibles of the Indian and African elephants were so different that they must be distinct species. He then showed that the mandible and teeth of the mammoth were so different from either of these species that it too must be a different species. Cuvier demonstrated again and again that fossil mammals were markedly different from any known living species. By 1814, he had described forty-nine species of quadrupeds (four-legged animals) from fossils that were not known from living species. He also integrated his fossil descriptions, comparative anatomy, and observations of living animals to generate descriptions of what the fossils were like when they were alive. He conceived of animals as integrated machines suited for specific lifestyles. Because of this integration, he claimed he could take even the smallest fragment of a fossil animal and reconstruct what the whole animal would have been like in life.

Jean-Baptiste de Lamarck (1744–1829), who was one of Cuvier's colleagues at the National Museum, was a champion of the alternative hypothesis, which was that these species had not gone extinct, but had instead changed over time into the species that were still alive. Lamarck argued that living species were just arbitrary points in a continuous history of transformation. A key part of Lamarck's theory was that this transformation unfolded too slowly to be apparent in a human lifetime. The large differences between fossilized and living organisms had therefore occurred over enormous time spans, and were a testament to how old the earth was. Lamarck recognized only anagenic change, or change within a lineage over time. A hallmark of Darwin's theory was cladogenesis, or the splitting of one lineage to form two or more descendant lineages.

Transmutation was a hard alternative to disprove, but Cuvier had two arguments against it. First, if living animals were highly integrated machines, then changing them would disrupt this integration. Second, when he studied animals in the fossil beds of the Paris Basin, he did not see any evidence of gradual change over time. Each species simply appeared, remained stable in character through time, then disappeared. Since he observed that species *did not* change in the fossil record and inferred from his study of living organisms that they were like integrated machines and *could not* change, he rejected Lamarck's alternative.

An important innovation followed soon after Cuvier's successful proof of extinction: the ability to weave information gleaned from fossils and the secondary rocks in which they were found into a chronological sequence that recounted the history of life. In the early 1800s, William Smith, a surveyor in England, was one of the first to recognize that fossils added information about the rocks in which they were found. One of his early occupations was to supervise the digging of canals for barges to transport coal and other products around the rapidly industrializing British Isles. Newly dug canals gave him access to sedimentary rocks over a large area and the opportunity to see fossils in the context of the sediments they were embedded in; most of his predecessors had studied fossils in the confines of museums. Smith's interest was in ordering sediments in a chronological sequence and in being able to relate the age of sediments in one part of the countryside to those found elsewhere. The structure of the rock in each layer provided some clues about age. The fossils found in the rocks changed over time, as some species disappeared and others replaced them, so they provided additional, critical information for determining the ages of sediments found in one location relative to those found elsewhere. By combining the information from the rocks and the fossils that they contained, he was able to construct a map of portions of the British Isles that defined the relative ages of the strata exposed on the surface in each of them. "Relative" means he could order them in a sequence from oldest to youngest; he did not know their actual age.

Smith was not a biologist and had little interest in what the fossils said about the history of life. His concerns were more in tune with the rising tide of capitalism that accompanied the industrial revolution. For example, coal was found in a particular sequence of sediments that became known as the Carboniferous period. He argued that if the surface rocks on an area of land were younger than this period, then it might be profitable to search for coal underneath them. If the surface rocks were older, then it would be fruitless to search for coal there because only older, non-coal-bearing sediments would be found underneath.

In 1808, Georges Cuvier worked in partnership with the geologist Alexandre Brongniart to duplicate Smith's discovery, but they went beyond Smith by recognizing the biological significance of the fossils. They argued that each layer of secondary rock was a different page in the history of life, revealed in chronological order. Each page revealed not only the organisms that lived at the time but also the nature of the communities in which they lived. Their initial papers were confined to surveys of the Paris Basin, but their results were soon extended by others throughout Europe and elsewhere.

Once this connection between the succession of fossils in sedimentary rocks and the history of life was made, there was a rapid series of discoveries that enabled scientists to reconstruct a history of life. This history is summarized in the "geologic timescale," which chronicles the succession of life-forms found over time from all known fossil-bearing strata (fig. 21). The chronology of sediments in different parts of the globe was established through the correlation of the fossil organisms they contained, creating a relative timescale that ranked rocks from oldest to youngest. We did not develop the ability to assign real ages, in terms of numbers of years, to these rocks until the early twentieth century, after the discovery of radioactive decay as a dating technique.

John Phillips, William Smith's nephew, recognized that chapters in the history of geological time could be grouped into three discrete eras—Paleozoic, Mesozoic, and Cenozoic, which stood for ancient, middle, and recent life. Each of these eras contained a distinct set of organisms; the boundary between eras was defined by the replacement of one group of organisms by another. Each era was subdivided into periods, and periods into epochs, each of which could be recognized wherever it was found by its characteristic flora and fauna. These subdivisions were often named for the regions where a given rock formation was first discovered and described. For example, Adam Sedgwick, who tutored Darwin in field geology, was the first to describe rocks from formations found in Wales with a characteristic fauna. "Cambria" is the Latin word for Wales, and therefore this time period would later be named the "Cambrian" period. Darwin did not adopt all of Phillips's nomenclature in the *Origin*. He used "Paleozoic" for the first era, but instead used the older terms "secondary" and "tertiary" to refer to the Mesozoic and Cenozoic eras.

The earliest layers of the Paleozoic era had only marine life, such as trilobites and brachiopods, with no trace of either vertebrates or land plants.

Formations	circa 1790	circa 1840		Modern	
Alluvium	Post-diluvial	Alluvium		Holocene	
Glacial deposits	Diluvial	Newer Pliocene	CAINOZOIC (Tertiary)	Pleistocene	CAINOZOIC (Tertiary)
Sicilian strata		Older Pliocene		Pliocene	
Subappenine strata	TERTIARY	Miocene		Miocene / Oligocene	
Parisian strata		Eocene		Eocene / Palaeocene	
Chalk					
		Cretaceous		Cretaceous	
Oolites			MESOZOIC (Secondary)		MESOZOIC
Lias		Jurassic		Jurassic	
Muschelkalk New Red Sandstone	SECONDARY	Triassic		Triassic	
Kupferschiefe		Permian		Permian	
Coal Measures					
Carboniferous (Mountain) Limestone		Carboniferous	PALAEOZOIC (Protozoic)	Carboniferous	PALAEOZOIC
Old Red Sandstone		Devonian		Devonian	
Wenlock Limestone				Silurian	
'Grauwacke'	TRANSITION	Silurian		Ordovician	
		Primordial (Cambrian)		Cambrian	
Longmynd strata	PRIMARY	AZOIC (Primary)		PRE-CAMBRIAN	
Scandinavian schists etc.					

Figure 21

The Geologic Timescale in 1790, 1840, and at Present

The timescale of 1790 predates the discovery of correlation (the equivalence of rock strata found in different formations all over the earth, based on the fossils they contain); so the 1790 time divisions correspond only approximately to those of the later periods. The two later timescales reflect the advent of correlation and increasing knowledge of the fossil record.

These earliest fossil-bearing strata overlay older strata that were described as "azoic," or devoid of life. Because these were the earliest fossil-bearing layers then known, scientists inferred that they represented the origin of life. Later layers recorded the replacement of old types of trilobites and brachiopods by new ones as well as the appearance and diversification of fish, followed by evidence of land plants. Overlaying these layers were strata that in some cases contained fossils of land vertebrates. Reptiles dominated the earlier of these layers, with mammals making an appearance only much later. There was little apparent overlap between the "age of reptiles" and the "age of mammals." Traces of humanity did not appear until the very end of the record.

Since each successive subdivision was represented by the disappearance of certain life-forms and the appearance of new ones, some scientists inferred that something had happened to end one era and begin a new one. Because species were clearly distinctive throughout each layer, showed no evidence of change over time, then disappeared and were replaced by new life-forms, Cuvier argued that the record revealed successive "revolutions" that drove some organisms to extinction, after which new organisms replaced them. He believed that extinctions of land animals were caused by continents subsiding under the ocean, because his investigations of the Paris Basin revealed an alternation between deposits from marine versus freshwater environments. He speculated that individual continents submerged under the sea, then reemerged and were recolonized by terrestrial species from other continents. He avoided the topic of where new species came from, since he felt it was futile to speculate on something that was so shrouded in mystery.

One of Cuvier's significant career accomplishments was the publication of his *Animal Kingdom* (*Regne Animal*, 1817, in Rudwick 1997), in which he presented an analysis of the comparative anatomy of all animals, including a reinterpretation of their classification. Prior to Cuvier, the diversity of animals living today and of those preserved in the fossil record was considered part of a single "scale of being," meaning they could be arrayed in a single linear hierarchy with simple, primitive life-forms at the bottom of the scale and progressively more advanced life-forms higher, and with humanity at the zenith. Lamarck advocated this "scale of being" and interpreted the temporal sequence of fossils as representing a progressive perfection of the animals. Cuvier's conclusions differed from Lamarck's and others in two critical ways. First, his scale was not strictly unilinear: he recognized four distinct "embranchments" of life that were as distinct from one another when they first appeared in the fossil record as they were in living organisms. These included the vertebrates, mollusks, Articulata, and Radiata. The first two classifications correspond well with the modern taxa of the same name, while the latter two are a composite of what we now recognize as a number of distinct groups of organisms. Nevertheless, his classification represents the emergence of the first well-defined recognition of what we would later refer to as "phyla," which subdivide the animal kingdom into distinct groups of organisms.

The earliest representatives of some of the embranchments were as anatomically complex as living organisms. For example, unusually well-preserved trilobite fossils found just above "azoic" rocks had compound eyes comparable in complexity to those of living insects. There was thus no empirical support for Lamarck's notion of a linear scale of being or of a progressive perfection of animals over time. These observations enhanced the mystery of the first appearance of life. If animals were so complex from the very beginning, where did they come from?

This brief history of geology is now overlapping in time with Charles Darwin's childhood. Charles Lyell, whose *Principles of Geology* (1830–1833) had such a great influence on Darwin, argued that the sequence of events revealed in secondary rocks could all be explained by the same processes that we can see in progress today. Whereas Cuvier argued that there had been successive revolutions in which one group of animals was replaced by another, Lyell argued instead that the fossil record could have been shaped by gradual extinction and replacement, one species at a time, much as we now see individual species declining in abundance, then going extinct. Lyell believed that discontinuities in the formation of fossil-bearing rocks explained the appearance of discontinuity in the organisms found embedded in the rocks (fig. 22). Recall that Lyell believed different regions of the earth were constantly changing in elevation, with portions of the earth alternately subsiding or becoming more elevated (introduction and chap. 10). Fossil-bearing sediments accumulated only in regions where the earth was subsiding, since only low-lying, subsiding regions could continue to accumulate sediments for a prolonged period of time. If subsidence ceased, then sedimentation would cease as well when the basin floor became level with the surrounding terrain. If the area then became more elevated, it would be exposed to erosion, which might wear away some fossil-bearing sediments. The breaks that were seen in the sediments deposited in any one region

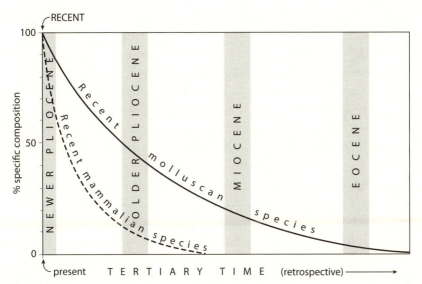

Figure 22
Lyell's Interpretation of Gaps in the Geological Record
The gray portions represent known fossil-bearing strata and their known order of appearance in the fossil record. The white portions are hypothetical intervals of time that separate these fossil-bearing strata. Fossil-bearing strata contain some fossils that are similar to animals alive today and others that are not; the latter must be species that are now extinct. The proportion of fossils that resemble animals still alive declines in older strata. Lyell postulated that extinction and replacement of species were continuous processes. "The use of a survivorship curve in this diagram is justified by Lyell's frequent reference to, and evident understanding of, the analysis of contemporary census returns" (Rudwick 1976, p. 184). Such curves assume a constant probability of extinction.

thus represented periods of nondeposition. Lyell hypothesized that discontinuities in the types of fossilized organisms the rocks contained were just artifacts of these periods of nondeposition, possibly exaggerated by erosion and loss of some sediments, rather than being products of "revolutions" of extinction and the creation of new life-forms. Although Lyell thus postulated the gradual extinction of some species and the gradual evolution of others, he interpreted the actual *origin* of new species, which might then evolve, as individual acts of special creation.

Those who studied the fossil record before Darwin defined the key questions that had to be answered. The first was what John Herschel defined as the "mystery of mysteries," or the origin of new species that appeared throughout the record. The second was to explain the presence in the fossil record of organisms that could not be found among the living organisms of the present day. A third was to explain why species, and even Cuvier's "embranchments of life," were distinct when they first appeared in the fossil record. The fossil record revealed a progression in the history of life in the sense that new types of organisms, like fish, then amphibians, then reptiles, then mammals, appeared in a chronological order; however, the earliest appearance of life in that record (as far as was known in 1859) was already structurally complex. The succession of life-forms, both in the appearance of distinct classes of organisms, then in the species found within each class, also begged the question of their relationships to each other. Finally, the juxtaposition of the "azoic," or apparently lifeless strata, with the strata that bore complex life-forms directly on top of them demanded an explanation.

Darwin has already argued, in previous chapters of the *Origin*, that his theory of evolution can explain the origin of species and extinction. In the following chapters, he argues that his theory can also answer the remainder of the questions derived from the fossil record and, as a consequence, explain the entire history of life. In doing so, he was telling the members of the growing discipline of geology that he had the right answers to all their questions and that most of their answers were wrong.

References

Rudwick, M.J.S. 1976. *The meaning of fossils: Episodes in the history of paleontology*, 2nd ed. New York: Neale Watson Academic Publications. (Pp. 179–91, Lyell; pp. 207–14, Owen.)

Rudwick, M.J.S. 1997. *Georges Cuvier, fossil bones, and geological catastrophes*. Chicago: University of Chicago Press. (Pp. 253–54, description of Cuvier's "embranchments of life.")

Rudwick, M.J.S. 2005. *Bursting the limits of time: The reconstruction of geohistory in the Age of Revolution*. Chicago: University of Chicago Press. (Pp. 194–555, source of historical detail prior to Charles Lyell's and Richard Owen's contributions.)

Chapter 17

Geology II: On the Imperfection of the Geological Record

In chapters 9 and 10 of the *Origin*, Darwin uses his theory to explain the geological record. He argues that the geological record contains the history of life as it evolved and diversified on a planet that is ancient beyond measure. If this is true, then the fossil record should reveal all the gradual transitions between species that his theory predicts have existed over time. It does not. It is episodic, with one group of organisms abruptly replacing another. This lack of evidence in the fossil record for transitions between species, Darwin concedes, is possibly "the most obvious and gravest objection which can be urged against my theory" (*Origin*, p. 280). If he is to convince the world that he is right, Darwin has to reconcile the expectations generated by his theory with the nature of the fossil record.

First Darwin argues that even if we found a fossil that was the common ancestor to different living species, we would not necessarily recognize it as such. We are inclined to think that the ancestor of two related species will be intermediate between them in appearance, but this is not likely to be the case. Darwin uses domestic pigeons to make the point. I will recast his example with dogs rather than pigeons since dogs are more familiar to most of us. Picture a dachshund and a French poodle, and then try to envision what their common ancestor might have looked like. We know from genetic analyses that it was the gray wolf, but you would never imagine something that looks like a gray wolf, because it is so different from a dachshund and poodle rather than being intermediate between them. We cannot see the relationship, because the dachshund and poodle are each the product of a long process of change in which their ancestors diverged from the gray wolf and from each other. We have no record of that change. Likewise, living species that are related to one another, such as the wolf, seal, and skunk—which are all in the order Carnivora—are derived from a common ancestor that did not look like any of them. Despite their diversity of outward appearance, these descendants still share some features of their anatomy that they inherited from this common ancestor.

Still, there must have been a continuity of change in the treelike pattern of succeeding generations that joins these animals together. When Darwin's theory is played out over a long timescale, it predicts that each step that joins a wolf, seal, and skunk to their common ancestor is as small as the differences that we see today between living populations of these animals. We would see this continuity in the record if it were complete. Darwin's argument, following Lyell's, is that we cannot see this continuity because the record is so fragmentary. Rather than containing the entire, branching tree of life, it contains only fragments of branches for some intervals of time. Most intervals of time are not even represented. Making sense of the relationships between animals alive today and their ancestors preserved in the record is like discerning the relationships between a wolf, poodle, and dachshund with all the steps between them having been lost.

From this point on, I will use Darwin's chapter subheadings.

On the Lapse of Time

Darwin's first goal is to establish the antiquity of the earth. If he is to be successful in arguing that all living organisms have been derived from a small number of common ancestors, and if his proposed process of natural selection is so slow that we cannot see it happening on the timescale of our own lives, then it follows that the amount of time necessary for evolution to account for the history of life must be unimaginably long. Darwin uses the total thickness of fossil strata found in the geological record to argue for the earth's antiquity.

Darwin cites the work of Professor Ramsay, who tried to envision what the dimensions of the geologic timescale would be if we took the thickest, most extensive deposits for each period in the earth's history from wherever they could be found, then stacked them all on top of one another in chronological order. The idea is that the amount of sediment found for one period in one place will be a function of how long that region spent as a low-lying basin that could accumulate sediment. In any one rock formation, a given time period may not be represented at all, because the rock formation was elevated at that time and exposed to erosion. It may be represented by a thin layer, because sediments were deposited for only a short time. Or it may contain thicker deposits representing a longer interval of sediment accumulation. When Professor Ramsay added up the thickest sediments that could be found in Great Britain for each time period, those from the Paleozoic summed to 57,154 feet, those for the Mesozoic (called "secondary" in the *Origin*) summed to 13,190 feet, and those for the Cenozoic (tertiary) summed to 2,240 feet. The entire timescale, if composed of the thickest sediments available for any one period, summed to 72,584 feet, or 13¾ miles. Some parts of the record that are thin in Great Britain are thousands of feet thick in Europe, so the true maximum thickness of strata that comprise the timescale is much greater. Darwin reasons that the thickest deposit for any given time period will come closest to representing the true duration of that period. He argues, given the rate at which we can observe sedimentary rock forming today, that a huge interval of time was required to deposit this total thickness of geological strata.

Darwin also applies Lyell's uniformitarian principles to estimate the age of the Sussex Weald, where subsidence of the land allowed the ocean to fill a valley, turning it into a bay. The tides eroded the exposed stratified rock along the margins of the bay. The intruding arm of the ocean eroded away 1,100 feet of sediment between the North and South Downs before the land was elevated, draining away the ocean and exposing the eroded land in between. Darwin estimates the rate of erosion per year, then calculates that 300 million years were required to remove 1,100 feet. His calculation was later shown to overestimate the true age by more than a factor of three. Nevertheless, his argument is still a valid appeal to the reader: Imagine how long it would take everyday processes, such as weathering by wind, rain, or tides, to shape the geological features of the globe. Such weathering must have taken many millions of years, and the earth must be very old.

Darwin's arguments stimulated his contemporaries' interest in the age of the earth. Before the *Origin*, geologists were satisfied to observe that the earth must be very old. After the *Origin*, it became relevant to ask, Exactly how old is the earth?—or, Is the earth old enough for Darwin's theory to account for the history of life and the diversity of living organisms? To some of Darwin's contemporaries, the answer was the earth was too young. The chief proponent of the argument that the earth was too young for Darwin's theory was William Thomson, later named Lord Kelvin. Thomson was one of the preeminent physicists of his time. Between the publication of the first and sixth editions of the *Origin*, Thomson developed a model of the earth that yielded estimates of the earth's age. First, it was known that the temperature of the earth increased as one descended into deep mine shafts. The existence of volcanoes showed that the core of the earth consisted of molten rock. Thomson argued that the earth began as a ball of molten rock, then cooled to form a solid crust on the surface with a molten core. He estimated the age of the earth with calculations of the likely rate of cooling. His upper limit for the age of the earth was 100 million years. He also developed a model for the rate of cooling of the sun. By that model, the sun initially radiated too much heat for the earth to support life. It was only after some interval of cooling that the origin of life was possible, which further constrained the amount of time available for Darwinian evolution. He suggested that the earth could have sustained life only for an interval of 20–40 million years, which was far too brief to be compatible with Darwin's theory. Because of Thomson's prominence as a physicist, and because we all tend to be overawed by the calculations of physicists who claim theirs to be the true science, his proposals held some sway in the debate about the *Origin*. Darwin modified his fifth edition of the *Origin* to try to reconcile Thomson's calculations with his theory, but he also maintained that even though he could not find fault with Thomson's argument, the details of the geological record were inconsistent with such a young earth or such a short duration of life on earth.

It was not until the early twentieth century that the discovery of radioactivity revealed the flaw in Thomson's model and at the same time provided a method for estimating the true age of the earth. The flaw was that the fission of radioactive elements in the mantle of the earth creates heat, as does the process of fusion in the core of the sun. Thomson's model of the earth and sun as globes of molten rock that dissipate some finite initial amount of heat is incorrect. Because new heat is constantly being generated, any model that estimates age from the amount of heat lost will underestimate the true age of the earth, apparently by a very large margin. The new knowledge also implied that the sun was not too hot for life to have existed on the very young earth but, instead, has been in a steady state very similar to its state today for a very long time. Furthermore, the rate of radioactive decay of elements found in ancient lava beds or volcanic ash made it possible to estimate their actual ages. The first estimates made in the early twentieth century suggested that the earth was more than one billion years old. Later refinements of the methods of absolute aging have revealed that the true age of the earth is around 4.6 billion years.

The important take-home message is that Darwin argued, correctly, that the earth is indeed ancient, and old enough for his proposed theory of evolution to account for the history of life.

On the Poorness of Our Paleontological Collections

If the fossil record contained the history of life, then Darwin had to explain why that history seemed so incomplete. In this section Darwin expands on Lyell's earlier argument for the imperfection of the fossil record.

In 1859, fossil-bearing rocks had been well explored only on a small subset of the globe. This incomplete sampling, combined with the high rate at which new fossil species were being discovered, suggested to Darwin that only a small fragment of the fossil record was known, and that some existing gaps would later be filled. We have now explored much more of the available fossil strata and have succeeded in filling many of the gaps in the record. We continue to discover new species at a high rate. This high rate of discovery implies that there are still many more fossil organisms to be found.

Modern researchers estimate that a complete inventory of all the species that have ever lived would number in the billions, but the total inventory of all species known from the fossil record numbers in the hundreds of thousands. (Estimates of the number that have lived are based on the number alive today and the rate of species replacement that we see in the fossil record.) This difference between the number of species estimated to have existed and the number found means that far fewer than 1% of the species that once lived have been found as fossils. This huge discrepancy is by itself a measure of the fossil record's imperfection.

A key source of imperfection in the record is the conditions required for fossilization. To form a fossil, a dead organism must be encased in sediments and preserved. The sediments form a mold around the carcass, after which the original components of the organism slowly dissolve away and are replaced by minerals, creating a cast. Whether this process will occur depends on the structure of the organism and its habitat.

Most fossils consist of what were "hard" or mineralized tissues, like bones or shells, in the living animal because these tissues are most resistant to decay. Soft tissues like skin, muscle, or internal organs decay much more rapidly and will be retained in fossils only if the animal dies in special circumstances. For example, the famous Burgess Shale fossils (see chap. 19 below) include many soft-bodied marine organisms. They were apparently buried by avalanches of mud that instantly enclosed them in an oxygen-free tomb, which inhibited the rate of bacterial decay. One consequence of the bias against soft-tissue preservation is that all the organisms that lack hard tissues will be poorly represented or absent from the record.

The likelihood of becoming a fossil also depends on the environment in which the animal lives. If the organism lives in a moist forest, then it is likely that the entire carcass, including the skeleton, will decay. If animals from this sort of environment are represented at all, then it is generally only by teeth, since teeth are the most durable part of the animal and hence the most likely to be preserved. Before he published the *Origin*, Darwin inferred from his barnacle studies that marine organisms living in the intertidal zone—that part of the shoreline that lies between high and low tide—are also unlikely to fossilize. The species of barnacles found in this zone today coat intertidal rocks around the world in huge numbers and are enclosed by rigid, mineral-rich shells that should readily fossilize. In Darwin's time, no living genus was known from the Cenozoic fossil record; however, one was discovered from earlier Mesozoic deposits shortly after Darwin published his barnacle monographs. This means that they must have lived throughout the Cenozoic, a period of 65 million years, without being preserved as fossils.

A second reason for the imperfection of the record is that it is intermittent. Fossils form only when the carcasses of dead organisms are embedded in sediments, but sediments accumulate only where the earth is subsiding. When we look at a formation of sedimentary rock, the layers that lie adjacent to one another may be separated by a prolonged period of nondeposition, a gap augmented, perhaps, by a time when that region was elevated and exposed to erosion, which erased some of its history. If we consider only the consecutive sediments found in one formation, then each layer is distinct with no evidence of a transition between layers. In some other region there may be thick deposits that represent the interval of time that separates the two layers. Piecing together different fragments of the record found in different regions provides a better representation of the history of life, but there remain many intervals of time that are not represented anywhere, because they have either been eroded away or remain buried under the surface of the earth.

Darwin's cites his observations of the west coast of South America as an example of conditions that can erase part of the record. There he found a rich coastal marine fauna, in terms of both abundance and diversity, plus a steady supply of sediment from the many streams draining the steep slopes of the Andes, yet very little in the way of recent fossil deposits. He thought that rich fossil beds were likely to have formed, but they were also likely to be quickly eroded away "by the slow and gradual rising of the land within the grinding action of the coastal waves" (*Origin*, p. 290).

Darwin cites the late E. Forbes, who argued that we see long sequences of high-quality, continuous deposits of marine fossils only when there has been a combination of special circumstances. Such deposits require a shallow basin. Shallow basins are productive environments that are host to large and diverse communities of organisms, many of which have mineralized tissues that readily fossilize. The basin must be continuously subsiding, and the rate of sedimentation must be closely balanced with the rate of subsidence. Such a basin would continuously sustain a rich fauna and would continuously preserve a sample of that fauna in the fossil record. This is a very rare combination of circumstances, and it pertains only to one type of community. It is found only sometimes in only some parts of the earth, yet this set of conditions is now known to be one of the few that record long, continuous intervals in the history of life.

The combination of all these limitations tells us that the fossil record will be incomplete, not only in the time intervals that are recorded, but also in the geographic area that is represented. The record is then further fragmented because, once such deposits are formed, they may later subside and be destroyed by the metamorphic processes of heat and pressure deep inside the earth, or they may be elevated and destroyed by erosion. The record that we usually see is the subset of the surviving deposits that are now on the surface of the earth and are currently exposed to erosion.

Darwin argues that the imperfections of the record will be further exaggerated by a disconnect between times when speciation is most likely to occur and conditions that favor fossil formation. As he said in a previous chapter (see chap. 10), the best conditions for speciation are on continents that oscillate in elevation and alternate between being elevated to form a continuous landmass and subsiding to form an archipelago. During the archipelago phase, species are subdivided into small, isolated populations, which will facilitate the evolution of small differences between them. When the land is once again elevated, these populations will be reunited, and a struggle for existence among these varieties will follow. Darwin predicted that this struggle would accelerate speciation and extinction. The transitions between species should thus be most common during these periods of elevation, but these are also the times when land is exposed to erosion, not deposition, so fossil formation is least likely to occur. The fossil record will thus tend to exclude intervals of rapid speciation and extinction, and thus not preserve transitions between species.

Darwin argues here that transitions in the fossil record are also rare because speciation is an extremely slow process that happens on a timescale longer than the duration of most continuous deposits of sediment. His contemporaries Bronn and Woodward claimed instead that the amount of time represented by a formation was often two to three times longer than the duration of a species. (They found that species were often absent at the bottom and the top of a sedimentary formation, so they concluded that the species had originated after the deposition of sediment began and had gone extinct before the deposition ended.) Darwin's counterargument is that the presence or absence of a species from a particular location is not a good measure of whether it was alive at the time. It could have been alive throughout the interval, but not always where sediments were forming.

We know now that species alive today have geographic ranges that can vary over time. In our current interval of global warming, many species have been observed to disappear from some regions or extend their ranges to new regions. Recall the blackcaps (chap. 15) that have recently extended their overwintering range to include Great Britain. Any fossil-bearing sediments that form there during that interval might record their sudden appearance, without any indication of their origin. Since the geographic range of a species can vary over time, it could easily have existed somewhere else for the entire interval represented in a given rock formation without ever being preserved there. A species can also disappear from an area and reoccupy it later, so a gap in its occurrence would appear in the fossil record of that one region. Such gaps in the record had been seen before publication of the *Origin*, and had been cited as evidence of a species having been created twice, but Darwin notes that a change in its distribution is a simpler explanation.

Darwin's model for speciation also argues that a new species will begin as a single population. It is only after a new species has become by some measure distinct that we should expect it to expand its range as it displaces its ancestors. The fossil record is not likely to include the site of origin and will instead capture the species only as it expands its range, so it will suddenly appear as a fossil without any record of its formation.

Finally, the differences between the way we recognize species today and the way they are perceived in the fossil record will cause us to fail to recognize some transitional forms even if we are lucky enough to find them. When we study living species and try to discriminate between what is a separate species versus a variety within a species, we consider how they vary across a geographical range. If we find populations in different places that are distinct, we are likely to call them varieties within a species if we can find areas where their ranges overlap and observe transitional forms between them. On the other hand, we will classify them as different species if we find a range of overlap and see no evidence of interbreeding (see salamander examples in chap. 9). The fossil record rarely incorporates this element of geographical variation; therefore, the same phenomena, when observed in fossils, will result in our simply naming morphologically distinct varieties from different regions as different species rather than seeing them as part of the transition in the formation of new species.

Given all these sources of discontinuity, the fossil record can hardly be expected to contain a complete history of life. In Darwin's words: "...we have no right to expect to find in geological formations, an infinite number of those fine transitional forms, which on my theory assuredly have connected all the past and present species of the same group into one long and branching chain of life. We ought only to look for a few links, some more closely and some more distantly related to each other; and these links, let them be ever so close, if found in different stages of the same formation, would, by most paleontologists, be ranked as distinct species" (*Origin*, pp. 301–2).

Darwin concludes that these many sources of imperfection in the geological record—its fragmentary representation of time and space, its selective representation of habitat and species, its susceptibility to destruction once formed, and its limited accessibility to exploration—account for the scarcity of evidence for species formation.

On the Sudden Appearance of Whole Groups of Allied Species

The difficulties presented by the geological record go well beyond those posed by individual species. Often we see whole groups of related species seeming to appear at once. For example, large orders of mammals, such as bats, seemed to pop out of the record with no evidence of their ancestry. "If numerous species, belonging to the same genera or families, have really started into life all at once, the fact would be fatal to the theory of descent with slow modification through natural selection" (*Origin*, p. 302).

Darwin proposes three aspects of imperfection in the geological record to account for such sudden appearances of groups of allied species. First, the earth is very large, but the fossil record (at his time) is known only from a small area, mostly in Europe and North America. This means that there are large gaps in the temporal sequence that may be filled with discoveries made elsewhere. Second, groups of species may have originated in some small area, only to disperse after they formed. Finally, there are huge gaps in the intervals of history recorded in the sediments. These gaps are easily large enough to obliterate not only the origin of an individual species but also the origin of some larger category in the taxonomic hierarchy. For all these reasons, whole taxa can appear at once in the fossil record without any evidence of transition. Darwin presents some examples to illustrate these arguments, but also cites some special properties of higher taxa that may account for such trends.

First Darwin presents examples of groups of animals that were once thought to have appeared suddenly in the fossil record, but were later found to be present in much earlier periods. The mammals were at first well known from the Cenozoic; it was only later that mammals were found in the Mesozoic, where they were apparently less abundant, smaller, and were represented by little more than teeth. (In the 1960s, the great anatomist Alfred Sherwood Romer speculated that all the known mammalian Mesozoic fossils in the world might fit inside a bowler hat. It has only been in the past two decades that we have found more complete skeletons from Mesozoic mammals and have begun to get a glimpse of their diversity. Perhaps scarcity, small body size, or habitat contributed to their poor representation in the fossil record.) Given these limitations, Darwin argues that to try to interpret the fossil record as a complete history of life would be "about as rash . . . as it would be for a naturalist to land for five minutes on some one barren point in Australia, and then to discuss the number and range of its productions" (*Origin*, p. 306).

Darwin also argues that there may be a special property of higher taxa that causes them to appear suddenly in the fossil record: "it might require a long succession of ages to adapt an organism to some new and peculiar line of life, for instance to fly through the air; but . . . when this had been affected, and a few species had thus acquired a great advantage over other organisms, a comparatively short time would be necessary to produce many divergent forms which would be able to spread rapidly and widely throughout the world" (*Origin*, p. 303). Higher taxa, such as order Chiroptera (bats) or birds, are often united by morphological novelties that enable them to use the environment in some new and different way. The perfection of such an adaptation may open the door to an adaptive radiation (chap. 1), since perfecting flight would make a diversity of new lifestyles suddenly available.

On the Sudden Appearance of Groups of Allied Species in the Lowest Known Fossiliferous Strata

Darwin faced a feature of the fossil record that was even more challenging to his theory. The lowest known fossil-bearing strata contained complex life, yet his theory predicted that some long interval of time must have been required for such complexity to evolve. How could he account for the sudden appearance of organisms as diverse and complex as the trilobites, which his theory predicted must have been derived from some earlier, more generalized arthropod? For his theory to be true, life must have existed for a very long interval of time before the appearance of these species, perhaps as long as the vast interval of time that was represented in the entire fossil record that postdated them. The alternative, championed by some of the most prominent geologists of Darwin's day, was that these earliest fossils represented the origin of life and that life was complex from the very beginning.

Darwin first makes recourse to the argument that only a small fraction of the available fossil-bearing strata have been explored. He predicts that researchers may well discover earlier forms of life and gain a better understanding of the history of life prior to the oldest-known fossil-bearing strata. He mentions some fossil-bearing strata predating the Silurian deposits that (it was thought in his time) may represent the oldest forms of life. There are also deposits of "phosphatic nodules and bituminous matter in some of the lowest azoic rocks" (*Origin*, p. 307), which suggest that some form of life existed long before fossils appear in the record, since such deposits in younger rocks are known to be of biological origin.

Darwin feels that it is not acceptable simply to argue that all the rocks that preceded the oldest-known fossil-bearing strata may have been eroded away or metamorphosed. The oldest-known fossil-bearing rocks (of his time) can be found "over immense territories in Russia and North America" (*Origin*, p. 308), so Darwin has no easy recourse to account for the absence of older fossil-bearing deposits. He does his best to imagine what kinds of changes must have occurred on the surface of the earth to account for either the absence of such strata or their metamorphosis to a condition that erased evidence of life. We need not recount his arguments here since this mystery has since been solved. We now have a reasonable record of earlier life (chap. 19 below) that supports Darwin's argument that there was a long history of life preceding the earliest-known fossils of his day.

Darwin concludes the chapter by acknowledging that "the most eminent" paleontologists of his day, "namely Cuvier, Owen, Agassiz, Barrande, Falconer, E. Forbes, &c., and all our greatest geologists, as Lyell, Murchison, Sedgwick, &c., have unanimously, often vehemently, maintained the immutability of species" (*Origin*, p. 310). This vehement opposition had often been expressed in their negative reviews of the anonymous *Vestiges of the Natural History of Creation* (1844, see introduction), so Darwin knew well what he was up against. Darwin had already made some progress in convincing Lyell of his views, but the others who were still alive when the *Origin* was published remained skeptics for the rest of their lives. Darwin's defense and counterargument lay in the nature of the record, which he, like Lyell before him, describes here as "a history of the world imperfectly kept, and written in a changing dialect; of this history we possess the last volume alone, relating only to two or three countries. Of this volume, only here and there a short chapter has been preserved; and of each page, only here and there a few lines. Each word of the slowly-changing language, in which the history is supposed to be written, being more or less different in the interrupted succession of chapters, may represent the apparently abruptly changed forms of life, entombed in our consecutive, but widely separated, formations. On this view, the difficulties above discussed are greatly diminished, or even disappear" (*Origin*, p. 310–11).

Darwin's argument in this chapter "on the imperfection of the geological record" contains some predictions that we have seen tested in the 150 years since the publication of the *Origin*. First, Darwin predicts that the history of life derived from the fossil record will become more and more continuous as our knowledge of the record improves. This prediction has been fulfilled (chap. 24). Darwin also predicts that some long history prior to the first fossil-bearing strata must have occurred, but he is at a loss to explain its absence from the record. While this prediction has also been fulfilled, there have been some unexpected twists to the story (chap. 19).

References

Browne, J. 2002. *Charles Darwin: The power of place*. Princeton, NJ: Princeton University Press. (Pp. 314–15, William Thomson's estimates of the age of the earth.)

Eicher, D. L. 1976. *Geologic time*, 2nd ed. Englewood Cliffs, NJ: Prentice-Hall. (Chaps. 1 and 6, estimates of the age of the earth.)

Wayne, R. K., and E. A. Ostrander. 2007. Lessons learned from the dog genome. *Trends in Genetics* 23:557–67. (Details on the origin of domestic dogs.)

Chapter 18

Geology III: On the Geological Succession of Organic Beings

Darwin's first chapter on the geological record is his preemptive strike against the way the record can be used to criticize his theory. In his next (tenth) chapter, Darwin takes the offensive. He argues that if we accept the record while keeping in mind that it is extremely imperfect, with more of life's history obscured by its gaps than revealed by known fossils, then we can find strong evidence there in support of his theory.

Consider first what Darwin's theory predicts about how the record should appear if it is indeed a history of the evolution of life. By Darwin's theory, there is no fixed law of replacement, whereby species should change by a given amount per unit of time, and there is no ironclad requirement that a change in any one species should necessarily be simultaneous with change in others. His theory instead predicts sporadic and unpredictable change among individual species. Whether or not a species evolves is a function of the availability of heritable variation for selection to act upon. Darwin proposed in previous chapters that variation is only sporadically available and that it is random with regard to whether or not it benefits the organism. Natural selection capitalizes on whatever variation is beneficial to the species, but how it evolves will also be influenced by whether its population is large or small, how mobile it is, perhaps by gradual changes in its physical environment, but most significantly by its interactions with other species.

Darwin's theory thus predicts that only some species will diversify into new species in any interval of time. Because the evolution of one species can alter its interactions with others in the community, we should expect that adaptation and diversification in one lineage will eventually cause evolution in those with which it interacts. We should also expect that species in the community "which do not change will become extinct" (*Origin*, p. 315) as those around it evolve and become more competitive. In this way, the entire community will gradually change over time.

These predictions are fulfilled, Darwin says, by successive fossil-bearing strata that represent close intervals of time. An individual stratum generally lacks a few species and has some new ones in comparison to an older stratum lying beneath it. If we consider strata separated by progressively longer intervals of time, the proportion of species that they have in common progressively declines (see fig. 22 in chap. 16).

If we now consider the history of individual species, we can predict that some will change little or not at all over long intervals of time, while others will quickly disappear from the record and be replaced by new species. Once a species disappears, it is almost never seen again. (The few exceptions that were known in Darwin's time can easily be accounted for by a change in geographic range, with the species disappearing from the specific site represented in the fossil strata, then later reinvading that region.) It seems that terrestrial animals evolved more quickly, as did animals that were more complex in organization, such as the vertebrates. Darwin speculates that this may be because they interacted with a greater diversity of other species, so they were more likely to encounter some change in biotic interactions that could induce variation and cause subsequent evolution.

The general features of the fossil record are thus what Darwin's theory predicts, particularly when a series of fossil-bearing strata are available that represent some long and nearly continuous interval of time. When patterns arise that seem inconsistent with his predictions, most often because there is a sudden change in the fauna in adjacent strata, they can generally be attributed to gaps in the fossil record caused by a break in deposition or erosion. "Each formation, on this view, does not mark a new and complete act of creation, but only an occasional scene, taken almost at hazard, in a slowly changing drama" (*Origin*, p. 315).

One of the mysteries of the fossil record that had been discussed before the *Origin* was that when a species was lost, it almost never reappeared. If species arise via special creation, then the same species should be able to reappear again and again throughout history to occupy the same ecological niche, but this almost never happened. Darwin's theory explains why species that disappear never reappear. A species may go extinct and be replaced by some new species that occupies the same ecological niche, but the new species does so with the burden of history. It will have been derived from some other branch on the tree of life, which means that it will always carry evidence of its ancestry even as it adapts to a new environment, which in turn means that it will never be the same as the extinct species that it replaced.

The rules that apply to individual species also apply to larger groups of related species, such as genera, families, and orders. Once they disappear, they do not reappear. If the fossil record was complete, rather than imperfect, then the history of these groups would be seen as including first an expansion in diversity, then decline, and all this could be shown to be gradual because there tends to be a gradual increase in the number of species in a group, then a gradual disappearance.

On Extinction

The causes of the disappearance of organisms from the fossil record remained a mystery to Darwin's contemporaries. As such, it provided an opportunity for Darwin to demonstrate the explanatory power of his theory. He has already offered a hypothesis for extinction earlier in the *Origin*, but he expands on it here in the context of the paleontological record.

Cuvier saw extinction as the product of a "revolution," or change of unknown cause that drove many species extinct at once. Darwin's contemporaries accepted that extinctions had occurred and that they were not necessarily a product of "revolutions," but still sought some special explanation for them. One proposal was that species, like individuals, had a finite life span and hence had some internal clock that predestined them to disappear, in the same way that individuals grow old and die. Darwin argues that we can understand extinction by looking at living organisms and recognizing that rarity is a precursor to extinction: "to admit that species generally become rare before they become extinct—to feel no surprise at the rarity of a species, and yet to marvel greatly when it ceases to exist, is much the same as to admit that sickness in the individual is the forerunner of death—to feel no surprise at sickness, but when the sick man dies, to wonder and suspect that he died by some unknown deed of violence" (p. 320).

By Darwin's theory, extinction is a by-product of evolution. As one species evolves and diversifies, it succeeds at the expense of others. Darwin sees the fossil record as fully consistent with this process. The details of the "tertiary" or recent fossil record show "that species and groups of species gradually disappeared, one after another, first from one spot, then another, and finally from the world" (pp. 317–18). There is no evidence of a fixed law that regulates the life span of species; each species or lineage retained the same individual properties in its disappearance as it did during its duration. The record shows that a lineage of species would gradually expand and multiply, then decline in diversity, range, and abundance. There are some apparent exceptions, such as the disappearance of the ammonites at the end of the "secondary period" (Mesozoic), which Darwin describes as "wonderfully sudden" (p. 318).

When we see that a living species is rare, we assume either that it has some special needs for survival that only some habitats can accommodate, or that there is something that limits its ability to multiply. All species, even whales and elephants, produce a sufficient number of offspring to sustain exponential population growth, so their not doing so means that something is restricting their abundance, even if we can not identify what it is. To Darwin, these "unperceived injurious agencies" (p. 319), be they competition, disease, predation, or something else, are a consequence of interactions with other species. All that is required for the checks on the population growth of a rare species to be converted into agents of extinction is for them to become gradually more severe. Such an increase in severity may be caused by any change in the many species with which a rare species interacts, which means that the actual cause may be difficult or impossible to identify.

Darwin's argument is thus an application of Lyellian uniformitarianism. Lyell argued that phenomena we see in our day-to-day lives, such as wind, rain, and tides, have shaped the face of the earth. Likewise, Darwin argues, all the everyday phenomena that contribute to population regulation also cause extinction.

Darwin's argument for the gradual extinction and replacement of one species by another applies as well to higher taxa. A new genus may replace an older genus from the same family, but a replacement may also come from a very different and unrelated group, sometimes with a few relicts hanging on. Darwin cites the ganoid fishes as an example of such relicts. These fish were a dominant, widely distributed group during the Paleozoic, but dwindled away and were later displaced by other bony fishes. Some ganoids hang on today, including the gar and bowfin that are found in freshwater rivers and estuaries in southeastern North America.

One possible wrinkle in Darwin's argument is the sudden disappearance of some large groups of species, such as the trilobites during the Paleozoic era or the ammonites at the end of the Mesozoic era. Both of these were abundant and diverse, then seem to have disappeared in an instant. Darwin suggests that each of these disappearances may be an artifact of a missing interval in the fossil record or may have been caused by some other group of organisms that replaced the previous group as it diversified and expanded its geographic range. Since ecologically equivalent organisms might be very different in appearance (e.g., both ants and birds feed on seeds), the cause of the disappearance may not be easy to discern from the record. (We now have a very different explanation for some of these sudden disappearances, which I discuss in chapter 19.)

Darwin concludes that extinction is not a mystery, even if we do not know the cause of a particular extinction: "We need not marvel at extinction; if we must marvel, let it be at our presumption in imagining for a moment that we understand the many complex contingencies, on which the existence of each species depends. If we forget for an instant, that each species tends to increase inordinately, and that some check is always in action, yet seldom perceived by us, the whole economy of nature will be utterly obscured. Whenever we can precisely say why this species is more abundant in individuals than that; why this species and not another can be naturalized in a given country; then, and not till then, we may justly feel surprise why we cannot account for the extinction of this particular species or group of species" (p. 322).

On the Forms of Life Changing Almost Simultaneously Throughout the World

The fossil record for marine organisms often reveals a seemingly simultaneous succession of life-forms throughout the world. For example, the European Chalk Formation, which was deposited in the Mesozoic era and is represented by the cliffs of Dover, can also be recognized "in North

America, in equatorial South America, in Tierra del Fuego, at the Cape of Good Hope, and in the peninsula of India" (*Origin*, pp. 322–23), because similar groups of organisms are found in all these places—sometimes in formations where chalk is not present. The fossilized species may not be the same in these different regions, but the genera and families are. Furthermore, the organisms that are found immediately above or below the chalk in Europe are different from those in the Chalk Formation, and this difference is repeated in the same sequence in these other places. It is this regularity of faunal succession that makes it possible to define the geologic timescale.

This worldwide, simultaneous succession of life-forms demands a special explanation, since it deviates from the more gradual pattern of replacement that Darwin's theory predicts. Darwin first qualifies what "simultaneous" actually means in the fossil record. Strata that seem contemporaneous because they contain similar organisms may actually be separated by millions of years. He bases his argument on different types of variation that we see among organisms. First, geographic variation: the living marine fauna we find today in Europe, South America, and Australia can belong to the same genera or families, but the species are often different. Second, variation over time: the Pleistocene fossils we find in Europe (up to 1.8 million years old) and the marine organisms alive in Europe today can also belong to the same genera or families, but the species are often different. So the variation we see among organisms living anywhere in the world at one point in time (today) is comparable to the variation we would see if we stayed in one place and watched those organisms evolve for two million years. This means that when an expert aligns two fossil strata from different parts of the world because they contain the same genera and families, we should expect a similar level of imprecision in defining how close to one another they are in age. Correlated strata from different parts of the world can represent deposits separated by millions of years.

It was generally accepted in Darwin's day that global changes in the marine fauna could not be accounted for by any physical aspect of the environment, such as "currents, climate or other physical conditions" (*Origin*, p. 325), because the replacements appeared worldwide and spanned a diversity of climates. Darwin's predecessors argued that there must be some larger, general law to account for such global changes. Darwin agrees and proposes evolution by natural selection as that general law.

Darwin's explanation for the global succession of species is his "principle of divergence" and extinction writ large. In chapter 9, I described how Darwin found that living genera of plants and insects that contained many species also tended to have species with many varieties, while genera with few species also tended to have species with few or no varieties. Here, in this chapter, Darwin infers that this pattern he found among living organisms is the single, last scene from a historical drama millions of years old. The large genera descended from some ancestral species that had gained some advantage in the "struggle for existence" and as a consequence diversified into many new species. Some of the descendant species retained this advantage and are now, in turn, diversifying into many varieties, which are incipient species. When a single species diversifies into many, it will become recognized first as a "species group" or "subgenus," then as a distinct genus. As multiple species within a genus diversify into new genera, the original genus will become recognized as a family, then as an order if this accelerated diversification continues. As the group multiplies and expands its range of habitats, its geographic range will expand as well. Such a phenomenal level of success will be rare, but when it occurs, it will be recorded in the fossil record as a wave of species replacements. This is what Darwin says we are seeing in the waves of global succession of marine fauna. As a successful lineage diversified and expanded its range and abundance, it did so at the expense of others, which were driven to extinction: "The forms which are beaten and which yield their places to the new and victorious forms, will generally be allied in groups, from inheriting some inferiority in common; and therefore as new and improved groups spread throughout the world, old groups will disappear from the world" (*Origin*, p. 327).

Some relics of these displaced forms of life may live on in protected pockets, like the freshwater ganoid fishes of North America; or the *Coelocanth*, found in separate deepwater populations off of the east coast of Africa and Indonesia; or the lungfish, found in freshwater habitats in South America, Africa, and Australia.

Darwin's argument is thus that the global successions of species that appear as sudden waves of replacement in the fossil record actually span millions of years. These replacements are consistent with the patterns of diversity that we see in living organisms, since we find that some lineages carry the signature of recent and continuing diversification, while others appear to be declining. The patterns and processes that we can see in action today are thus the same as those that have occurred throughout the history of life and have shaped the fossil record.

On the Affinities of Extinct Species
to Each Other, and to Living Forms

Georges Cuvier described many Pleistocene fossils, such as mammoths, mastodons, or rhinoceroses, that lived during what we now recognize as the recent ice ages. All were clearly different from species alive today, but were also clearly related to living species. Cuvier found that older strata contained fossils of extinct organisms that were quite different from anything alive today but, by applying his skills in comparative anatomy, he could often show that they were related to and could be classified with living species. One odd fossil was *Paleotherium* (ancient beast), which was from the Eocene epoch (55–33 million years before the present). He found that the complex, articulating facets on the bones of the ankle and foot bore critical similarities to what he had classified as two different orders: the "pachyderms" (thick-skinned animals such as elephants) and ruminants, including cattle. The animal he reconstructed looked like a tapir.

Cuvier's findings revealed two general trends in the fossil record. First, the fossils found in progressively older strata are progressively more different from those alive today. Second, although it may be possible to classify the fossil organisms in a living genus or family or order, based on details of their anatomy, sometimes classification is problematic because they have properties that seem intermediate between what are now two distantly related lineages, as was the case with *Paleotherium*. Darwin argues that his theory of evolution can explain both of these trends. To see how, he refers us again to his "tree of life" (see fig. 12 [p. 175] in chap. 10).

First, think of the figure as the fossil record; each horizontal division represents a different geological stratum, and the whole sequence, from the bottom to top, is a chronological series of strata. Think of all the living descendants of species A as comprising an order that includes three families. If a^{14}, q^{14}, and p^{14} each represent a different genus, then they would be classified together in one family; b^{14} and f^{14} are a second family, and o^{14}, e^{14}, and m^{14} are a third family. Darwin originally used this figure to illustrate how the principle of divergence would cause each surviving lineage to diverge from other survivors and drive closely related species extinct. In this fashion, the surviving lineages would become more and more different from one another over time. Now we are instead comparing living species with those found in the fossil record. The living species are the twigs at the top of the tree of life, and the fossils are the underlying branches. A different consequence of the principle of divergence is that, as we go back in time, fossil species will tend to become progressively more different from those alive today. For example, a^{10} might be classifiable as an extinct species in genus a^{14}, but a^5 or a^6 may share fewer, and only more general, similarities with living species that define them as a member of the same family or order.

Imagine finding a fossil of species B in some ancient stratum. It was in the original genus but went extinct and left no living descendants. It was closely related to and probably very similar to species A. It may well have some characters that place it in the order of organisms that descended from species A, but otherwise may possess some anatomical features found only in the family (a^{14}, q^{14}, p^{14}), others found only in the family (b^{14}, f^{14}), and yet others that are not characteristic of either of these families. In this regard, it may easily be seen as in some way intermediate to or forming a bridge between living families. This is true even though species B went extinct without leaving any living descendants. Darwin's figure shows that the large majority of species that have existed over time represent such dead ends.

Darwin's hypothetical species B is like Cuvier's *Paleotherium*, which had general features of its anatomy that placed it between pachyderms and ruminants. By the time of the *Origin*, Richard Owen had described so many fossils that seemed allied to the pachyderm and ruminant orders but did not fit into either of them, plus others that blurred the distinction between these orders, that he discarded Cuvier's classifications and created new orders to replace them. *Paleotherium* became part of a new order, the Perisodactlya, which includes the horses, tapirs, and rhinoceroses.

People often use the term "missing link" to describe fossils that seem to be the ancestor of species alive today or seem to bridge two living groups of organisms, then imagine that the fossil in hand is indeed that common ancestor. It is important to realize that this will almost never be true. Any such "link" is much more likely to be just one representative of what was a diversity of species related to a common ancestor of those living today. So, though we now know that *Paleotherium* is from the horse lineage, we still cannot assume that horses are directly descended from it. It is just one representative of what was a diverse group of related species alive at the time, most of which went extinct without leaving any descendants.

What we actually see when we compare living organisms with the fossil record or successive layers in the record can be far more complicated than what might be imagined from looking at Darwin's figure. There will be gaps in the record caused by intervals of nondeposition and gaps in the representation of individual lineages, possibly because of a shifting geographic range that sometimes did not include the area where fossils were forming (chap. 17). Lineages evolve at different rates and persist for different amounts of time. For all these reasons, only fragments of the tree will be present, and the amount of change seen in different lineages over time will vary. "All that we have a right to expect, is that those groups, which have within known geological periods undergone much modification, should in older formations make some slight approach to each other; so that the older members should differ less from each other in some of their characters than do the existing members of the same groups; and this by the concurrent evidence of our best paleontologists seems frequently to be the case" (*Origin*, p. 333).

Darwin thus argues that these two prevalent features of the fossil record—the tendency for organisms found in progressively older strata to be progressively more different from those alive today, and their tendency to sometimes have traits intermediate between existing taxonomic categories—are both easily explained by his theory of evolution. The combination of the slow transmutation of species and the imperfections of the record means that adjacent strata may contain similar species, but the fossils in ancient strata may bear only some general resemblance to younger fossils or animals alive today.

On the State of Development of Ancient Forms

Lamarck thought that the more ancient an organism was, the less perfect it was. Each successive age contained new and improved designs for life, such that all organisms could be arranged on a single scale of being. Agassiz thought that fossil organisms tended to look like the early developmental stages of the species that were descended from them. Darwin followed individuals like Pictet and Huxley in not being convinced that there is any evidence for general progression of any kind in the fossil record. Very ancient forms like trilobites were as complex in their adaptations as modern species, for example. Darwin did think, however—and here argues—that his theory predicts one type of inexorable progress. Because evolution is driven by the struggle for existence, each successive community of organisms must have beaten out its predecessors and driven them to extinction. "If under a nearly similar climate, the Eocene inhabitants of one quarter of the world were put into competition with the existing inhabitants of the same or some other quarter, the Eocene fauna or flora would certainly be beaten and exterminated; as would a secondary [Mesozoic] fauna by an Eocene and a Paleozoic fauna by a secondary fauna" (*Origin*, p. 337).

Darwin has already argued (chap. 4 above) that we can see this same process of competitive displacement in action today. Organisms from Europe, which represent the products of a larger landmass—where population sizes will be larger, new varieties and species will be generated more quickly, and the ensuing struggle for existence will be more intense—were known to be multiplying prodigiously on the smaller, isolated landmass of New Zealand. They would predictably drive some of New Zealand's endemic fauna to extinction. However, if we had fossils of both faunas, it would not be possible to conclude simply by comparing them that one was superior to the other. In the same fashion, Darwin does not think we can see evidence in the fossil record of the unyielding progress through natural selection that his theory describes.

On the Succession of the Same Types within the Same Areas, during the Later Tertiary [Cenozoic] Periods

Here Darwin addresses patterns seen in recent fossils of terrestrial animals. Whereas marine faunas display a worldwide replacement of one fauna by another, the terrestrial faunas often display regional differences. In the recent fossils from South America, we see a succession of mammals in the order Edentata. Darwin found giant, extinct species of this order, as later identified and described by Richard Owen, while he was on the voyage of the *Beagle* (introduction). His finds included a specimen of the slothlike *Megatherium*, originally described by Cuvier, and a glyptodont, which was a hippopotamus-size animal that was armored like an armadillo. The living edentates, which include armadillos, sloths, and anteaters, are found in the same place. Likewise, the fossil fauna of Australia is dominated by marsupials closely related to the living marsupials still found there. The general result is that the recent terrestrial fossils from any region tend to be closely related to the living species found there.

Darwin is so impressed with this pattern that he names it the "law of the succession of types" and argues that it is explained by his theory. All these continents and islands are geographically isolated, so their terrestrial inhabitants have evolved independently of those on other landmasses. The fossil record retains the imprint of this historical relationship: "On the theory of descent with modification, the great law of the long enduring, but not immutable, succession of the same types in the same areas, is at once explained; for the inhabitants of each quarter of the world will obviously tend to leave in that quarter, during the next succeeding period of time, closely allied though in some degree modified descendants" (*Origin*, p. 340). The fauna of marine environments do not show such dramatic regional differences, because marine habitat is more continuously distributed on the face of the earth, so there can be greater migration between and less differentiation among organisms in different regions.

These regional affinities do not necessarily hold in the more ancient strata, because the ranges of some types of organisms have changed over time. For example, in earlier times a diversity of marsupials could be found in the Northern Hemisphere. Cuvier discovered the first marsupial fossils in Europe. He recognized his first specimen as being from this group based on details of its anatomy that were revealed in early phases of excavating it out of its rock matrix. He showed his aptitude as an impresario and demonstrated the predictive power of comparative anatomy by arranging for an audience of experts to witness his completion of the preparation. He predicted that the completed preparation of the pelvic girdle would reveal bones that distinguished the marsupials from other mammals, then completed the preparation of the fossil in front of witnesses to reveal the predicted bones.

If we combine Darwin's law of succession of types with his argument that the struggle for existence will cause an inexorable increase in competitiveness, then there seems to be a problem with what we see in these regionally distinct terrestrial faunas. The recent fossil record shows again and again that the giants of the past are represented by miniaturized relatives of today. Can we really accept that the diminutive sloths, anteaters, and armadillos alive in South America today outcompeted their giant ancestors and drove them to extinction? Darwin refers again to his figure to explain why this is an inappropriate question. His figure shows that, of all the species that are alive at any one point in time, most will go extinct without leaving any descendants. Of the initial eleven species at the bottom of his figure, only three have surviving descendants at the top of the figure. The known fossil fauna of South America includes edentates of various sizes. It is the smaller lineages from the past that are the likely ancestors of the small species found living there today. It is interesting that the large ones have disappeared, but this does not imply that they did so as a result of competition with the smaller species from a different lineage. Why we often see a selective loss of larger species remains an open question.

Darwin concludes by arguing that his theory of evolution by natural selection explains all that we see in the fossil record. Accepting his argument means accepting that the fossil record is highly fragmented and represents only a tiny fraction of time or of the species that once existed. Our modern assessment of the record, which includes estimates of the absolute ages of many strata, supports his interpretation.

If we accept the premise of imperfection, then Darwin is able to offer answers to all the questions raised by prior geologists and paleontologists. These include explaining the origin of species, extinction, why species and even higher taxa can suddenly appear in the record fully distinct, and why we see a particular succession of life-forms over time. He also explains a diversity of other mysteries, like the existence of living fossils. Darwin brings all these phenomena under a single, explanatory framework when interpreting the fossil record as a history of life as governed by the process of evolution by natural selection: "all of the chief laws of paleontology plainly proclaim . . . that species have been produced by ordinary generation: old forms have been supplanted by new and improved forms of life, produced by the laws of variation still acting round us, and preserved by Natural Selection" (*Origin*, p. 345).

References

Long, J. A. 1995. *The rise of fishes*. Baltimore: Johns Hopkins University Press. (Pp. 138–48, details on the ganoid fishes.)

Rudwick, M.J.S. 1976. *The meaning of fossils*. New York: Neale Watson Academic Publications. (Pp. 207–14, details on Owen.)

Rudwick, M.J.S. 2005. *Bursting the limits of time: The reconstruction of geohistory in the Age of Revolution*. Chicago: University of Chicago Press. (Pp. 399–409, details on Cuvier, including the public preparation of a marsupial fossil.)

Chapter 19

Geology IV: Evolution Today

The Expansion of the Geologic Record and a Brief History of Life

When the first edition of the *Origin* was written, the Silurian was the earliest period of the Paleozoic era. Beneath it lay what were thought to be azoic (lifeless) rocks. By the time of the sixth edition (1872), sufficient diversity had been found within the Silurian to subdivide it into an older Cambrian period, followed by the younger Silurian period. Shortly thereafter a third period, the Ordovician, was inserted between the Cambrian and Silurian periods. Each of these periods was defined by a characteristic fauna. This progressive subdivision of the former Silurian period was a product of the growing knowledge of the geological record.

Today, the Cambrian, Silurian, and Devonian periods are part of the Paleozoic era, which is the earliest of the three eras (later ones being the Mesozoic and Cenozoic eras) of what we now call the Phanerozoic eon. "Phanerozoic" means "visible life." The Phanerozoic eon is preceded by the Archean and Proterozoic eons. The Archean eon begins with the origin of the earth, approximately 4,600 million years (myr) ago, and ends at 2,500 myr ago. The Proterozoic spans the time between 2,500 and 540 myr ago.

To Darwin and his contemporaries, the "beginning" occurred when the fossils of animal bodies first appeared around 510–520 myr ago. This means that the fossil record known to Darwin and the geological timescale presented in figure 21 (chap. 16) represents only around 12% of the history of the earth. Although some of Darwin's contemporaries believed this to be the beginning of life, it was instead just the beginning of easily recognized life. Today, the starting point of the Phanerozoic (543 myr ago) is defined by the appearance of multicellular organisms that left traces of their meanderings on the surface of soft sediments.

Evidence for life on earth is almost as old as the oldest rocks that could possibly contain such evidence. The earliest evidence comes in the form of chemical traces in rocks that are 3,700–3,800 myr old. The earliest fossils of real organisms are still debated, but there is general agreement that there are fossils in the 3,500 myr-old Warrawoona Formation in Western Australia and the 3,400 myr-old Fig Tree Formation in southern Africa. Both "fossils" are stromatolites, or dome-shaped rock formations created from layers of bacteria and sediment. Stromatolites can be found alive and well in a few places today. They give us some sense of what life was like 3,500 myr ago. Living stromatolites have a community of bacteria living in a thin film on the upper surface. The stromatolite is built up one thin layer at a time when sediments are washed over the upper surface of bacterial slime, then the bacteria overgrow the sediments. The chemistry of life in the bacterial mat adds precipitated minerals that solidify the sediments into a rocky pedestal. The stromatolites that we see today are the relics of a life-form that was once widely distributed.

The transition from the Archean to Proterozoic eons is marked by geological changes that include a cooling and thickening of the earth's crust and the formation of extensive, shallow seas where stromatolites proliferated and other forms of bacterial life thrived. Some of these were photosynthetic cyanobacteria. Their oxygen production was at first absorbed by chemical reactions, but an "oxygen revolution," or the accumulation of free oxygen in the atmosphere, began around 2,200–2,300 myr ago. The presence of free oxygen represented an important transition in the history of life. To us, oxygen is essential for life, but it is actually a toxin because it oxidizes, or degrades, macromolecules that are essential for life. The advent of oxygen in the atmosphere and oceans selected for organisms that had the capacity to defend themselves against it and drove those that could not defend themselves into recesses where oxygen remained scarce. It also created the conditions for the evolution of oxidative metabolism, which is a more efficient way of capturing chemical energy from food and set the stage for the origin of eukaryotic cells, which is another great transition in the history of life.

Animals, plants, fungi, and protozoans are all made of eukaryotic cells. All of them inherited this type of cell from a single common ancestor. A hallmark of the living eukaryotes is that their cells have a membrane-bound nucleus that contains DNA in the form of chromosomes. Prokaryotes (bacteria) instead have a circular strand of DNA that is not enclosed by a membrane. Eukaryotic cells, with only a few exceptions, also contain mitochondria. These are membrane-bound organelles that have their own DNA and have oxidative metabolism. The mitochondria were originally an independent bacterium that was incorporated into another bacterium. The eukaryotic cell is thus a partnership, or endosymbiosis, between what had been at least two different types of free-living bacteria that joined forces to create a new life-form. The eukaryotic cell represents a form of evolution called "reticulate evolution." Reticulate evolution is defined as the exchange of genetic material between different species to create a descendant species that is a blend of its ancestors. Reticulate evolution is also the source of the photosynthetic chloroplast found in plant cells; these cells are the product of the fusion of a mitochondria-bearing eukaryotic cell and a cyanobacterium. This kind of evolution through the joining together of separate life-forms is one of the unexpected turns in the history of life that Darwin could not have anticipated. Endosymbiosis is a rare phenomenon, but has clearly played a major role in defining life as we know it.

The earliest fossil evidence of eukaryotic cells date to approximately 1,800 myr ago. These are referred to collectively as "acritarchs" and can be recognized as eukaryotes by their large size and the structural complexity of their surface. Eukaryotic cells are generally much larger than prokaryotic cells. There is chemical evidence that suggests that eukaryotes may have originated a few hundred million years earlier.

By 575 million years ago, following an interval of more than 120 million years of severe ice ages, we see the emergence of multicellular life. Darwin envisioned a gradual and prolonged ramping up of the complexity of life. But, in a second unexpected twist, what we actually see is that although life is very old, complex, multicellular life is relatively young. Our earliest record of multicellular life is only tens of millions of years older than the earliest fossils known to Darwin.

The Ediacaran fauna, first found in Ediacara Gorge in the Flinders Ranges of Australia, is the oldest assemblage of multicellular fossils. It consists of soft-bodied organisms preserved under conditions that retained details of their anatomy. It at first seemed that there was a gap of millions of years between the disappearance of this fauna and the appearance of younger multicellular organisms. The Ediacaran fauna was so strange that some paleontologists argued it represented an independent, failed experiment in the evolution of multicellular life, and that multicellular life started anew tens of millions of years later. This time gap has now been filled with similar fossils from dozens of deposits that range from 575 to 543 myr ago, with increasing diversity in the latter half of this interval. A few Ediacaran fossils have even been found in the Cambrian. Our current interpretation is that these fossils include two phyla that are still with us today: Porifera (sponges) and Cnidaria (including jellyfish). Other phyla may also have been there, but some organisms in the Ediacara may be unrelated to anything alive today.

The beginning of the Cambrian is defined by trace fossils identified as *Treptichnus pedum*. These "fossils" are actually trackways made by an unseen, soft-bodied organism (fig. 23). This animal had a new and efficient way

Figure 23
Treptichnus pedum
The first appearance of this fossil is used to define the beginning of the Cambrian period (Paleozoic era; see figure 21). *Treptichnus pedum* was chosen for this status because its fossils are widespread and relatively abundant, so they are useful for correlating the ages of strata from different parts of the earth. This is a "trace fossil," which means that you are looking at burrows made by the animal, rather than the animal itself. The nature of the burrows suggests an animal that was motile and capable of complex probing and feeding activities. Each "stitch" of the fossil represents a probing of the sediment, followed by the animal withdrawing, then probing in a different direction. These tunnels later filled with sediment that was different in composition from the surroundings; then both the filled tunnels and the surrounding sediments lithified, or became solid rock.

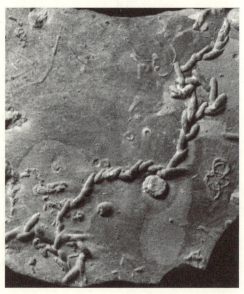

of mining the sediment for food by extending the anterior tip of its body forward, then withdrawing its body and probing in a different direction. These trace fossils are joined some 10 million years later by an assemblage of small, shelly fossils. These are often sponges that had mineral skeletons, but also early cnidarians, mollusks, and some organisms of unknown affinity. Then, beginning around 520 myr ago, we see the abrupt appearance of the large, complex life-forms familiar to Darwin. The dominant fossils are trilobites, from the phylum Arthropoda, and echinoderms and brachiopods, each of which is a phylum of its own.

Many of these creatures had mineralized tissues that fossilized easily, which is the source of the name "Phanerozoic" ("visible life") because fossilized life suddenly becomes much more visible in rocks dating to the start of this eon. Darwin's contemporaries described older rocks as lifeless because the life they contained was either microscopic or did not fulfill their search image, which was for shells and skeletons.

One type of hard tissue that evolved was a skeleton, either external or internal, which was attached to muscles. Muscles plus skeletons enabled animals to function like machines, with jointed legs that facilitated more efficient movement, whether scurrying along solid substrates, swimming through the water column, or burrowing in sediments. Improved mobility meant being able to expand into new territories. Having hard tissues also meant being able to have jaws that grasped and processed food, allowing these animals to expand their diets. This interval includes the earliest fossils of what are clearly specialized predators. If there were predators, then there was selection in favor of the evolution of defense against them. The ensuing arms race helps to explain why life suddenly becomes more visible at this point in the fossil record; this sudden appearance is called the "Cambrian Event."

The diversity of life we see around us today includes many soft-bodied organisms that will rarely make their way into the fossil record. Even animals with hard tissues that do fossilize consist mostly of soft tissues, like skin and internal organs. These soft parts play an important role in defining what the animal looks like and how it works, but leave little or no trace in fossils. The fossil record usually contains only fragments of hard tissues from a small subset of all the organisms that were alive at the time and rarely includes soft-bodied animals or the soft parts of animals with mineralized tissues. The rare exceptions include what we refer to as "lagerstätten," or mother lodes where whole communities have been fossilized in a fashion that preserves images of soft tissues.

The Burgess Shale fauna is a lagerstätten that yields a photograph of the type of community that existed during the Cambrian Event. It was first described from the Burgess Shale in the Canadian Rockies, but is now known from a few dozen sites around the world. The original Burgess Shale fossils were preserved remarkably intact and retained impressions of soft body parts that normally do not fossilize. Fossilization began with the sudden immersion of entire communities in anaerobic, fine-grained sediments that inhibited bacterial decay and retained impressions of the entire, intact bodies. Because of the remarkable detail, it is possible to reconstruct what the preserved organisms were like in life. This fauna includes eight phyla that are well represented in the fossil record that follows, including Porfera, Cnidaria, Brachiopoda, Mollusca, Arthropoda, Echinodermata, Hemichordata, and Cnidaria. The more than twenty other phyla alive today are soft bodied, so they have left little trace in the fossil record, but even some of these are seen in the Burgess Shale.

This brief history leaves us with the same hard question Darwin faced: How can we explain the apparently sudden appearance of complex life? One possible answer is that complex multicellular life had a long but invisible evolutionary history that predates its appearance in the fossil record. This proposal is based on a feature shared by the larvae of all the more simple multicellular organisms. The larvae, which are tiny free-living plankton that feed in the open water, may consist of just a few thousand cells. Each larva has a small group of cells that multiply and diversify as they metamorphose into later life stages. It is these late-developing cells that become the adult structures that define the different phyla. If the larvae of the living phyla are like the adults of their ancestors, then we do not see them in the fossil record because they are very unlikely ever to be preserved. By this hypothesis, the "Cambrian Event" was the origin of larger, more readily preserved body plans rather than the origin of the phyla themselves. We do see what might be the eggs and embryos of animal phyla preserved in phosphatic nodules that are as old as 600 myr.

The additional hypotheses that have been proposed all share the property of accepting the suddenness of the Cambrian Event but offering some biological explanation for it. One hypothesis is that the event was driven by the evolution of predation. This idea was inspired by a combination of the anatomy of the Cambrian fauna and experimental work in modern ecology. Robert Paine, an ecologist who studied the intertidal zone, found that removing predators from an ecosystem could cause a crash in biodiversity. He removed starfish, which are predators on mussels, from experimental plots. In the absence of starfish, the mussels proliferated and coated the intertidal rocks to the exclusion of most other organisms. When starfish were present, they fed on the mussels and opened up space for other organisms. Because the Burgess shale fauna includes the first appearance of large-bodied predators, it has been proposed that the sudden burgeoning of multicellular life was caused by the advent of a predator-driven increase in species diversity.

A second hypothesis is that the explosion was triggered by an increase in the availability of oxygen. Being big and having mitochondria-driven aerobic metabolism requires free oxygen, so an increase in the abundance of oxygen would facilitate the evolution of increased body size. There is evidence that an increase in oxygen in the atmosphere occurred at this time.

None of these explanations overcome the quirkiness of what looks like a sudden ramping up of the diversity of multicellular organisms. The event may be understood best as a convergence of events: the evolution of genetic novelties that facilitated the origin of multicellularity and large body size; the progressively changing environment of the earth after the end of the long period of glaciation; an increased availability of oxygen; and a ramping up of the complexity of biotic interactions. For those of us who study the process of evolution in real time and have seen how rapid it can be, the few tens of millions of years over which these changes occurred seems more than adequate for Darwin's mechanism of evolution by natural selection to explain all that we see. So, while it is interesting that complex life began this way, the Cambrian Event does not run counter to our concept of evolution by natural selection or beg for some other explanation.

Extinction Today

Although Darwin proposed a new explanation for extinction, the formal study of extinction did not develop until late in the modern synthesis period. It then took another thirty years for the study of extinction to really take off.

The first protagonist in the modern study of extinction was George Gaylord Simpson, who is credited with bringing the field of paleontology into the modern synthesis era. In chapter 8 I introduced Ernst Mayr and Theodosius Dobzhansky, who were two of the founders of the "modern synthesis" of evolutionary biology. I also mentioned that New York City was the epicenter of the modern synthesis, since Dobzhansky was a professor at Colombia University and Mayr was a curator of ornithology at the American Museum of Natural History when they wrote their key books in 1937 and 1942. G. G. Simpson continued this tradition with his book *Tempo and Mode in Evolution*, published in 1945 as part of the same series that included the earlier books by Dobzhansky and Mayr. Simpson was then a curator of paleontology at the American Museum of Natural History.

Simpson applied ideas from population genetics, mostly those proposed by Sewall Wright during the 1930s, to the interpretation of the different rates of evolution we see in the fossil record. He argued that many features of the record were consistent with Darwin's proposal of evolution by natural selection, as modified by Wright's updating of the concept. He also elaborated on how Darwin's ideas could be expanded on to account for persistent patterns in the record that seemed inconsistent with evolution as a prolonged, gradual process. One such phenomenon was the sudden appearance of mammalian orders with little evidence for their origin. Darwin thought such suddenness was an artifact of imperfections in the fossil record. Simpson observed that more than eighty years of research had done little to resolve such discontinuities through the discovery of new fossils. In spite of the absence of gradual transitions in such cases, he argued that much of the fossil record was consistent with Darwin's theory, when viewed through the prism of the modern synthesis. One key to Simpson's explanation is that evolution varied in tempo, or how quickly it happened. Under special circumstances, it could happen much more quickly than Darwin imagined. The sudden appearance of mammalian orders in the fossil record could be the product of episodes of accelerated evolution.

Simpson departed from Darwin on one issue: the nature of extinctions. Darwin argued that extinction was a by-product of evolution; species that became more successful in the struggle for existence drove others to extinction. If extinction occurred this way, then we should see a complementary process; as organisms disappeared, they should be replaced by their superior descendants. Simpson found instead that extinctions often involved the disappearance of whole groups of organisms, followed by an interval of reduced species diversity. The organisms that took their place did not begin to diversify until after this interval of reduced diversity. Those that inherited the future seemed to be just some random taxon that was lucky enough to survive an extinction event rather than being a group of organisms that had caused the extinction of their predecessors.

One example of this pattern of extinction and replacement known to Simpson was the abrupt disappearance of dinosaurs on land, pterosaurs in the air, and plesiosaurs, ichthyosaurs, and mososaurs in the oceans at the end of the Mesozoic. Mammals later filled all niches vacated by these giant reptiles, but only after a long interval when there was a scarcity of large animals in all these habitats. Prior to the end of the Mesozoic, the mammals were small bodied and apparently not too abundant. It is impossible to look at the diminutive Mesozoic mammals and predict their future glory.

This pattern of delayed replacement does not support Darwin's idea of a perpetual struggle for existence. Simpson argued that, if anything, extinction seemed to be an independent event that created opportunities for the lucky survivors. In the first chapter of *Tempo and Mode*, titled "Rates of Evolution," Simpson presented analyses that became the model for later studies of extinction. His approach was similar to the one used by actuaries to evaluate birth and death rates in humans. For Simpson the "individual" was a genus. "Birth" was its first appearance in the fossil record, and "death" was its last appearance. Its presumed "age of death" was the difference between these two events.

Jack Sepkoski was a key figure in further developing the study of extinction, beginning in the 1970s. He was a practitioner of a different kind of paleontology. Rather than excavate fossils, he made the library his territory for collecting data. He mined the world literature for reported occurrences of all marine organisms at the level of family and genus, then followed Simpson's lead by compiling a record of the "birth" and "death" of each taxon. Sepkoski concentrated on marine taxa because of the abundance of rock formations that contained marine fossils and because their hard tissues meant that they were likely to be represented as fossils.

Sepkoski's more complete record of life confirmed that there were sometimes dramatic declines in diversity and then each time, after an interval of low diversity, a gradual increase marked by the origin of new genera, families, and orders. This pattern of replacement of one group of organisms by another was what the nineteenth-century paleontologists were often seeing when they used fossils to define the divisions of the geological timescale. The divisions appeared as unusually sharp "boundaries" between different rock strata, with an older group of fossils in the lower stratum and a newer group in the upper stratum.

David Raup and Sepkoski performed a formal statistical analysis of Sepkoski's data and concluded that they revealed five "mass extinctions," or five intervals in the fossil record when 65% or more of the known species disappeared. Two of these events corresponded to the boundary between the Paleozoic and Mesozoic eras (the Permeo-Triassic or P-T event) and the boundary between the Mesozoic and Cenozoic eras (the Cretaceous-Tertiary or K-T event). The other events corresponded to boundaries between periods within the Paleozoic and Mesozoic eras. It now appears that Cuvier, Phillips, and others were partly right; the history of life is indeed punctuated by periodic "revolutions" in which many organisms disappeared, seemingly at once, to be replaced later by others. Darwin's observation that the disappearance of the ammonites at the end of the Mesozoic seemed to be "wonderfully sudden" is also right. Contrary to Darwin's argument that the suddenness was an artifact of an imperfect fossil record, their disappearance really was sudden, in geological terms.

The P-T event is described as the closest that metazoans have come to extinction in the history of life. Approximately 95% of marine species disappeared. There was a prolonged cessation of coal production, which implies a severe curtailment of photosynthesis, which would have cause inevitable cascading effects through the ecosystem. The K-T event is less dramatic in terms of the proportion of marine species that disappeared, but it was a sad time for those of us who love dinosaurs, since it is marked by their disappearance, along with a diversity of other organisms.

We now know that the "big five" extinctions are all correlated with unusual events. There were major asteroid impacts and major volcanic eruptions near the time intervals corresponding to the P-T and K-T boundaries. The massive volcanic eruptions that date to the P-T boundary represent the largest cluster of eruptions known in the history of the earth. They left behind the Siberian Traps, or flood basalts, that cover 4 million square kilometers of land with 2 to 3 million cubic kilometers of basalt. The K-T boundary is associated with the Deccan Traps, which are similarly huge flood basalts in India. The other three mass-extinction events are also associated with some combination of asteroid impacts, clusters of volcanic eruptions, ice ages, and/or indications of large changes in ocean chemistry.

These "catastrophes," which may appear instantaneous in the fossil record, are likely to have been more prolonged. The massive volcanic eruptions would have occurred repeatedly over hundreds of thousand of years. The environmental changes that they or the asteroid impacts caused would have been locally immediate, but more gradual and prolonged over the whole earth. More intensive study of the fossil record around the temporal boundaries of the mass extinctions has often shown that some groups of organisms persisted for some time after the event, then slowly disappeared.

There is an element of redemption to this view of extinction. Going extinct is not a brand of inferiority if it is caused by some catastrophe rather than by competition from a superior descendant. For those of us who were impressed by the might of the dinosaurs, it was painful to see the word "dinosaur" used as a brand of failure for an outmoded idea or a design that was overdue for replacement. Now we can think of dinosaurs as capable of forever being masters of their universe and never relinquishing their territory to mammals, were it not for some intervening catastrophe.

One useful perspective for envisioning what "sudden" means in geology is to think about how the world is changing today. We are in the midst of the sixth mass extinction. One hundred million years from now, the fossil record of our time will reveal dramatic evidence of the dispersal of humans out of Africa 100,000 years ago, followed by the spread of agriculture beginning around 10,000 years ago, the advent of the industrial revolution, then the super-exponential growth of the human population. The current extinction event began during the Pleistocene with the beginning of the decline of the mammalian megafauna, including many of the animals in Cuvier's menagerie of the disappeared, such as mammoths, mastodons, and cave bears. In fact, what had been interpreted by his contemporaries as extinctions caused by the biblical flood were instead consequences of the expansion of humanity. Then there was a global decline of forest, expansion of deserts and grasslands, accumulation of industrial wastes, and an accelerating rate of extinction. From our perspective, the change may not seem sudden, nor the rate of species loss dramatic. The reason we do not sense cataclysm, even though the geological record is certain to preserve it this way, is because of the difference in the time frame of our lives versus the time frame of the geological record. To us, 100 years is a long time. In the fossil record, 100,000 or even a million years can appear as an instant.

An important feature of the current extinction event that distinguishes it from the big five is that this one is Darwinian. The current wave of extinction is caused by the way humans evolved as they adapted to Africa. Their adaptations enabled them to become superior predators and competitors and to expand their range throughout the globe. In chapter 12, I reported on studies that show that some organisms are evolving rapidly and are apparently keeping pace with human-generated environmental change. Perhaps these are the candidates who will survive this mass extinction, and will then expand and diversify after it ends.

Raup later observed that, however dramatic the five mass-extinction events may be, they account for only 4% of all extinctions during the Phanerozoic. He refined his analysis to look at how extinctions were distributed throughout the Phanerozoic and found that the "big five" are actually not in a class by themselves. They instead are the tail of a curve. Extinctions tend to occur in clusters, with the big five being the largest of these clusters. Progressively smaller and more abundant clusters of extinctions appear throughout the record.

Some of the smaller extinction events are certainly smaller versions of the big five—that is, triggered by catastrophe—but they also admit the possibility of being caused by the sorts of biotic interactions that were proposed by Darwin. For example, the fossil record seemed to show that the dinosaurs appeared suddenly toward the end of the Triassic period, after the abrupt disappearance of other groups of large land-dwelling reptiles. One recent paper reports on new finds that show there was instead a prolonged period of overlap between the first dinosaurs and the reptiles that they replaced, so it is plausible that dinosaurs outcompeted and displaced the earlier reptiles in a Darwinian fashion.

It is also true that the Darwinian mechanism of extinction, which may well have operated more or less continually, just as he envisioned, would often be invisible in the record. The Darwinian model of extinction invokes a gradual process of decline that is driven by the diversity of biotic interactions responsible for the regulation of population size. These interactions might include competition between closely related species but also might include predation, disease, or anything else that can contribute to population regulation. The resulting pattern of extinctions could include the disappearance of individual species here and there with no apparent rise of other species to replace them.

The last thirty years of research on extinctions makes it clear that extinction is a product of different causes. Darwin's mechanism cannot account for all extinctions. It is also clear that catastrophic extinctions have at times reshaped the history of life by wiping out some groups of organisms and creating vacancies later filled by others. Their disappearance seems to have been caused by changes in the physical environment that outran the ability of some organisms to adapt to them. Darwin's concept of extinctions driven by biotic interactions remains viable, but it also remains hard to see and prove. At the present time, we cannot even offer a conjecture about the relative importance of Darwin's mechanism of extinction.

References

Cowen, R. 2005. *The history of life*, 4th ed. Malden, MA: Blackwell Publishing. (Pp. 1–83, early history of life, origin of eukaryotes, extinction.)

Eicher, D. L. 1976. *Geologic time*, 2nd ed. Englewood Cliffs, NJ: Prentice-Hall. (Pp. 137–40, subdivision of Precambrian time and the age of the earth.)

Gould, S. J. 1989. *Wonderful life: The Burgess Shale and the nature of history*. New York: W. W. Norton & Co. (Pp. 53–78, details on the Burgess Shale fauna.)

Irmis, R. B., S. J. Nesbitt, K. Padian, N. D. Smith, A. H. Turner, D. Woody, and A. Downs. 2008. A Late Triassic dinosauromorph assemblage from New Mexico and the rise of dinosaurs. *Science* 317:358–61.

Raup, D. M. 1991. A kill curve for Phanerozoic marine species. *Paleobiology* 17:37–48. (Demonstration that mass extinctions are the tail of a curve.)

Raup, D. M., and J. Sepkoski. 1982. Mass extinctions in the marine fossil record. *Science* 215:1501–3. ("Discovery" of mass extinctions.)

Simpson, G. G. 1945. *Tempo and mode in evolution*. New York: Columbia University Press.

Chapter 20

Geographical Distribution

Biogeography was an emergent rather than established science at the time of the *Origin*. It was a by-product of this age of exploration and the accumulating knowledge of Earth's flora and fauna. Many foreign outings, be they military adventures such as Napoleon's expedition to Egypt, cruises with a purpose such as the mapping of the coast of South America by the *Beagle*, or voyages of discovery such as the *Endeavor* voyage to the South Pacific, included official naturalists, often doubling as the ships' surgeons, who cataloged the plants and animals of the places that they visited. Darwin's unofficial position as the naturalist on the *Beagle* was part of this tradition.

Joseph Hooker was a member of this fraternity of wandering naturalists and was Darwin's closest correspondent as he developed his ideas about biogeography. Hooker was a veteran of James Clark Ross's Antarctic expedition (1839–1844), during which he assembled botanical collections from New Zealand, Tasmania, many other Southern Hemisphere islands, and the southern tips of Africa and South America. He later mounted an expedition to the Himalayas (1847–1850), sponsored by Kew Gardens. He was the first European to assemble plant collections from places like northern India, Tibet, Sikkim, Bhutan, and Assam. Hooker identified Darwin's plant collections from the *Beagle* voyage. Darwin shared his ideas about natural selection and species transmutation with Hooker in 1844 and many times thereafter. It was Hooker, along with Charles Lyell, who arranged to have some of Darwin's papers presented alongside Wallace's paper in 1858. By the time of the *Origin*, Hooker had published extensive floras that covered the regions he traveled and was one of the world's leading experts on the geographical distribution of plants. He was also one of Darwin's earliest converts.

The products of these explorations were syntheses of the global distribution of plants and animals, which first appeared shortly before the *Origin*. Alphonse de Candolle published his *Géographie Botanique Raisonnée* in 1855. Philip Sclater published an article on the global distribution of birds in the *Biological Journal of the Linnaean Society* in 1858. This same journal, in the same year, published the joint Wallace/Darwin papers.

In the two "geographical" chapters of the *Origin*, Darwin argues that the geographic distribution of life-forms can be explained by his theory of evolution. His theory predicts that organisms always carry footprints of their history, be it in their anatomy, their distribution over time as revealed in the fossil record, or their distribution over the face of the earth. While Darwin was the first to view biogeography from an evolutionary perspective, there is poetic justice in Alfred Russel Wallace being recognized as the founder of modern biogeography. Wallace's status in biogeography should provide solace to those who feel that history cheated him of due credit for the discovery of natural selection. Darwin and Wallace gained priority in their respective fields for the same reason. The scope and depth of the *Origin of Species* so overshadows Wallace's essay of 1858 that Darwin's name would dominate our thinking as the discoverer of evolution by natural selection even if he had not published alongside Wallace's essay. Comparing Wallace's essay to the *Origin* is like comparing Beethoven's *Für Elise* with his portfolio of piano concertos. Likewise, Wallace's two-volume biogeographic classic, *The Geographical Distribution of Animals; with a Study of the Relations of Living and Extinct Faunas as Elucidating the Past Changes of the Earth's Surface* (1876) overshadows chapters 11 and 12 of the *Origin*.

It might seem strange for Darwin to devote two of the fourteen chapters of the *Origin* to biogeography; however, the distribution of plants and animals played a key role in how Darwin developed his theory. There were three features of biogeography that captured Darwin's imagination as he first contemplated the transmutation of species; all three were well articulated in his *Voyage of the* Beagle.

The first feature pertained to distributions of plants and animals, which he saw firsthand during the Beagle voyage. One insight came as he traveled over the Andes in March 1835. On March 23, after he had traversed the barren crest of a pass through the mountains, then descended the eastern escarpment, he made the following observation:

I was much struck with the marked difference between the vegetation of these eastern valleys and those on the Chilean side: yet the climate, as well as the kind of soil, is nearly the same, and the difference of longitude very trifling. The same remark holds good with the quadrupeds, and in a lesser degree with the birds and insects. I may instance the mice, of which I obtained thirteen species on the shores of the Atlantic, and five on the Pacific, and not one of them is identical. . . . This fact is in perfect accordance with the geological history of the Andes; for these mountains have existed as a great barrier since the present races of animals have appeared; . . . we ought not to expect any closer similarity between the organic beings on the opposite sides of the Andes than on the opposite shores of the ocean. (Darwin [1860] 1962, p. 328)

The next two features of biogeography that captured his attention were a consequence of his exploration of the Galapagos. One was the presence of so many species that were unique to the islands. The other was that the closest relatives of these species could be found on the nearest mainland. His impressions of the Galapagos include the following description:

It was most striking to be surrounded by new birds, new reptiles, new shells, new insects, new plants, and yet by innumerable trifling details of structure, and even by the tones of voice and plumage of the birds, to have the temperate plains of Patagonia, or rather the hot dry deserts of Northern Chile, vividly brought before my eyes. Why, on these small points of land, which within a late geological period must have been covered by the ocean, which are formed by basaltic lava, and therefore differ in geological character from the American continent, and which are placed under a peculiar climate,—why were their aboriginal inhabitants, associated, I may add, in different proportions both in kind and number from those on the continent, and therefore acting on each other in a different manner—why were they created on American types of organization? It is probable that the islands of the Cape de Verd group resemble, in all their physical conditions, far more closely the Galapagos Islands, than these latter physically resemble the coast of America, yet the aboriginal inhabitants of the two groups are totally unlike; those of the Cape de Verd Islands bearing the impress of Africa, as the inhabitants of the Galapagos Archipelago are stamped with that of America. (Darwin [1860] 1962, p. 393)

The more remarkable feature of the islands was that there were sometimes different, closely related species on the different islands. The tortoises
from many of the islands were so distinct that Darwin was told it was possible to tell their island of origin from the shape of their shells. Many plants,
insects, and birds were found on only one or a few islands, but there was
often a very close relative on a different island. Different islands sometimes
had distinct races of iguanas. Darwin did not imagine that animals might
differ among islands that were often within sight of one another, so he did
not even bother to record the island of origin of many of the finches that he
collected. Their accurate identification was based on collections made by
others on the voyage who did record the island of origin. "Reviewing the
facts here given, one is astonished at the amount of creative force, if such
an expression may be used, displayed on these small, barren, and rocky islands; and still more so, at its diverse yet analogous action on points so near
each other" (Darwin [1860] 1962, p. 398).

I now follow Darwin's original subheadings from the *Origin*.

Geographic Distribution/Single Centers of Creation

Darwin states here that he has been impressed by "three great facts" that
characterize the geographical distribution of organisms. While it might be
reasonable to think that the physical environment of a region determines
what animals and plants will be there, this has often been contrary to what
he has observed. The east and west slopes of the Andes have similar soils
and environment, but the plants and animals are different. The Galapagos
and the deserts of Chile have very different environments, but the animals
are often related. On a larger scale, South America, Africa, and Australia all
contain some regions with very similar climates, but there is no affinity between the organisms found in a desert or tropical forest or Mediterranean
shrubland of Australia and those found in the same habitats of the other
continents. Clearly, the global distribution of any particular species is not
simply a matter of its filling all the places that have a suitable environment.

The second great fact is that the distribution of organisms seems to be
most often defined by barriers, such as mountains on land or great depths
in the oceans. It was the presence of the Andes, rather than differences in
climate, that had caused the differences between plants and animals on either slope. It was the presence of the deep Pacific, not climate, that had

caused the differences between shallow-water marine faunas on either side of the ocean.

The third "great fact" is that groups of related organisms are often confined to specific regions. South America, Africa, and Australia each have characteristic faunas and floras. The closest relative of any species of plant or animal is most likely to be found nearby, but often in a different climate. The mammalian order Edentata (sloths, anteaters, armadillos) is found almost exclusively in South and Central America, while Australia is dominated by marsupials. There is thus an affinity among the animals found within each of these continents, regardless of habitat. "We see in these facts some deep organic bond, prevailing throughout space and time, over the same areas of land and water, and independent of their physical conditions. The naturalist must feel little curiosity who is not led to inquire what this bond is. This bond, on my theory, is simply inheritance" (*Origin*, p. 350).

Darwin's explanation for these great facts of biogeography is that each taxon of organisms is ultimately derived from a single population that evolved in a single place to the point of being a distinct species. It then dispersed and perhaps diversified into new species, then genera, then families, and so on. This single site of origin, followed by a pattern of dispersal and differentiation, led to regions of the world filling up with groups of related organisms, each adapted to its local environment. Dispersal was restricted by whatever natural barriers the organisms encountered, be they oceans, mountain ranges, deserts, or rivers.

If each species did, in fact, originate in one place and then migrated outward, the predictions of how their descendants would be distributed, according to Darwin's theory of evolution, exactly match what these "three great facts" reveal: species will diversify into genera and perhaps even families or higher levels in the taxonomic hierarchy, but may be confined to one geographic region. Their distribution will be a function of how well they can disperse and the barriers to dispersal they encounter along the way. If species were instead the product of some form of special creation, then there would be no reason to expect such patterns, since the creation of each species would have been an independent event. Under special creation, it would be reasonable for a species to have been created in more than one place and to occupy a given type of environment wherever that environment was found, so that its distribution could be independent of barriers. There would also be no reason to expect the distribution of a given species to be related to that of others with which it was classified. Linnaeus and those who followed used anatomy, rather than biogeography, to classify organisms.

While Darwin's evolutionary interpretation of biogeography can account for many of the distributions that we see, there are also many exceptions in the form of species, genera, and higher taxa whose distributions are broken up into widely separated regions. For example, the same species of plant or animal may be found only on mountains isolated from one another by a "sea" of unsuitable, lowland habitat. Some are found on widely separated islands or on islands and the mainland, again separated by a sea (in these cases, literally). Such disjunct distributions make Darwin's assumption that related groups of organisms originated in one place, then emigrated to all the places where they now occur, seem impossible. So how did they get to such far-flung locations, and how can Darwin reconcile these many apparent exceptions to his theory? To an advocate of special creation, such disjointed distributions would suggest that the same species was created multiple times. To Darwin, rejecting the idea of a single point of origin "rejects the *vera causa* of ordinary generation with subsequent migration, and calls in the agency of a miracle" (*Origin*, p. 352).

Darwin offers two general explanations for the quirks we see in animal and plant distributions. One is that organisms vary in their ability to emigrate to new locations because they vary in their ability to surpass barriers. The second is that the species distributions we see today are a product of geographic ranges that have changed over time as species were shifted around by a changing environment.

Before Darwin addresses these difficulties, he reminds his readers of three properties of evolution. First he describes the different ways in which species originate. To say that a species has a single origin does not mean that it sprang from a single hermaphrodite or a single male or female. New species emerge initially as a population of individuals that represent a local variety. Some varieties later expand their range and displace their ancestor. In this fashion, a very small subset of local varieties will emerge as distinct species.

Second, evolution is determined by how much variation is present within a population and by changes in its environment. Darwin thinks (contrary to our current thinking) that a change in environment, such as when plants or animals are domesticated, can induce variation. The equivalent change in nature would be a change in any of the organisms with which the population in question is interacting. If a whole community of organisms migrates to some new locality, then these interactions will remain intact, and there may be no evolution in any of the species. If a single species migrates into a new community, then it will inevitably interact with a different suite of species, which may cause a bout of natural selection and evolution both in the migrant and in some of the species with which it interacts. When Darwin describes the Galapagos, he emphasizes that the plants and animals he found there were "in different proportions both in kind and number from those on the continent, and therefore acting on each other in a different manner." His point is that changes in biotic interactions are what cause evolution, rather than changes in the physical environment.

Third, and last, is Darwin's assertion that there are no fixed rules of change. The fact that organisms can evolve does not mean that they will evolve. The amount of evolution we see is neither a function of time nor of movement to a new geographic location. Some organisms, like the "living fossils," show little or no change over long intervals of time, while others have evolved relatively quickly. The amount of evolution that we see is a function of the amount of change experienced by the organism.

Means of Dispersal

Some types of organisms tend to have coherent distributions. Mammals, for example, rarely have disjunct populations. When they do, such as when the same species is found in Great Britain and Europe, Darwin suspects that its range dates to times when Great Britain was joined to Europe by a land bridge. There are often pronounced regional differences in mammal faunas; edentates (sloths, armadillos, anteaters) are mostly confined to South America. Many groups of marsupials are found only in Australia. Terrestrial mammals, as opposed to aquatic ones like seals, are absent from oceanic islands. However, their absence from islands or restriction to certain regions does not mean they cannot survive there. Darwin cites many cases in which mammals introduced to islands or to different continents thrived, often at the expense of the residents. Rabbits, for example, became a pestilence in Australia, as did the foxes that were imported to eat them. The typical patterns of distribution suggest that terrestrial mammals are capable of walking from place to place but are not well suited for crossing oceans.

Plants, on the other hand, are notoriously ill behaved in their distributions. One of my favorite examples is the genus of fortnight lilies (*Dietes*), which are familiar to gardeners in many parts of the world. Three of the four species are from South Africa. The fourth is from Lord Howe Island, which is 600 kilometers east of Australia, meaning that it is on the opposite side of Australia from Africa. Plants present us with many such conundrums regarding their distribution.

The favored hypothesis for mode of dispersal in Darwin's time was via land bridges. These could come and go while continents alternately emerged or submerged as the earth's surface oscillated in a Lyellian landscape. In the *Origin*, Darwin politely defers to the advocates of this mode of dispersal, including Lyell, his late friend Edward Forbes, and his colleague Joseph Hooker, all of whom invoked oscillating land bridges to resolve what seemed like the most improbably disjunct distributions of organisms. While Darwin is polite in the *Origin*, he was more direct in his private correspondence, such as in the following quote from a letter to Lyell: "Here poor Forbes made a continent to North America and another (or the same) to the Gulf weed; Hooker makes one from New Zealand to South America and round the world to Kerguelen Land. Here is Wollaston speaking of Madeira and Porto Santo as the sure and certain witnesses of a former continent. . . . If you do not stop this, if there be a lower region for the punishment of geologists, I believe, my great master, you will go there" (as quoted in Browne 1995, p. 521).

Darwin was convinced from his own observations that there was much more to dispersal than land bridges. The Galapagos and other oceanic islands were a key inspiration for him. Under Lyell's theory of land bridges, such islands were the summits of submerged mountains once found on what are now submerged continents. If this were true, then "some at least of the islands would have been formed, like other mountain-summits, of granite, metamorphic schists, old fossiliferous or other such rocks, instead of consisting of mere piles of volcanic matter" (*Origin*, p. 358).

Darwin had seen many of these islands during his travels. After the Galapagos, he visited Tahiti and hiked through its vertiginous landscape of volcanic peaks split by nearly vertical-sided valleys produced by the eroding action of mountain streams. Such islands could never have been joined to some distant continent; they were the product of volcanic eruptions and had emerged de novo in the middle of the ocean. The plants and animals that inhabited them had to have arrived by some means other than travel across a land bridge.

In the *Origin*, Darwin proposes that there must on occasion be long-distance dispersal across the ocean. This is not a wild guess. On one of the last legs of his journey home, he had visited and made plant collections on Keeling Island, a coral atoll in the Indian Ocean. He succeeded in collecting almost every plant known from the island, save a type of tree that was known from a single individual found "near the beach, where, without doubt, the one seed was thrown up by the waves" (Darwin [1860] 1962, p. 453). He also recounted the observations of other travelers who had seen seeds wash ashore on Keeling Island, including seeds from trees native to Sumatra, from other parts of the Malay Archipelago, or seeds trapped in the roots of eucalyptus logs from Australia. Many of the seeds were viable.

After his barnacle years, Darwin developed a research program to characterize the potential for oversea transport, which he now describes in this chapter. Since he was most impressed by the wildly disjunct distributions of plants, and since he knew of observations of the oversea transport of seeds, he concentrated on the properties of seeds that could facilitate ocean transport. He immersed seeds of different kinds in jars of seawater for different intervals of time, then measured their ability to germinate. Darwin found that 64 out of 87 species of plants he tested had seeds that could germinate after 28 days in seawater. Some survived for 137 days.

He noticed, with Hooker's prompting, that many of the seeds had become waterlogged and sank, even though they could still germinate. If they sank, then survival times were of little use because the seeds were not going anywhere. Darwin repeated the studies and noted how long the seeds could stay afloat. He also worked with seeds and fruits that were first dried before immersion, since drying might often occur before a seed was washed out to sea. Dried seeds often stayed afloat much longer. Hazelnuts, for example, quickly sank when green, but remained afloat for 90 days when dried. In all, 18 out of 94 tested species floated for at least 28 days. Johnston's *Physical Atlas* reported on ocean currents in different parts of the world, so Darwin simply multiplied the number of days of survival by the distance traveled per day to estimate how far a seed could be transported and still survive. He found that on the order of 10–20% of the plants he had tested could travel nearly 1,000 miles from one landfall to the next, then germinate and establish a new colony. Some plants could go much farther. In fact, he inferred that some of the seeds reported to have washed up on Keeling Island "must have traveled between 1800 and 2400 miles" (Darwin [1860] 1962, p. 454).

Darwin also tells readers of the *Origin* that the soil contains a bank of viable, dormant seeds. He once extracted soil that was trapped behind a rock entwined in the roots of a recently felled tree that was more than fifty years old. He found seeds in this soil that readily germinated. It follows that anything that can transport bits of soil from place to place can bring along hitchhiking seeds. He recalls that old tree trunks sometimes washed up on the shores of remote Polynesian islands bearing rocks still encased in their roots. Since some of these islands are composed entirely of coral and have no native stone, the rocks are greatly valued as sharpening tools. If rocks could get there, so could seeds that were compacted in any soil encased behind the rocks. Icebergs have been seen carrying branches, rocks, and soil to distant landfalls, so Darwin speculates that seeds could have been transported long distances on icebergs during the periods of glaciation.

Darwin tells us here that he also studied birds as a source of transport. He had his servant shoot some partridges after an interval of rainfall and found sufficient soil and pebbles adhering to their feet to bear seeds, so he reasoned that birds could transport seeds, perhaps great distances, as they migrated from place to place. He then embarked, with the help of his children, on ever more fanciful studies of potential accidental transport. After hanging the severed feet of ducks in snail-infested pond water, they found that the snails could remain attached and viable for up to 20 hours. Darwin collected bird droppings from the garden and found that these could contain viable seeds. Seed-eating birds often hold seeds in their crop for 12–18 hours before swallowing them; these remain fully viable because a bird's crop is just like a purse for storing seeds before they are processed by the gizzard and gut. The Darwin team found that a dead pigeon could float on seawater for 30 days with a crop full of viable seeds, so a carcass could convey seeds to a distant shore. A colleague told him of an acquaintance who had given up flying pigeons from France to England because they were so easily caught by birds of prey when they arrived, exhausted after the long flight over water; so the Darwin team fed carcasses stuffed with seeds to birds of prey at the London Zoo, and found that the predators regurgitated pellets of hair and bone that also contained viable seeds. Darwin reasoned that the accidental transport of a bird by a storm could thus transport hitchhiking seeds, perhaps long distances. Some of these seeds would remain viable after passage through the gut, but even more would remain viable if the bird were snatched by a hawk or eagle while it still held seeds in its crop. Any such accidental transport of a bird could become the accidental transport of a seed, which could then introduce that plant to some new territory.

Birds can be blown long distances by occasional storms and in that way disperse around the world. Every bird fancier knows of those rare occasions when an unusual species shows up thousands of miles from where it is normally found as a consequence of such accidental transport. In the same fashion, Darwin reasons, birds can colonize distant landmasses and are regular inhabitants of oceanic islands.

Animals other than birds, Darwin says, can be transported across the ocean on flotsam, which would explain the origin of the South American fauna that he saw on the Galapagos. Such transport would act like a filter, selecting against animals that were sensitive to exposure to saltwater or that could not survive long without food or water. This is probably why reptiles are frequent colonists of oceanic islands (e.g., the two species of iguanas, tortoises, and a snake on the Galapagos), whereas no native amphibians or terrestrial mammals are found there.

All such events may be rare and most landfalls may be unsuccessful, particularly if a seed or animal is transported from one continent to another that is already well stocked with plants and animals, leaving little room for an invader to fit in. However, migrants will have a higher probability of success if they land on a new piece of land, like a volcanic island, that is not yet well stocked with animals and has many vacancies yet to be filled. Given large intervals of time, Darwin reasons, such rare forms of transport can account for the stocking of oceanic islands and the occasional movement of plants and animals between isolated landmasses.

Because he suspects that coral atolls have grown on the tops of subsiding volcanoes, Darwin also infers that over time there must have been other volcanic islands, which later receded into the ocean, that served as stepping stones for the overseas migration of terrestrial plants and animals. This means that some distributions we see as improbable today may have been facilitated by intermediate landfalls that have since disappeared. (We have since found that the Galapagos archipelago and some islands in the Indian Ocean once contained such stepping stones that were closer to continents.)

Darwin concludes that there are probably many more ways for plants and animals to get around than those he considers here. Each added mechanism, no matter how improbable in itself, enhances the overall chances of colonizing different parts of the world and in turn can help explain the existence of disjunct distributions, even if we cannot explain individual oddities.

We have now witnessed some examples of long-distance migration that tell us that the improbable can occur. Censky and colleagues (1998) documented the landfall of a small population of green iguanas on the island of

Anguilla in September 1995. Three males and five females made it to shore. An adult female, possibly carrying developing eggs, was found in 1998. The iguanas arrived on a large raft of flotsam following a sequence of hurricanes. The flotsam raft included logs up to 30 feet long with mats of roots. Their likely origin, based on the distribution of this species in the Caribbean and the direction of the prevailing winds, was Guadeloupe, around 250 kilometers to the southeast.

Platyrrhine monkeys, caviomorph rodents, and some lizards appear to have rafted from Africa to South America around 30 million years ago. Molecular-clock dating suggests that the South American and African members of each lineage shared a common ancestor at a time that corresponds to the oldest-known fossils for each lineage in South America. If these organisms had taken a land route from Africa to South America, then some fossils should be found in Eurasia or North America, but none have been seen. Today we see large rafts of flotsam washed out to sea from the mouths of the Congo and Senegal rivers, then picked up by prevailing east-to-west currents and conveyed across the Atlantic. Estimated transit times are too slow to allow a mammal to survive the trip, but the Atlantic was once narrower (chap. 21 below), and unusual circumstances may have created a faster transatlantic crossing (Renner 2004).

Dispersal during the Glacial Period

Darwin's colleague and correspondent Asa Gray, from Harvard University, reported that plant species found in Labrador could also be found in the White Mountains in the United States, but not over much of the territory in between. Some of these same species were also found in the "loftiest mountains of Europe" (*Origin*, p. 365). Other plants could be found high in the Pyrenees and Alps, then at lower elevations in northern Scandinavia, but were absent from the large expanse of land in between. Stranger affinities of plants had been reported from the cooler parts of the Southern and Northern hemispheres. Joseph Hooker found that most of the flowering plants of Tierra del Fuego, on the southern tip of South America, were in the same genera as species found in North America and even some in Europe. He also found plants on isolated mountaintops in the Himalayas that were most closely allied with ones found on mountaintops in India, on Ceylon, and on volcanic cones in Java and Europe, but were not seen in any of the lowlands that lie in between. He even found affinities between plants found on mountaintops in Australia, New Zealand, and Europe.

I was skeptical of these reported affinities between plants in New Zealand and Europe, since similar claims had been made for the birds of Australia and Europe but were later revealed, with the aid of DNA sequences, to be untrue. I followed Darwin's lead to find an answer, which was to find an expert and ask him. Darwin probably had to wait months for a response to one of his inquiries; I have access to the Internet and had an answer within ten hours. Steve Wagstaff, from the Allan Herbarium in Lincoln, New Zealand, confirmed that Hooker was right. While there is some dispute about whether the species in New Zealand and western Europe are in the same genus or in two closely related genera, they are certainly more closely related to each other than to any other species. Two genera that have such a distribution and that may be familiar to gardeners are *Ranunculus* and *Veronica*.

Similar patterns are seen in marine life. Hooker found twenty-five species of marine algae common to New Zealand and Europe, but nowhere in between. Darwin quotes a Professor Dana as saying "it is certainly a wonderful fact that New Zealand should have a closer resemblance in its crustacea to Great Britain, its antipode, than to any other part of the world" (*Origin*, p. 376). Darwin reports that "even as long ago as 1747, such facts led Gmelin to conclude that the same species must have been independently created at several distinct points" (*Origin*, p. 365). Darwin is not willing to subscribe to miracles, but allows that such distributions are a serious challenge to his proposal that each species originated in a single place, then migrated elsewhere.

Darwin proposes that these odd distributions are a consequence of the waxing and waning of glaciers caused by climates that alternately cooled down and warmed up. It was Louis Agassiz who convinced the world that there had been a recent, widespread, and dramatic ice age. In fact, Agassiz overturned one of Darwin's conclusions in the process. In a paper published in 1839, Darwin argued that the three parallel roadlike benches that lined the valley of Glen Roy in Scotland were beaches created at a time when the land had subsided and seawater filled the valley (fig. 24). Agassiz instead argued that the "roads" were the shoreline of a lake that had filled the valley when it was impounded by a glacial dam. Agassiz was right. In Europe, it was clear that glaciers had covered vast areas of land and flowed down from mountains, carving out valleys and pushing large boulders in their wake. By 1859, there was also evidence from North and South America, Asia, and

Figure 24
The Parallel "Roads" of Glen Roy
The three parallel, roadlike benches that line the valley of Glen Roy in Scotland
were the shoreline of a lake that filled the valley, impounded by a glacial dam. Darwin
instead interpreted the "roads" as evidence that the land had once subsided so that
the valley was filled by an arm of the sea.

New Zealand that there had been intervals of dramatic glaciation that were
certainly accompanied by a cooling of the environment.

In retrospect, Darwin realized that he had seen such evidence of glaciation while on the voyage of the *Beagle*. In western Chile, he was "astonished
at the structure of a vast mound of detritus, about 800 feet in height, crossing a valley of the Andes; and this, I now feel convinced was a gigantic mo-
raine" (*Origin*, p. 373) left behind by a retreating glacier. The erratic, granite
boulders that Darwin saw strewn across the plains of Patagonia had also
been deposited there by glaciers that had swept them out of the Andes. All
these glaciations took place during the same geological period and were of
long duration. Darwin allows that they may not have been strictly simultaneous, but they would at least have been overlapping in their duration,
which also means that the climate would have been cooler. Now we know
that they do indeed represent a global phenomenon that was accompanied
by a worldwide decline in temperature.

Darwin invites us to envision how the world would change if such an ice
age began today. There would be a progressive cooling of the environment
and an expansion of polar ice sheets and glaciers in higher latitudes north

and south, or on the tops of mountains, even those found near the equator. In central North America during the last glaciation, the Laurentide ice sheet extended all the way across Canada into the Midwest as far south as Illinois. Other ice sheets covered large areas of North America and Eurasia. The expansion of ice and the cooling climate would push artic communities to lower latitudes and alpine communities to lower elevations. The current inhabitants of the more temperate zones would likewise shift to the south. Those of the tropics would be squeezed into an ever smaller band of habitat near the equator. Many would go extinct. During a later phase of climate warming, the reverse would occur. The Arctic plant communities would be displaced northward or would climb the slopes of mountains as the ice caps and glaciers retreated. As the climate warmed further, the cold-adapted plants would disappear from the lowlands, leaving the alpine populations stranded, as if they were on an island in the ocean.

Many plant and animal species in the higher latitudes of the Northern Hemisphere have a circumpolar distribution, meaning that they are found across North America, Europe, and Asia. This occurs because at times there have been land connections between these regions, like the Bering Land Bridge that joined Siberia to Alaska. If you look at the North Pole on a globe, you will see that, even now, there is near-continuous, circumpolar land in the subpolar latitudes. Because some species have long had circumpolar distributions, the end product of the migrations of plant communities toward the south in front of advancing ice sheets, then north again behind retreating ice sheets, would be the disjunct distributions that we see today. Arctic plants would have become stranded on mountaintops at lower latitudes and would have resumed their former distributions at northern latitudes, filling in the land behind the retreating glaciers. This explanation predicts that species we find on mountaintops of North America, Europe, and Asia will have affinities with those found in northern latitudes, but not with species in the lower-elevation habitats in between. The land in between was reclaimed instead by temperate plant communities as they migrated north behind the retreating ice.

In northern latitudes, the advance and retreat of glaciers may simply have caused a wholesale shifting of intact communities south and then north again, but in the mountains of temperate and tropical regions there might instead have been a mixing of Arctic plants with the local alpine species that were present before the glaciations. Because evolution is driven by changes in biotic interactions, we may not expect to see much change in the Arctic flora throughout this cycle because they always remained in the same community of plants. On the other hand, there should have been a mixing of Arctic and temperate-zone alpine species as the climate warmed and temperate-zone plants reclaimed the mountaintops in temperate regions. Here, Darwin's theory predicts that the displaced Arctic species will be exposed to new biotic interactions as they become isolated on mountaintops and will tend to evolve into new varieties or species as a consequence. In fact, the plants isolated on the tops of temperate-zone mountains are now often represented by distinct varieties or even species whose closest relatives are found only in the Arctic.

Such a glacial cycle may explain distributions of plants found at different latitudes in the Northern Hemisphere, but what of the similarities that we see between the Northern and Southern hemispheres, such as between the crustaceans of New Zealand and those found in Great Britain? Darwin reasons that a period of extreme cooling could have forced some temperate species into equatorial regions, where they would have intermingled with tropical species. When the climate again warmed up, many of these species would have migrate back north, but some may just as well have followed a warming climate to the south and thus ended up in temperate zones south of the equator.

Northern Hemisphere migrants have been more likely to move south than southern forms north. Darwin suggests that this is because there is much more land to the north and hence more species, each of which consists of larger populations, which also means that they are honed to a sharper competitive edge than those from the south. He draws an analogy between this biogeographical pattern and the way that species introduced from Great Britain or Europe to Australia or to oceanic islands are displacing the native species. We have rarely seen successful colonization in the reverse direction. Likewise, the tops of mountains in tropical latitudes tend to be dominated by temperate plant species derived from the Northern Hemisphere. These mountaintops are as much habitat islands as the oceanic islands are, so the same rules apply.

Darwin acknowledges that these ideas cannot account for all the known biogeographic anomalies. One that he finds most disturbing was reported by Joseph Hooker for species restricted to the Southern Hemisphere, but found only in places as remote from one another as Kerguelen Land (islands 48°S, between Australia and Africa), Australia, New Zealand, and Tierra del Fuego. One such plant is the southern beech tree (genus *Nothofagus*), which is found in all these places except Kerguelen Land. Hooker suggested that there was once a land bridge that connected all these southern islands and that the land bridge even included Africa. To Darwin, this is the silliest of all land bridge proposals.

References

Browne, J. 1995. *Charles Darwin: Voyaging*. Princeton, NJ: Princeton University Press. (Pp. 431–33, Agassiz, glaciers, and Glen Roy; pp. 451–53, Hooker's introduction to transmutation; pp. 516–21, Darwin's dispersal experiments.)

Browne, J. 2002. *Charles Darwin: The power of place*. Princeton, NJ: Princeton University Press. (Pp. 33–42, Lyell and Hooker's arrangement for the co publication of Darwin's letters and Wallace's essay.)

Censky, E. J., K. Hodge, and J. Dudley. 1998. Over-water dispersal of lizards due to hurricanes. *Nature* 395:556.

Darwin, C. [1860] 1962. *The voyage of the* Beagle. Annotated and with an introduction by Leonard Engel. Garden City, NY: Doubleday and Company.

Renner, S. 2004. Plant dispersal across the tropical Atlantic by wind and sea currents. *International Journal of Plant Sciences* 4 (supplement): S23–33.

Chapter 21

Geographical Distribution, Continued

The Galapagos Islands are remarkable for their endemic species, but not all island groups have abundant endemics, so we have to wonder what else is special about the Galapagos. If Darwin's theory is to have general value, then it must explain why we see the evolution of endemics on some archipelagoes but not others. Also, some types of organisms (birds, reptiles, and insects, for example) are much more likely to be found on islands than others (mammals and amphibians). These patterns must be explained as well. The ever-present alternative, in Darwin's day, was that each species was the product of an independent act of special creation. Darwin's counter is that, if species are the product of independent acts of creation, then we should not see such coherent patterns in species distribution, yet we do. In this chapter, Darwin extends his theory to account for oddities in the geographical distribution of plants and animals, such as the irregular evolution of endemics and the skewed representation of different types of organisms on islands. From here I follow Darwin's original subheadings.

Fresh-Water Productions

Ponds and lakes are islands of water isolated by dry land. Rivers are isolated from other bodies of freshwater by dry land along their courses, then by saltwater if they empty into the sea. The sharpness of the subdivision of freshwater environments leads us to expect that there will be considerable differences in the kinds of animals and plants found in each of them, as there often are among those found on islands in the sea. The opposite turns out to be true. Freshwater organisms of all sorts are often widely distributed and show little differentiation among "islands," even among ones that are well isolated by wide expanses of hostile territory. We even see closely related species on different continents. Darwin says he witnessed such patterns while on the voyage of the *Beagle*: "I well remember, when first collecting the fresh waters of Brazil, feeling much surprise at the similarity of the fresh-water insects, shells, &c., and at the dissimilarity of the surrounding terrestrial beings, compared with those of Britain" (*Origin*, p. 383).

Darwin proposes that the inhabitants of freshwater ponds and streams evolved adaptations that enabled them to disperse short distances to new locations. Such adaptations are a necessity for persistence because ponds and streams are often ephemeral, at least on the timescale of millennia. They almost inevitably disappear as they fill with sediment and are converted first into marshes, then dry land. New ponds and lakes can form when land subsides, creating low-lying basins. Ponds and lakes were created by the thousands as ice sheets melted and retreated to higher latitudes and meltwaters accumulated in natural depressions. Similarly, streams disappear and new ones form as the landscape erodes or elevations change, thus changing the patterns of drainage. Organisms that occupy such variable habitats without evolving the ability to escape deteriorating environments by colonizing new ones will soon go extinct. Those found in such habitats today were successful in evolving the ability to disperse—for example, the ability of larval snails to adhere to the feet of migrating ducks (see chap. 20).

Explaining fish distributions represents a greater challenge because fish so rarely have life stages that allow them to persist for extended periods out of water, making it difficult to imagine how they could have moved between isolated bodies of freshwater. Darwin observes that the widely dispersed fishes are often from ancient lineages, so the passage of extremely long periods of time may account for their having eventually hit upon ways to colonize widely dispersed habitats by whatever means were available. The living lungfish are part of a subclass that dates to the Paleozoic era, which makes them one of the most ancient of all living fish lineages. Different genera of lungfish are found in South America, Africa, and Australia—but this still begs the question of how they got there. Some fish can be transported by "accidental means." In India, live fish or fish eggs have been transported by whirlwinds, but such windstorms are rare and are not well suited to long-distance dispersal. Many fish groups lineages include closely related marine and freshwater species, so related freshwater species in different regions

could have evolved independently from the same or related marine ancestors. One example of such independent invasions of freshwater habitats
by a marine family of fish is seen in the sculpins (family Cottidae). These
fish are like the aquatic equivalent of a toad—they rest on the bottom and
are small, squat, and cryptic. Most members of the family live in shallow
marine waters, but many lineages, particularly in the genus *Cottus*, have
invaded freshwater rivers. Many species are found only in freshwater, but
with nonoverlapping ranges in very different parts of the world. For example, one species (*Cottus klamenthis*) is found just in rivers on the West Coast
of North America. Another (*Cottus poecilopus*) is found just in Scandinavia
and Russia.

Darwin knows of no individual fish species that is found on different
continents, but many species have wide distributions among ponds and rivers within a single continent. He speculates that the most common mechanism of dispersal within a continent would be changes in the elevation
of the land and in the distribution of freshwater environments over time,
plus the occasional intervention of flooding. Another form of exchange
not mentioned by Darwin is that different river basins can often exchange
tributaries via "stream capture," which occurs when an increase in elevation or erosion causes a tributary of one drainage system to shift course
and join a different drainage. The dynamics of lakes and rivers can also be
linked; lakes may at times be joined to one another by running water and
at other times isolated. Rivers may end their course in lakes, then shift to
flowing into the sea or shift from flowing into the sea back to flowing into
a lake. All these changes in the distribution of aquatic habitats, caused by
a dynamic, oscillating, and eroding landscape, will cause freshwater organisms to migrate among basins. The isolation of freshwater habitats that we
see today is thus not a permanent condition. Many of these habitats have
been connected in the past.

Smaller freshwater organisms can be transported between bodies of water in many different ways. Darwin had shown that freshwater snails can attach themselves to the feet of ducks and remain viable out of water for 12–20
hours. He found that the tiny hatchlings were the best suited for transport
because they tightly adhered to the ducks' feet, whereas the larger ones fell
off. Their 12- to 20-hour period of viability outside water would be sufficient
time for a duck to fly hundreds of miles. An ideal setting for such dispersal
would be the annual migrations of ducks: they routinely fly long distances,
then settle on ponds to feed and rest before the next leg of their journey.

When the ducks fly away, they must often inadvertently carry snails and other organisms with them.

Freshwater plants are low in diversity and often very widely spread, so Darwin infers that they too must be capable dispersers. Seeds are probably the most easily dispersed life stage, since they are often small and packaged in a protective coating that resists drying. They can remain viable for a very long time, sometimes more than 100 years, before germinating. Darwin showed that seeds could accumulate in a "seed bank" in the mud at the bottom of ponds. He recounts his scooping three tablespoons of mud from the bottom of one of his ponds: After drying it to the consistency of soil, he left it covered in his study for six months. He inspected it regularly and pulled up each seeding that appeared. He found that 537 seedlings of many kinds of plants germinated from this tiny amount of mud. He reasons that any bird that flies away from a pond with mud on its feet inevitably carries seeds with it.

Many of the smaller organisms that live in ponds, such as the water fleas in the genus *Daphnia*, produce eggs that can remain dormant for decades and accumulate in the mud in the same fashion as seeds, so they too can be dispersed in mud adhered to birds' feet. Because birds are so itinerant and because some seek out similar aquatic habitats wherever they go, dispersal among these habitats will be common. Such frequent dispersal can cause an organism to be spread far over the landscape and to remain uniform in character across a wide range.

Darwin also speculates that the seeds of aquatic plants can be transported by many of the other means he has described for terrestrial plants. Fish sometimes eat seeds that can remain viable in their guts. Birds eat fish, then can either fly or be transported by storms to distant places and later defecate or regurgitate viable seeds into a new home. Darwin recalls puzzling over Alph. de Candolle's observation that water lilies in the genus *Nelumbium* were distributed over North and South America in widely separated freshwater lakes and ponds. These lilies produce very large seeds that do not seem to be amenable to long-distance transport, but James Audubon once found one in the stomach of a heron.

The formation of a new pond is the emergence of a new island of habitat, but one that is not nearly so isolated as an oceanic island. Visiting birds will introduce the eggs, seeds, and juveniles of freshwater organisms. These new colonists will find themselves in an empty or sparsely inhabited environment with little competition, making them likely to survive, reproduce, and establish new populations. Species diversity tends to be low, which takes the edge off the struggle for existence. Because most pond-dwelling organisms share the ability to be readily dispersed and because birds as dispersal agents so frequently move from pond to pond, similar communities will be assembled in each pond. As Darwin observes, "Nature, like a careful gardener, thus takes her seeds from a bed of a particular nature, and drops them in another equally well fitted for them" (*Origin*, p. 388).

On the Inhabitants of Oceanic Islands

The biotas of oceanic islands, meaning those islands that have never been joined to the mainland, offer a special opportunity to discriminate between Darwin's theory and special creation. Darwin argues that the peculiarities of the flora and fauna of oceanic islands are dictated by their history of colonization; everything found on these islands must have penetrated the selective filter of overseas transport. Special creation suffers no such constraints.

The flora and fauna of oceanic islands have other peculiarities that belie their history. They are species-poor in comparison to equal areas on continents. They are very susceptible to invasion by species that arrive from continents with human help. The continental invaders often drive the natives to extinction. By the doctrine of special creation, there is no reason why islands should be stocked with either fewer species or competitively inferior species than continents, but Darwin's theory predicts both of these trends. Competitiveness and speciation are products of the struggle for existence. This struggle is more intense on continents than on small islands because more species are present on continents and they occupy larger areas of land, so each species tends to be represented by more individuals and hence will contain more variation for selection to act upon. The products of the more intense struggle on continents will be competitively superior to island endemics.

Some oceanic islands have very few endemic species and instead share the same species as the nearest continent. If Darwin's theory is true, then it must also explain why not all invaders evolve into endemic species. His theory predicts that whether an immigrant evolves into a new species depends on how readily it disperses and on whether it finds itself in a community composed of organisms different from those on the mainland. Almost all land birds on the Galapagos are endemics and are part of communities of plants and animals that are quite different from those on the mainland. In contrast, none of the marine birds are endemic species. The marine birds are much more mobile, so there will be a constant exchange of individuals between populations on the islands and others on the mainland. They also inhabit a coastal environment with organisms similar to those of mainland South America. Because there is little change in their biotic environment, there is weak selection for local adaptation.

Darwin's argument for sea birds also applies to the land bird fauna of some islands. We see very few endemic species of land birds on Bermuda, in the western Atlantic Ocean, even though it is as far from North America as are the Galapagos from South America. We also see very few endemic birds on Madeira, which lies in the eastern Atlantic Ocean, closer to Africa. In contrast to the Galapagos, Bermuda and Madeira are stopovers for birds that migrate between nesting grounds in the north and overwintering habitat to the south, so there is regular contact with large numbers of land birds from the neighboring continent. Those birds that take up permanent residence on each of these islands tend to co-occur with one another on the mainland, so there is not a large change in their biotic interactions. This combination of constant exchange with the mainland and little change in biotic interactions means that there is weaker selection for the formation of local varieties than on the Galapagos.

Oceanic islands often lack certain large classes of animals, particularly mammals and amphibians. Very different animals have sometimes evolved to occupy the ecological niches left vacant by their absence. Some of the most spectacular examples of this phenomenon were found on New Zealand. The only mammals found there before the arrival of humans were bats. There were small, flightless birds that appeared to have filled niches normally occupied by rodents, and gigantic, grazing birds whose role would have overlapped with that of mammalian grazers and browsers like deer. The ecological niches most often filled by mammals were thus occupied by birds.

One of Richard Owen's early claims to fame was to describe a species of moa, an endemic bird from New Zealand, that filled the mammalian browser niche. He did so on the basis of a fragment of a single bone. Owen applied Cuvier's discipline of reconstructing the whole animal from a fragment and concluded that the one in question was from a flightless bird larger and more massive than an ostrich. Since it was bone rather than fossil, he concluded that it was either recently extinct or might even still be living somewhere in New Zealand. The editors of the journal that published the paper did so with some reluctance. If such birds were around, then someone should have noticed them. Owen was right, and it appears that Europeans just missed seeing live moas. Several species existed when the Maoris arrived in New Zealand, around AD 1300. The largest was estimated to have been 3.6 meters tall and to have weighed up to 250 kilograms! The Maoris rapidly ate them into extinction.

Darwin reports his finding that terrestrial mammals are unknown from any island situated more than "300 miles from a continent or great continental island; and many islands situated at a much less distance are equally barren" (*Origin*, p. 393). Amphibians have similar distributions. Amphibians and mammals are not excluded from these islands because the habitat is unsuitable. Some have been introduced to islands and have multiplied to the point that they are a nuisance. What amphibians and mammals have in common is that they are ill suited for long-distance rafting over the ocean. Saltwater is toxic to amphibians, and both classes have a greater need for a supply of freshwater to survive than do other animals. Bats and marine mammals, such as seals, are the only mammals in remote areas because they are able to disperse across ocean barriers.

The composition of the plant communities on islands can also be very different from that of the communities on the nearest continents. Darwin observes that trees and woody shrubs found on islands are often most closely related to herbs on the mainland. It seems that herbs are often among the earliest colonists and, once on the islands, evolve to occupy niches occupied by shrubs and trees on the mainland. One such example is plants in the genus *Echium*, which includes the 'Pride of Madeira' shrubs that are popular garden plants in California. They produce striking, conical stalks of purple/blue flowers in late winter and early spring. Twenty-seven species of *Echium* are endemic to the Canary Islands, Madeira, and the Cape Verde archipelago—all are woody shrubs or rosette trees. In contrast, all continental species of *Echium* are herbaceous. Recent research has shown that all the island endemics of *Echium* are derived from a single, herbaceous ancestor that first colonized the Canary Islands less than 20 million years ago.

There is sometimes a relationship between the depth of the ocean that separates an island from the mainland and the similarity of the flora and fauna. Ocean depths have fluctuated as ice ages waxed and waned, alternately tying up water in polar ice caps or releasing it into the seas. Islands that are separated from the mainland today by shallow channels may well have been joined to the mainland at a time when the sea level was lower; at those times, they would have been much more accessible to all organisms.

By the time of the sixth edition of the *Origin*, Darwin was able to report on "Wallace's line" as an example of a deepwater channel that was an effective barrier to dispersal because it persisted throughout the glacial cycles. Wallace had cataloged the faunas found on the Malay Archipelago, which extends from Southeast Asia to Australia but is broken by a deep ocean channel between Borneo and Celebes. He discovered that the mammalian faunas on all islands west of the divide were characteristic of Asia while all east of the divide were characteristic of Australia. During glacial maxima, when ocean levels were lower, land bridges emerged on either side. The bridges allowed Australian mammals to migrate to the eastern shore of the channel and the Southeast Asian mammals to migrate to the western shore, but none crossed in either direction.

Darwin concludes: "All the foregoing remarks on the inhabitants of oceanic islands,—namely, the scarcity of kinds—the richness in endemic forms . . .—the absence of whole groups, as of batrachians (amphibians), and of terrestrial mammals notwithstanding the presence of aerial bats—the singular proportions of certain orders of plants,—herbaceous forms having been developed into trees, &c.,—seem to me to accord better with the view of occasional means of transport having been largely efficient with the long course of time, than with the view of all our oceanic islands having been formerly connected . . . with the nearest continent" (*Origin*, p. 396).

Darwin recognizes that there are still some distributions that are hard to explain, even given the strength of his arguments. He can only suggest that there may yet be means of dispersal that are unknown to us. One is that the isolation of some islands today might have been less severe in the past because there might have been island stepping stones that have since disappeared beneath the sea (indeed, such islands have since been found). Another is that some organisms may have as yet unknown biological properties that allow them to disperse farther than previously imagined. For example, land snails can be found in some of the most remote parts of the Pacific, yet are very susceptible to saltwater. Darwin reports his discovery that adults can enter a resting stage when exposed to certain environmental stresses. They withdraw their bodies deep into their shells and seal the entrance with a hardened disc of tissue called an "opercula." Adults that were in such a resting phase proved to be as durable to exposure to seawater as many plant seeds.

Different but closely related endemic species can be found on different islands within an archipelago, as Darwin found in the Galapagos. How can his theory explain their origin? Such islands can have the same climate and geology, plus a very similar set of original colonizers from the mainland. One key to the explanation is that the different species of tortoise, mocking thrush, and finches on different islands of the Galapagos are more closely related to each other than they are to any other species found elsewhere. This pattern suggests that there was only one successful migration from the mainland, followed by multiple speciation events on the islands. Darwin argues that the causes of different endemics on individual islands are the same as those that establish the island-mainland differences, writ small. Each island may have been independently colonized by the same species from the mainland, but may also have received its initial migrants from other islands. Each of the islands is well separated from the others, in spite of their proximity, by strong ocean currents that are driven by the steady trade winds, so successful island-to-island transport may be a rare event. Each island also has a unique biota because of chance differences in the species that initially colonized it. This means that a species that migrated from one island to another was likely to find itself in a different community and hence to experience strong selection to adapt to those differences: "a plant, for instance, would find the best-fitted ground more perfectly occupied by distinct plants in one island than in another, and it would be exposed to the attacks of somewhat different enemies . . . natural selection would probably favour different varieties in the different islands" (*Origin*, p. 401).

The key to endemism on these islands, but not in freshwater ponds, is in the supply and the source of the immigrants. In freshwater ponds, immigration is common, and the immigrants come from very similar, simple communities whose species are all well adapted to dispersal. In the Galapagos, immigration is rare, and the source of immigrants is either the mainland or another island that has different communities.

If we accept that a single species can first colonize one island, then another, and possibly evolve into two or more species, then we must ask why the species that evolves on one island does not then migrate to the others. Conversely, why are there so many species that occupy multiple islands without diversifying into different varieties and species on each of them? One general answer Darwin proposes is that colonization is like buying lottery tickets. A very large majority of tickets are losers. When we look at the species that are well established on an island, we are seeing only the winners. Why did they win? For a colonist to be successful, it must find the necessary resources to sustain itself, which means it must be different enough from the species that are already there to be able to fill a vacancy and persist. Alternatively, it might be superior to a current occupant and might then displace it. What we see on the Galapagos today is the product of the dynamic equilibrium of this constant casting of lottery tickets, most of which lose, some of which are successful and add new species to an island, and some of which are successful in replacing one species with another.

Darwin then argues that "the principle which determines the general character of the fauna and flora of oceanic islands . . . is of the widest application throughout nature" (*Origin*, p. 403). Oceanic islands are not special in their isolation. The species found on isolated mountaintops that were recolonized after the glaciers receded, as well as those in every isolated freshwater habitat, are related to organisms from the surrounding habitat because that is where the original colonists came from: "thus we have in South America, Alpine humming-birds, Alpine rodents, Alpine plants, &c., all of strictly American forms" (*Origin*, p. 403) that colonized from the surrounding lowlands. Likewise, the animals that we find in the caves of Kentucky are most closely related to those found in the surrounding countryside, as are those from caves in Europe of European origin (chap. 6 above). If we ever find many species in one region being closely related to species in another, however distant the regions are from each other, then Darwin predicts we will also find that at some time in the past there was free migration between them. Today, some species will be the same in both places, some will be represented by different varieties, and some by different but closely related species. The fate of each ancestral species will have been determined by how easily it moved between the "islands" and by the differences in biotic interactions that it experienced when it colonized an island.

Darwin also argues that the ability to disperse is a heritable character. If we consider a genus that has an unusually large range, then it almost certainly will include one or more species that also have a very large range. All species in the genus must have been derived from a common ancestor that ranged widely, then diversified. At least some of the descendant species will have thus inherited the ancestor's ability to disperse. The underlying cause of having a large range is multifaceted, since it requires both mobility and the ability to persist in a new location.

If species were products of independent acts of creation, then we would not expect to see such regular patterns in their distribution. There would be no reason for widely ranging genera to tend to be made up of widely ranging species; for the inhabitants of any habitat in any region to tend to be more closely related to those from neighboring habitats regardless of how different the environments may be; for the inhabitants of oceanic islands to be most similar to those from the nearest continent; nor for any of the other regular patterns that we see in how organisms are distributed. All these patterns are "utterly inexplicable on the ordinary view of the independent creation of each species, but are explicable on the view of colonization from the nearest and readiest source, together with the subsequent modification and better adaptation of the colonists to their new homes" (*Origin*, p. 406).

Summary of the Last and Present Chapter

In his summary, Darwin refers to his late colleague Edward Forbes, who observed a "striking parallel in the laws of life throughout time and space" (*Origin*, p. 409). Both the time and the space inhabited by any species are continuous. The exceptions we see in time can be attributed to the imperfection of the geologic record, while those we see in space can be attributed to a species having disappeared from parts of its former range or to its occasional dispersal to new environments. Whether we look at lineages as they have changed over time or as they have migrated to new places, they are joined by the same "bond of ordinary generation." Those found closer together in either time or space tend to be more similar to one another than those found more distantly separated in time or space; "in both cases, the laws of variation have been the same, and the modifications . . . have been accumulated by the same power of natural selection" (*Origin*, p. 410).

Evolution Today: The Silliest of All Land Bridges

Darwin was consistent in two stands that he took on biogeography. One was that where the continents are today is where they have always been, at least for the time period that is represented in the fossil record. Continents had oscillated up and down, but they had not moved. His second stand was that his colleagues were far too reliant on land bridges to explain geographic distributions. Darwin's alternative was that organisms had various means of dispersing across ocean barriers.

Joseph Hooker's proposed land connections between South America, Africa, Australia, New Zealand, and islands found between these larger landmasses had been a last straw for Darwin, yet Darwin still puzzled over the similarities in plants found on these widely separated landmasses. He argued that if plants were distributed as seeds floating on the ocean or by hitchhiking on icebergs or flotsam, then what we find on a given island should be most similar to what is found on the nearest continent. This logic worked well for the flora and fauna of the Galapagos and Cape Verde Islands, but Hooker's Southern Hemisphere plants challenged Darwin's logic. The strong affinity between the flora of New Zealand and Australia was consistent with his hypothesis because they were each other's closest neighbors, but how could he explain the strong affinity between New Zealand and South America? It strains credibility to think that seeds might drift between these two landmasses. Darwin's answer was to postulate a source of migrants that was once proximate to all of these Southern Hemisphere landmasses. He proposed, as others had before him, that the Antarctic once had a much warmer climate and was home to a diverse community of organisms. Plants could have dispersed from Antarctica radially outward to all the other southern landmasses on drifting icebergs. Even this mechanism seemed to fall short of explaining some of the affinities between plants on widely separated landmasses in the Southern Hemisphere, "but this affinity is confined to the plants," he pointed out, "and will, I do not doubt, be some day explained" (*Origin*, p. 399).

In some ways, both Darwin and Hooker were right. We now know that the Antarctic was a warmer and livelier place even within the Cenozoic era. We have a fossil record from parts of the Antarctic that provide a historical record of its role in the origin of a diverse flora and fauna.

More surprisingly, Hooker's proposed land connections actually existed. There was another mechanism at work, unknown to Darwin, slower than Lyell's proposed oscillations of the earth's surface, slower than the rise and fall of sea level during the ice ages, that vindicated Hooker's silliest of all land bridges. Darwin and his contemporaries were wrong about the constancy of continents. Continents have moved and have moved quite far over time spans of hundreds of millions of years. They continue to move today. The theory of plate tectonics encapsulates the explanation for this phenomenon. This theory, which solidified during the 1970s, now has the status of being a unifying concept in the geological sciences that is on par with evolution as the unifying concept of the biological sciences. It provides a mechanistic explanation for the origin of the major topographic features of the surface of the earth, including mountain ranges and ocean basins, the distribution of different types of rocks in the crust of the earth, the structure of the earth's core, earthquakes, volcanoes, the changes in elevation that were the basis of Lyell's geological theories, and much more.

Like evolution by natural selection, the theory of plate tectonics has its roots in much earlier proposals, such as Alfred Wegener's book *The Origin of Continents and Oceans* (1915). Wegener recognized some of the evidence for the existence of moving continental plates, including features of now-distant continents with matching contours, like puzzle pieces, and similar matches in the distribution of living and fossil organisms, which suggested that landmasses now distant from one another had once been adjacent but then pulled apart. However, no mechanism that could cause such "continental drift" was known, so Wegener's ideas were not widely accepted. Acceptance of his theory came with the discovery of such a mechanism; it is called plate tectonics. My use of the word "theory" here parallels its application to evolution, which is that a theory is a statement of general principles that has wide explanatory value. It is an embodiment of facts, rather than speculation.

I offer here a brief explanation for the complex phenomenon of plate tectonics, then some examples of its relevance to biogeography. In chapter 17, I described the discovery of radioactivity, its role in disproving Lord Kelvin's conclusion that the earth was much younger than geologists were proposing, and how the measurable rate of radioactive decay provided a reliable tool for estimating the true age of the earth. The key to disproving Kelvin's estimate for the age of the earth, based on his assumption of steady heat loss, was that nuclear fission in the interior of the earth provides a continuing source of heat that propagates to the earth's surface. This same source causes the interior of the earth to be persistently hot. This core heat, in turn, causes the asthenosphere, or layer that underlies the solid outer layer (the lithosphere), to be plastic and slowly circulate. The lithosphere is divided into discrete plates that are constantly moving, driven by the circulation of the asthenosphere. The continents and continental islands are lighter material that is "floating" on the surface of these plates. The lava from the volcanoes that created the oceanic islands that so influenced Darwin is derived from the asthenosphere. Most volcanoes form along the borders between adjacent plates.

The plates are in constant motion because in some places the force of the circulating asthenosphere draws lithosphere into this inner layer, where it is destroyed. As some lithosphere is destroyed, new lithosphere is created elsewhere as plates are pulled apart then lava rises to the surface, and cools. Continents are moved about, as if they were on a giant conveyor belt, as lithosphere is destroyed in some places and replaced elsewhere. In other places, like the famous San Andreas Fault of California, the plates are moving laterally to one another. Darwin was one of the first people to recognize the possibility of a common mechanism causing volcanoes, earthquakes, and changes in the elevation of land after he observed their association while on the voyage of the *Beagle* (chap. 10 above): "From the intimate and complicated manner in which the elevatory and eruptive forces were shown to be connected during this train of phenomena, we may confidently come to the conclusion, that the forces which slowly and by little starts uplift continents, and those which at successive periods pour forth volcanic matter from open orifices, are identical" (Darwin [1860] 1962, p. 314).

The first clue to continental drift was the way the east coast of South America seemed to complement the west coast of Africa, like two pieces in a jigsaw puzzle that can be locked together. The explanation we have from plate tectonics is that South America and Africa are on two different plates and were once fused as a single landmass. Patterns of circulation in the underlying mantle caused a tension that pulled the two plates apart, resulting in eruptions along a line of volcanoes in the stressed junction between them. These eruptions added new lithosphere, in the form of cooling lava, to the margins of each plate. The growth of the lithosphere along this line of volcanoes was matched by the consumption of lithosphere at subduction zones elsewhere on the globe. The subduction of crust elsewhere has continued to pull the plates apart and split the single landmass in two, and then, like a widening wedge, new crust formed between them, creating the basin filled by the Atlantic Ocean. These events began early in the Cenozoic era (more than 100 myr ago) and continue today.

The split between Africa and South America follows a much longer history of the movement, fission, and fusion of the continents. Further back in time, South America and Africa had been joined to Australia, Antarctica, and New Zealand to form a supercontinent called Gondwana. Gondwana was then fragmented by plate movements similar to the more recent splitting of South America and Africa. As the landmasses separated, they also drifted northward.

The affinities between plants of the Southern Hemisphere discovered by Hooker date to a time when the now widely scattered southern landmasses were part of a single continent. As this supercontinent broke up into separate landmasses that drifted apart, the ancestors of the organisms that Hooker studied were carried along, and their descendants, though now separated by wide oceans, are still closely related.

The discovery of the movement of the continents represents one of the key missing links of biogeography, making it possible to define when land bridges that joined what are now well-isolated landmasses existed, as well as when the barriers between them developed. This added information has resolved many of the problems raised by disjunct distributions of plants and animals that troubled Darwin.

References

Bohle, U.-R., H. H. Hilger, and W. F. Martin. 1996. Island colonization and the evolution of insular woody habit in *Echium* L. (Boraginaceae). *Proceedings of the National Academy of Sciences* 93:11740–45.

Browne, J. 1995. *Charles Darwin: Voyaging*. Princeton, NJ: Princeton University Press. (P. 521, Darwin's response to Hooker's silliest of all land bridges.)

Darwin, C. [1860] 1962. *The voyage of the* Beagle. Annotated and with an introduction by Leonard Engel. Garden City, NY: Doubleday and Company.

Flannery, T. 1994. *The future eaters*. Chatswood NSW: Reed Books. (Pp. 52–66, fauna of New Zealand before the arrival of humans.)

Grotzinger, J., T. H. Jordon, F. Press, and R. Siever. 2007. *Understanding Earth*. New York: W. H. Freeman. (Details on plate tectonics and continental drift.)

Owen, R. 1839. *Proceedings of the Zoological Society* No. LXXXIII, pp. 169–71. (Description of a Moa.)

Chapter 22

Mutual Affinities of Organic Beings:

Morphology: Embryology: Rudimentary Organs

In my view, Darwin saved the best for last. His chapter 13 is a giant in defining his theory as the unifying concept of the biological sciences. It is in this chapter that Darwin concentrates his arguments for how the process of evolution has left an imprint on all aspects of living organisms. He takes on three well-developed disciplines that together comprised the bulk of the life-science research of his day: the classification of organisms, comparative anatomy, and comparative embryology. Each of them had a long history prior to the *Origin*, and each was accompanied by well-established paradigms that Darwin sought to overturn. Comparative anatomy was Richard Owen's home turf, so we may well imagine that Darwin's invasion of this territory was the last straw for Owen. Owen would have had due cause to be irritated with the *Origin* up to this point, but Darwin's reinterpretation of comparative anatomy, which overturned one of the crown jewels of Owen's career, must have been enough to blow the top off of Owen's head, figuratively speaking. Rudimentary organs were a fourth topic that had been noticed and discussed by scientists, but remained unexplained. Darwin argues that such organs are easily explained by his theory. The unifying feature of his arguments is that these disciplines can be understood if the differences between species reflect their descent from a common ancestor. He goes further by showing how his theory unites these diverse disciplines under a single explanatory framework.

Darwin addressed these topics in his earlier, unpublished essay of 1844, but the intervening barnacle years gave him personal research experience in all these disciplines. The theoretical nature of Darwin's presentation in chapter 13 of the *Origin* is thus backed by a wealth of personal experience.

Classification—Background: "God created, Linnaeus organized"

Linnaeus's motto reflected his view that the "natural system" of classification that he had invented reflected the Creator's plan for life; many of Darwin's contemporaries shared this view. Linnaeus's *Systema Naturae*, in which he defined and applied his system of classification, was first published as an eleven-page pamphlet in 1735, then expanded and revised nine times to become a multivolume work that included the classification of thousands of species of plants and animals. One measure of his success is that his system of classification remains in use today. The Linnaean system named each species with a Latin binomial, which was a convention that had been proposed two hundred years earlier. The first part of the name was the genus and the second was the species. For example, lions and tigers are in the genus *Panthera* and carry the specific epithets of *leo* and *tigris*, respectively. A more significant quality of Linnaeus's system was to have a series of taxa above the level of genus and to arrange the classification in a hierarchical fashion. Linnaeus had five levels in his hierarchy, but by Darwin's time the hierarchy had expanded to seven levels. We still use this seven-level hierarchy—kingdom, phylum, class, order, family, genus, species—plus subheadings within these big seven. The complete classification of a lion, without subheadings, is kingdom Animalia, phylum Chordata, class Mammalia, order Carnivora, family Felidae, genus *Panthera*, species *leo*.

Organisms were classified together on the basis of similarities in their appearance. Linnaeus's classifications for plants were based largely on the morphological details of the different components of flowers. His narrow focus on flowers led to some unnatural groupings. For example, he grouped a diversity of "plants" into a class he called "Cryptogamia," which means literally "hidden marriage," since they all shared the property of not having flowers so that their mode of reproduction was hidden. This grouping included fungi, algae, mosses, lichens, and ferns. Fungi are now considered to be a separate kingdom on par with animals and plants; algae are not plants; and lichens are a symbiosis of a fungus with a photosynthetic partner, usually a green alga or a cyanobacterium. However, in general many of Linnaeus's classifications were good, and many are still valid today.

A major preoccupation of systematists after Linnaeus was to identify the characters that were of value in classification and to refine classifications so that they were "natural," meaning that the organisms that were grouped together really did share some affinity with one another, even if the cause of that affinity was unknown. The range of characters used for classification was expanded to include any anatomical detail of the organism. The systematists also took a closer look at development, since some organisms that look very similar in early life stages can look very different later in life, or vice versa. For example, Cuvier classified barnacles as mollusks, based on their appearance as adults. Their larvae were later discovered to be classically crustacean in appearance; it was the larvae that revealed their true identity and placed them in a different phylum.

Today we can add genetic characters to the mix. Sequences of DNA provide a vast quantity of information because organisms have many thousands of genes that vary in how rapidly they evolve and hence the information they convey. They also have extensive DNA that is not part of their genes but can carry information about relatedness. Rapidly evolving segments are useful for evaluating recent events, ranging from identifying the parents of an individual to resolving the relationships between populations within a species or closely related species. Less rapidly evolving segments are useful for resolving details of events that occurred in the more distant past, such as defining the relationships between more distantly related species and evaluating how they group together in higher levels of the taxonomic hierarchy. Some very stable segments of DNA enable us to reach all the way back to the early diversification of the prokaryotes, perhaps more than 3 billion years ago.

Classification—The Origin

Darwin's goal in this section is to ask if we can see any evidence of a plan of creation in how organisms are arrayed in the Linnaean hierarchy. Science is built on discriminating between alternative theories, and in the nineteenth century special creation was the prevailing alternative to evolution, which is why Darwin, in the *Origin*, always comes back to comparing those two alternatives. He concludes that there is no evidence of a plan. Instead, classification reveals the evolutionary history of the organisms, meaning their descent with modification from common ancestors. Darwin's alternative of evolution by natural selection predicts a nested hierarchy, but otherwise makes no predictions about any sense of order within the hierarchy. Darwin's success in defining evolution as the cause of hierarchical classification has been so complete that it may be difficult for a modern reader to identify with the argument that follows. To appreciate the impact of Darwin's argument, it is necessary to put yourself in the historical context, which was a scientific community that was seeking evidence for some plan to the workings of nature, directed perhaps by the hands of a creator.

To ask whether there is evidence for a plan in nature, Darwin first considers the characters that have proven useful for categorizing organisms into what was accepted as a "natural" scheme. His first observation is that organisms classified together often have very different habitats and lifestyles. For example, species in the class Mammalia can be sustained fliers, aquatic, terrestrial, fossorial, or arboreal. The mammalian order Carnivora includes both aquatic (seals) and terrestrial species. Those that are terrestrial include species that are adept at burrowing, like badgers, or at living in trees, like many of the cats. In spite of a name that implies a life of eating meat, the order includes the panda, which subsists on bamboo, the aardwolf, which has reduced teeth and feeds on termites, and many omnivores.

Conversely, habit is not a good indicator of relationships. For example, sharks, cetaceans (whales and porpoises), and the extinct ichthyosaurs (reptiles) are similar in shape, habitat, and locomotion, but are from three different classes. It is their independent adaptation to a lifestyle of sustained swimming that shapes their appearance, not their relationship to one another. The existence of such close similarities between distantly related organisms was recognized long before Darwin and was known as something to avoid considering when classifying organisms if one wanted a "natural" classification.

Darwin argues that the most reliable characters for classification often have little adaptive value. In fact, the best characters are often of seemingly "trifling importance." Their most important quality is that they are stable in their expression in many species. Their stability means that they can be used to define relationships between different species, even if these species have very different lifestyles. For example, in the mammals, both living and fossil, the marsupials can almost invariably be distinguished from the placentals because the marsupials have an inflection of bone on the inner side of their lower jaw that placentals lack, yet this inflection serves no known purpose.

The importance of any character also is a function of how well it is correlated with other characters. Darwin cites the embryonic leaves, or cotyledons, of plants as an example. The two major divisions of the angiosperms,
or flowering plants, recognized at the time of the *Origin* were defined by
whether the seedling had one or two of these leaves. The major divisions
were thus named for those plants that had one (Monocotyledons) versus
two (Dicotyledons) cotyledons (fig. 25). It has since been discovered that

**Figure 25
Morphological
Characters that
Distinguish between
Monocotyledons and
Dicotyledons**

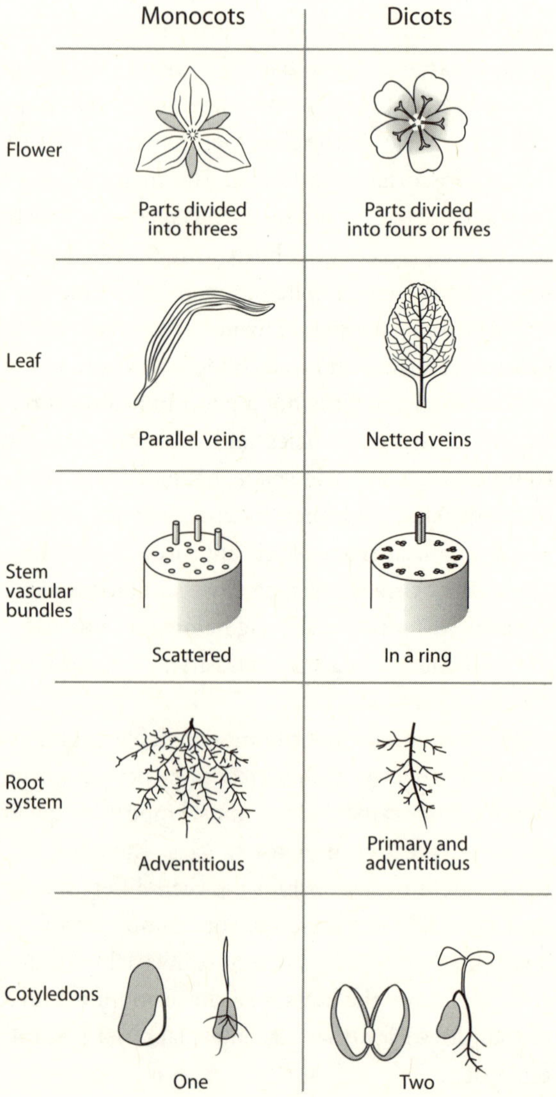

	Monocots	Dicots
Flower	Parts divided into threes	Parts divided into fours or fives
Leaf	Parallel veins	Netted veins
Stem vascular bundles	Scattered	In a ring
Root system	Adventitious	Primary and adventitious
Cotyledons	One	Two

the "dicotyledons" of Darwin's time included species that did not all share a single common ancestor. If a few oddballs are excluded, then we are left with a group that does share a single common ancestor and contains the large majority of Darwin's "dicotyledons." This new grouping is sometimes referred to as the eudicotyledons. With this small refinement, the remainder of Darwin's presentation is intact.

This seemingly trivial difference of having one versus two cotyledons was significant for classification because it was correlated with many other features of flowering plants. Monocots tend to have leaves with parallel veins, as in a blade of grass, while dicots have leaves with veins that are branching and reticulate, as seen in the leaves of trees. Monocots have trimerous flowers, which means that structures like petals come in multiples of three. Dicots instead have tetramerous flowers, with structures that come instead in multiples of four or five. There are also differences between the groups in the structure of the pollen, stems, and roots. It is this association between the number of embryonic leaves and many other characters that makes this seemingly trivial character of great significance for classification.

Traits of embryos, such as the larvae of barnacles, can also be very useful for classification. Louis Agassiz and other contemporaries of Darwin argued that the anatomical features of embryos are the most reliable for classification because the early life stages are more stereotyped than the adults in appearance. Rudimentary organs, ones that serve no apparent function, can be useful as well. The Perissodactyla, an order of mammals that includes horses, tapirs, and rhinos, provides an example of the combined use of embryonic traits and rudimentary organs in classification. Perissodactyls are referred to as the "odd-toed" ungulates (hoofed animals), as opposed to the artiodactyls, or even-toed ungulates, which include deer and antelope. The recent evolution of both groups included a reduction in the number of digits, from five to either three or one in some perissodactyls, and from five to two in some artiodactyls. At birth, horses have only a single digit—their hooves are the expanded toenail of their third digit. We can see their affinity to the other perissodactyls in their development and in the skeleton of the adults. Early horse embryos have five digits but lose all traces of the first and fifth early in development. As the embryo grows, the growth of third digit keeps pace with the rest of the body, while the second and fourth digits have reduced growth rates, so they end up as internal, thin splints of bones, or rudimentary digits, in the legs of adult horses. With rare exceptions, no trace of these digits can be seen in the external anatomy of the horse. The presence of the first and fifth digits early in development and of the rudiments of the second and fourth digits in adult horses helps to establish their relationship to tapirs and rhinos (fig. 26).

The actual subdivision of organisms into different levels of the taxonomic hierarchy is seemingly arbitrary because the number and diversity of species we see in a taxon of plants or animals is quite variable. There are families and even orders that contain only one or a few species, all of which are quite similar in appearance, then other taxa of the same rank that have hundreds to thousands of species that can be quite variable in appearance. The frog family Ascaphidae, for example, contains only one genus and two species of frogs, which are found in the steep, fast-running streams of the northern Rocky Mountains and mountains of the Pacific Northwest. These two species are so similar that they were classified as one until 2001. In contrast, the frog family Leptodactylidae has roughly fifty genera and 1,100 species. It includes the genus *Eleutherodactylus*, the largest of all vertebrate genera, with more than 700 species. It continues to grow with the discovery of new species. In fact, I helped catch what are now recognized as two new species during a single night of collecting in a steep-sided ravine in Darien Province, Panama. The genus *Eleutherodactylus* is divided into subgenera, some of which are as different from one another as different families of

Figure 26
Developing Forelimb of an Embryonic Cow
Adult cows, like horses, have hooves that appear to be only a single digit, but their cloven hooves are actually the fingernails of the third and fourth digits. Pictured here is the appearance of the forelimb early in development. It has five digits, plus all the bone homologies illustrated in figure 17, including the humerus (H), radius (R), ulna (U), the bones that define the wrist and body of the hand, and five digits (I–V). Only digits III and IV will be retained in the limb of the adult. Ca3 and ca4 are the carpals, which are homologous to the bones in our wrist. In cows, they will fuse to form a single cannon bone. The development of horses begins with the same five-digit arrangement as illustrated here, but differs in the pattern of digit loss and fusion (see text).

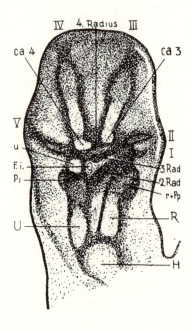

other frogs. They range from robust terrestrial frogs that look like Jabba the Hutt of *Star Wars* fame to slender, arboreal frogs that breed in bromeliads that line the branches of trees, far above the forest floor.

In summary, Darwin argues that classification reveals no evidence of a plan or design. If there were some plan, then there should be some observable regularity to the rules of classification or in the composition of each division of the taxonomic hierarchy, but there is not. To Darwin, the so-called "rules" of classification "are explained, if I do not greatly deceive myself, on the view that the natural system is founded on descent with modification; that the characters which naturalists consider as showing true affinity between any two or more species, are those which have been inherited from a common parent, and, in so far, all true classification is genealogical; that community of descent is the hidden bond which naturalists have been unconsciously seeking, and not some unknown plan of creation or the enunciation of general propositions, and the mere putting together and separating objects more or less alike" (*Origin*, p. 420).

Darwin argues that the reason "natural" classifications take the form they do—groups nested within groups; each group united by some odd selection of stable traits; very little pattern in the number of species within any one group; and differences both among and between groups being quite variable—is that their classification represents a mapping of their evolutionary history. All these patterns are consistent with what Darwin's theory predicts about the history of life.

The next step in Darwin's argument is to show how evolution by natural selection causes the hierarchical nature of taxonomic classification, with groups nested within groups. He begins this argument in the second and fourth chapters of the *Origin* (my chapters 3, 5, 9, 10), starting with a description of how he made a table containing the genera of plants, arranged by how many species each genus contained. He could see from this table that genera with many species tended to have species with many varieties. The individual species in large genera also tended to have large geographic ranges. He inferred that a genus with many species was derived from a single species that had the ability to expand in population size and range and to occupy diverse environments, and that this ancestral species had, as a consequence, diversified into multiple species. This expanding cluster of species would first be recognized as a "species group" within the genus and later as a separate genus. Some species that comprised this new genus would inherit whatever the traits were that had enabled their parent species to diversify, so they too would diversify into multiple varieties, then species. Darwin argues here that it is this progressive diversification of single, successful species into multiple descendants that results in a nesting of species within genera, then genera within families, and so on.

Darwin's versatile figure, the only one he included in the *Origin* (see fig. 12 in chap. 10 above), shows how this process looks if diagramed as a tree of life. He originally used this figure in his fourth chapter to envision the fate of eleven species in a genus. The diagram was interpreted from the bottom up as some of these species diversified into many descendants while others went extinct. Here, he instead considers the figure from the top down and treats the original eleven and all numbered nodes as genera rather than species. The top of the figure represents a large group of related organisms as we see them today, divided into discrete genera, families, and orders. The question is, how did they become what they are today?

If we begin at the upper left side of the tree and trace downward, then we first see a cluster of three genera (a^{14}, q^{14}, p^{14}) derived from a single common ancestor (a^{10}); the relationships between these three genera are reflected in their greater similarity to each other than to species found outside this cluster. To the right lies a second cluster of genera (b^{14}, f^{14}) that trace their ancestry back to a different common ancestor (f^{10}); they too bear similarities to each other that reflect their close relationship. The f^{10} and a^{10} ancestors in turn share a^5 as a common ancestor, so the two clusters of living genera will carry evidence of this more distant relationship, although they will share less in common than do the genera within each of the two clusters. We can think of these two different clusters of genera as forming two subfamilies, and all five genera within the two clusters as forming a distinct family.

This interpretation of the figure shows that species cluster together into a genus because they share a common ancestor. Likewise, genera cluster within a family because they share a more distant common ancestor. In this fashion, Darwin provides a biological explanation for the hierarchical relationship among organisms that was invented by Linnaeus for classification, then adopted by all who came after him because it seemed "natural." The taxonomic hierarchy is indeed "natural," Darwin argues, because it maps out a family tree of living species.

The order Carnivora provides an example of this hierarchy. It includes organisms as diverse as lions, bears, wolves, skunks, and seals, each of which represents a different family. All these families are derived from a single common ancestor that lived during the Eocene, approximately 50 million years ago. The order split into the "catlike" and "doglike" suborders approximately 42 million years ago, then into the recognized families. The living Felidae are derived from a common ancestor that lived approximately 11 million years ago and later diversified into several genera, including *Panthera* and *Felis*. The genus *Panthera* includes the lion, tiger, leopard, and jaguar, while the genus *Felis* includes the domestic cat, its closest relative, the wildcat (*Felis sylvestris*), plus at least five other species of similar, small cats from Africa, Europe, and Asia. More generally, as you go up the tree from order to family to genus, you progress to groups of animals that share a more recent common ancestor and are progressively more similar to one another in appearance.

Darwin's figure also illustrates why he thinks taxa differ in the number of species they contain and in how diverse those species are. The genera alive in period 14 represent two diverse orders descended from single species in the A and I lineages; however, the living genera also include f^{14}, which are the descendants of the original genus F that persisted because it occupied some environment where it was sheltered from interactions with the descendants of A and I. The low rate of diversification in F may mean that f^{14} is classified in the same genus as the original parent species. This genus is thus much older than the living genera in the A and I lineages. We see examples of such stability in living species. The brachiopod genus *Lingula* includes species alive today along with ones that lived during the Paleozoic, hundreds of millions of years ago. "Thus, on the view which I hold, the natural system is genealogical in its arrangement, like a pedigree; but the degrees of modification which the different groups have undergone" vary, and this variation is represented by their occupying different ranks in the taxonomic hierarchy (*Origin*, p. 422). The descendants of F may represent a small and not very diverse genus, while those of A and I represent large and diverse orders. It is these differences between lineages in their rate of diversification and their rate of formation of new species that account for the haphazard nature of what we now recognize as different levels of the taxonomic hierarchy.

The process of expansion of some lineages at the expense of others explains why, as we progress up the hierarchy from genus to family to order, and so on, the number within each rank declines: "The larger and more dominant groups . . . tend to go on increasing in size; and they consequently supplant many smaller and feebler groups. Thus we can account for the fact that all organisms, recent and extinct, are included under a few great orders, under still fewer classes" (*Origin*, pp. 428–29).

Many biologists today are impressed that all living animals are classified into only approximately thirty phyla. Each phylum can be characterized by a distinct body plan, and many of them date to the earliest-known fossils of multicellular animals. It has been inferred by some that this small number of phyla represents some mysterious constraint on the number of body plans available to animals. Darwin would likely say instead that the limited number and distinctness of phyla tell us nothing about constraints. They simply represent the few dominant lineages that have continued to expand and diversify, driving others to extinction. The only reason their body plans are so distinct is that their origin is in the very distant past, so divergence has driven them far apart as extinction erased whatever connections once existed between them.

Darwin's theory also explains why some characters are not useful for classification. Adaptations to a particular lifestyle are generally not good characters for classification because, as a lineage expands and occupies different habitats, there is rapid evolution of adaptations specific to new environments. These adaptations will not be shared by closely related species that adapted to different habitats. Their relationships to one another may instead be more easily recognized by "seemingly trifling" or rudimentary characters that are not closely tied to local adaptation and hence will be more stable and better indicators of shared ancestry, like the vestigial toes of horses.

Convergent evolution represents a second reason that characters tightly linked to adaptation to a particular environment are often not useful for classification. The similar body shapes and "fins" of whales, seals, manatees, and fish represent independently evolved adaptations to their aquatic lifestyle. Their similarity reflects their common function rather than shared ancestry. We view convergent evolution as the systematists' bane because it is a source of anatomical similarity, sometimes to an extreme degree, that can confuse rather than promote an understanding of relatedness. I illustrate one remarkable example of convergence in figure 27.

Darwin argues that the naturalists of his day were already intuitively using genealogy when they classified organisms, so his argument for evolution as the cause of the taxonomic hierarchy goes only a few short steps beyond their accepted principles. For example, common sense dictated that males, females, and juveniles be classified as the same species regardless of how different they were in anatomy or lifestyle. The juvenile life stage of many organisms is often a larva that is separated from the later life stages by

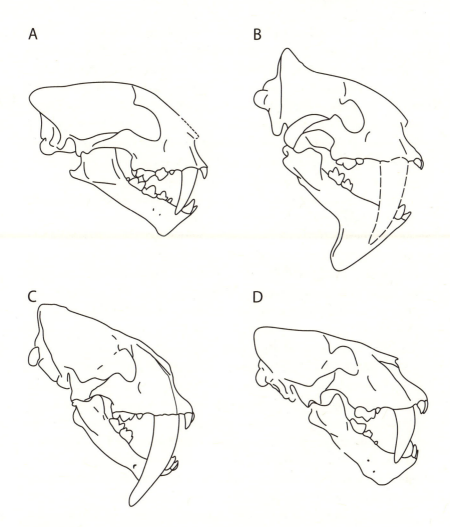

Figure 27
Convergence in Saber-Toothed Predators
Pictured here are representatives of two lineages of predators in the order Carnivora, one from the family Nimravidae (A = *Dinictis felina*, B = *Eusmilus sicariu*) and one from the family Felidae (C = *Smilodon fatali*, D = *Homotherium serus*). Each lineage independently evolved saber-toothed morphology. This morphology evolved at least two additional times, one or more times in the order Creodonta and once in the marsupial order Sparassodonta. All these animals shared the similarity of having enlarged upper canines (the "sabers") that were specialized for inflicting shallow, slashing wounds, but these animals also shared integrated changes to the morphology and biomechanics of the remainder of the skull that enhanced their ability to kill and process prey. One such change that is visible in all the skulls pictured here is the incisors that are in front of the sabers. These enabled the predators to tear apart subdued prey without having to engage the saberlike canines.

a metamorphosis that completely changes anatomy and habitat; for example, legless, aquatic tadpoles metamorphose into terrestrial frogs. When tadpoles and frogs are classified, the developmental link between larvae and adults takes precedence over any anatomical difference between them, however dramatic. Likewise, males and females can be dramatically different from one another (e.g., see fig. 8, ch.5), yet are classified as the same species.

Darwin's contemporaries also routinely recognized varieties within species and classified them as such. They made the distinction between calling two forms "varieties within a species" versus different species by studying their geographic ranges to see if there was any evidence of overlap and interbreeding between them. This practice is again an intuitive acceptance of genealogy, the ability to interbreed, as a criterion for classification. Darwin argues that systematists should accept his theory that a species can diversify into different varieties and that those varieties can become different species, because this "same element of descent (has) been unconsciously used in the grouping of species under genera and genera under higher groups" (*Origin*, p. 425).

Finally, the emphasis on using multiple, correlated characters for classification, rather than individual characters (e.g., classifying plants by embryonic leaves plus flower structure plus leaf venation), is also an unconscious mapping of genealogical relationships, which is why the system seems "natural." "We may err . . . in regard to single points of structure, but when several characters, let them be ever so trifling, occur together throughout a large group of beings having different habits, we may feel almost sure, on the theory of descent, that these characters have been inherited from a common ancestor" (*Origin*, p. 426).

Darwin concludes that his theory accounts not only for the hierarchical classification of organisms but also for the rules that systematists were using (at that time) to construct the hierarchy. His theory explains why correlated characters or characters that are rudimentary or trifling can be of value in classification and why similarities that are products of convergent evolution are not of value for classification. Combining classification as practiced in Darwin's time with Darwin's theory of evolution also opens up new possibilities: "We shall never, probably, disentangle the inextricable web of affinities between the members of any one class; but when we have a distinct object in view, and do not look to some unknown plan of creation, we may hope to make sure but slow progress" in defining the tree of life (*Origin*, p. 434).

Morphology—Background

Morphology was one of the largest divisions of the biological sciences of the nineteenth century. The general endeavor was to describe and understand how organisms were put together and how they worked. By the time of the *Origin*, the growth of knowledge of comparative anatomy had resulted in a synthesis and transformation of the study of morphology similar to what was seen in biogeography. This growth of knowledge made it possible for Cuvier to develop comparative anatomy as a predictive science, as when he predicted the structure of the pelvis of a fossil marsupial before it was extracted from its rock matrix or when he was able to reconstruct a whole animal from just a few fossil fragments of its skeleton. This synthesis also led Cuvier to envision and define the embranchments of life, as opposed to the idea that all life was part of a single scale of being, as argued by Lamarck and others.

Richard Owen began his career in this tradition, such as when he accurately described the moa from a fragment of one of the bird's leg bones. He then extended the synthesis to a more general concept of the structure of organisms' body plans. One of the hallmarks of his career was his interpretation of the comparative anatomy of the vertebrates and other organisms as variations on the theme of an archetype, or ideal. This ideal is what I referred to earlier (chap. 14) as the "unity of type." Comparative anatomists of the day recognized, for example, that all quadrupeds, or vertebrates with four limbs, have the same basic arrangement of bones in their limbs, regardless of lifestyle (see fig. 17, chap. 14, and fig. 26, this chap.). The arrangement of bones is so stereotyped that it is possible to recognize and apply the same name to each bone in the limbs of all quadrupeds. Quadrupeds with limbs highly modified for flight (birds, bats, pterosaurs), swimming (seals, whales, ichthyosaurs), or running (horses, deer) all have the same arrangement of bones, regardless of the extent to which the limb has been modified for a special function. Some bones are lost and some may be fused, but the basic topography of the limb remains the same, hence the descriptor "unity of type."

One concept that Owen formalized from this work was that of "homology," which he defined as "the same organ in different animals under every variety of form and function." In the case of the limbs, the fingers and toes could be defined as homologous in contexts as diverse as the single, enlarged toe of a horse limb, a contributor to the fin of a porpoise, or as a strut supporting the elastic flight surface of a bat's wing.

Owen argued that each of the embranchments of life, as originally defined by Cuvier, could be represented by a distinct archetype. The archetype represented a conceptual ideal for the design of an organism, while individual species represented variations on this central theme. He introduced this concept to Darwin as Darwin was contemplating his taxonomic revision of the barnacles in the late 1840s. He urged Darwin to study the barnacles by first characterizing the archetype, then conceptualizing each species and genus and order as a variation on this central theme by defining the homologies among all of them. Darwin made good use of this advice, but arrived at a different interpretation of what homologies represent.

Morphology—the Origin

Over the course of his career, Owen concocted different interpretations of his archetypes, but the prevalent one was that the archetype was an imprint of the vision of the Creator. Here, Darwin proposes an alternative interpretation. He argues that the homologies are products of evolution; the reason that all vertebrate limbs follow a common design is that all vertebrates are descended from a single common ancestor that evolved limbs with that same arrangement of bones. As the quadruped lineages multiplied and adapted to an increasing diversity of environments, they inherited the same arrangement of bones from their common ancestor, but the structure of individual bones was modified by natural selection. (Recall that evolution is a process of tinkering, which means that it is most often a modification of existing structures for new functions.)

Darwin deals next with a phenomenon called "serial homology," or segmentation. The body plans of many organisms, including vertebrates, consist of serially repeated units that are all very similar in structure. Envision a centipede, with its many pairs of legs. Each pair of legs is part of a defined segment, which you can see as a separate ring of cuticle girdling the upper surface of the body. The body as a whole is like a horizontal stack of coins, or a series of segments lined up along the length of the body, all very similar in structure. Many groups of animals are similar in being composed of repeated units. If you take a close look at an earthworm, you will see that it too consists of a series of rings, all of which are very similar in structure. When you look in a mirror, you might find it hard to see anything in common between yourself and a worm or centipede, but that is because our segments are hidden under our skin and only part of the human body is segmented. Our segmentation extends from the top of our neck to the base of our spine. Each vertebra is part of such a unit, with the other parts including nerves and muscles. Vertebrate limbs are not part of this segmentation; they were added to a segmented trunk. The head is actually a composite of what was the anterior tip of a segmented body attached to some unsegmented elements, which now include the skull and much of the brain, laminated around the segmented core.

One of Owen's accomplishments was to characterize the segmented body plan that was his archetype for all vertebrates, including those lacking limbs, such as the fishes (fig. 28). From a Darwinian perspective, Owen's archetype defines the anatomy of the common ancestor of all vertebrates. When a body plan consists of such repeated units, it is possible for natural selection to modify some independently of others as the organism adapts to new environments. Some segments may become specialized for a particular function while others retain the same function they had in the ancestor. In the case of the vertebrates, the vertebrae of fish are all very similar in struc-

Figure 28
Richard Owen's Vertebrate Archetype
This diagram represents Owen's vision of the ideal around which all vertebrates were created. He argued, from his vast knowledge of comparative anatomy, that components in this diagram are homologies that can be found in all vertebrates, so that all of them can be considered to be variations on this unifying theme. Many of the components of this archetype are part of your vertebrae. The vertebrae of adult humans, and of most vertebrates, are a single block of bone, but they actually represent the fusion of a number of bones depicted here that are distinct early in embryonic development. Owen envisioned the entire skeleton of all vertebrates as being derived from these same building blocks.

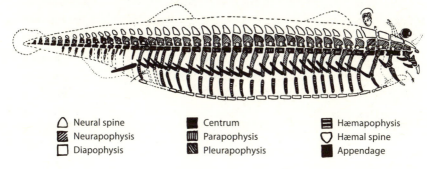

△ Neural spine	■ Centrum	▤ Hæmapophysis
▨ Neurapophysis	▥ Parapophysis	▽ Hæmal spine
☐ Diapophysis	▧ Pleurapophysis	■ Appendage

ture and function. They serve as a horizontal, central axis for the waves of muscular contractions that propel fish through the water. With the transition to life on land, the vertebral column became converted into a central axis from which the body weight of the animal was suspended. Some vertebrae became fused to the pelvic girdle and provided an anchor point for the hind limbs to transmit force to the rest of the body. Some became modified into the neck, which facilitated the ability of land animals to move their head independently of the rest of the body.

In arthropods, the primitive body plan consisted of many segments, each of which had a pair of appendages. Each appendage had two branches, one of which served as a gill, the other as a walking leg. As the number of species of arthropods multiplied and diversified, this basic body plan was modified as segments fused to become distinct body parts and the appendages associated with each segment became modified for special functions, or perhaps lost. In the insects, for example, segments fused to form a body that consists of three divisions—head, thorax, and abdomen. The thorax retains three pairs of legs derived from what were three distinct segments in the arthropods' common ancestor. The mouthparts are modified appendages derived from some of the fused anterior segments. The huge diversity of morphologies that we see in the arthropods can be interpreted as the product of this same ancestral body plan, as modified by the fusion and specialization of the different segments. Darwin asks: "Why should one crustacean, which has an extremely complex mouth formed of many parts, consequently always have fewer legs; or conversely, those with many legs have simpler mouths?" (*Origin*, p. 437). The answer is that they are derived from a common ancestor that had a given number of segments in the anterior part of the body; each lineage then differed in how this finite number of segments was allocated to the head or body.

Darwin presents one misconception here, retained through the sixth edition of the *Origin*, that demands correction. Richard Owen and others thought that the skull of vertebrates consisted of modified vertebrae. In one of his many victorious confrontations with Owen, Thomas Huxley argued that the skull was not derived from modified vertebrae. Huxley was right. Rather than being composed of modified vertebrae, the skull is something new that was added on top of the segmented core of the common ancestor of the vertebrates.

Darwin here interprets the similarities we see in comparative anatomy as reflections of traits inherited from a common ancestor, then modified as descendant species adapted to different environments. This may seem like a small step. If you see it this way, it is because Darwin was so successful that we now take this interpretation for granted. His legacy, from a modern point of view, is that here he took on one of the largest endeavors in nineteenth-century biological science and gave a unifying explanation for all the patterns described by prior investigators. His explanation also redefined the goals for all anatomical research done after the *Origin*. It was now possible to interpret similarities in structure from an evolutionary perspective. They could be used to make inferences about how different groups of organisms were related to one another and how anatomy had been shaped and transformed by natural selection as organisms adapted to different environments. If he had not already done so, then it was here that Darwin turned Richard Owen into a lifelong adversary.

Embryology—Background

Embryology, like morphology, was one of the largest areas of endeavor in the nineteenth-century biological sciences. While anatomists concentrated on the anatomy of free-living life stages, embryologists instead studied how anatomy unfolds during development in the egg or womb of the mother. Long before the Origin, naturalists had seen a relationship between how organisms develop and how they are related to one another. A common observation was that organisms that are very different in appearance as adults can be very similar in appearance early in development. Louis Agassiz recognized this relationship when he advocated embryonic traits as being of great value in classification. Karl Ernst von Baer described this similarity in his epic work on development, Entwickelungsgeschichte der Thiere (1828), where he stated that

> the embryos of mammalia, of birds, lizards, and snakes, probably also of chelonia [turtles], are in their earliest states exceedingly like one another, both as a whole and in the mode of development of their parts; so much so, in fact, that we can often distinguish the embryos only by their size. In my possession are two little embryos in spirit, whose names I have omitted to attach, and at present I am quite unable to say to what class they belong. They may be lizards or small birds, or very young mammalia, so complete is the similarity in the

mode of formation of the head and trunk in these animals. The ex-
tremities, however, are still absent in these embryos. But even if they
had existed in the earliest stage of their development we should learn
nothing, for the feet of lizards and mammals, the wings and feet of
birds, no less than the hands and feet of man, all arise from the same
fundamental form." (quoted in *Origin*, 6th ed., p. 388)

It was this similarity of early life stages that made embryonic and larval
characters so useful for classification, although there were competing ideas
about what these similarities in appearance meant. One theory was embodied in a school of thought named *Naturphilosophie*, which had its origins in
Germany in the late eighteenth century. An early adherent to this philosophy was Johann Wolfgang von Goethe, who is perhaps better known for
his contributions to literature. Prominent nineteenth-century contributors
include J. F. Meckel and E. Serres. Under this school of thought, the stages
that embryos went through during development represented a recapitulation of their creation. Before the *Origin*, recapitulation was tied to the idea
that all organisms could be arrayed on a single scale of being. Each step up
the scale represented an improved version of life that was attained by tack-
ing a new stage of development onto the last, adult stage of the organism
that occupied the previous tier.

Von Baer opposed the recapitulationist interpretation of development.
That interpretation implied that the development of every kind of organism would precisely retrace the same climb up the scale of being until each
type reached its own level in the hierarchy. Thus, an embryo should proceed through the last developmental stage of whatever type of organism
occupied the tier below it, then add some new stage at the end. Von Baer
disagreed with this interpretation of embryology for two reasons. First, his
observations were not consistent with its predictions. He argued that what
we first see early in development is the definition of the large group of animals that an embryo belongs to. More specific characters appear as development proceeds. So, for example, we might first recognize an embryo as
a vertebrate, then as a quadruped, and only later as a mammal. This was
why he could not tell what class his unlabeled embryos belonged to. This
progression from general to specific traits meant that one could never see
an embryo pass through the adult stage of some other organism on its path
toward the completion of development.

Second, von Baer, like Cuvier, interpreted the relationships between animals as consisting of multiple, independent embranchments, rather than being organized into a unilinear scale of being. He claimed to have discovered embranchments independently of Cuvier, through his study of comparative embryology. Cuvier's interpretation was instead derived from a study of the comparative anatomy of the adults. Von Baer found that he could recognize what embranchment of life an organism belonged to in the earliest stages of development. Milne-Edwards, whom Darwin also cited in the *Origin*, summarized von Baer's ideas: "The metamorphoses of embryonic organization, considered in the entire animal kingdom, do not constitute a single, linear series of zoological phenomena. There are a multitude of these series. . . . They are united in a bundle at their base and separate from each other in secondary, tertiary, and quaternary bundles, since in rising to approach the end of embryonic life, they depart from each other and assume distinctive characteristics" (1844, p. 72, as cited in Gould 1977, p. 57).

Gould (1977) noted that this interpretation of development as a treelike branching process predates Darwin in envisioning the organization of life as a tree, even if it does not do so from an evolutionary perspective. It would be natural to think this sort of perspective would make von Baer an early adherent to evolution, but some of his last essays were devoted to opposing Darwin's theory.

Louis Agassiz's contribution to this argument was to synthesize observations from development and the fossil record: "Agassiz insists that ancient animals resemble to a certain extent the embryos of recent animals of the same classes; or that the geological succession of extinct forms is in some degree parallel to the embryological development of recent forms. I must follow Pictet and Huxley in thinking that the truth of this doctrine is very far from proved" (*Origin*, p. 338).

Embryology—the *Origin*

Before Darwin, the recapitulationists saw strict rules that governed the relationship between embryology and classification: just as embryos seemed to follow a linear scheme of development, so should taxonomy be organized on a linear scale of being. Here, in this section of his thirteenth chapter, Darwin enters the fray with von Baer's perspective, which is to view development as a process during which the general properties of a group of organisms appear first, followed by the appearance of the more refined characteristics of a given taxon. Because general traits appear early in development, they often reveal how organisms are related to one another. Beyond this, Darwin argues there are no fixed rules for the relationship between development and evolution. Development is the path that is followed to define the adult organism, so it makes sense that if natural selection can modify the anatomy of the adult, then it will do so by modifying the developmental processes that lead to adulthood. In addition, many organisms go through discrete life stages on the path to adulthood; first they are embryos, then they may be free-swimming larvae, then they metamorphose into adults. Larval stages may be added, such that multiple, discrete larval stages are seen between the embryo and adulthood. Or, larval stages may be deleted, so that development goes directly from the embryo to a juvenile that has a morphology and lifestyle similar to those of an adult. Natural selection can act on each life stage somewhat independently of the others, although it is also possible for selection on one life stage to cause correlated changes in other life stages. Darwin argues that there is good cause to expect natural selection to have a smaller impact on early development than on the morphology of the adult. However, he concludes that there are no deeper rules dictating the patterns of development, no matter how heavily embryologists that preceded—and many that postdate—the *Origin* have speculated to the contrary. Development may record some aspects of the history of evolution, but in Darwin's view, as we shall see, it will be a quirky and imperfect history at best.

One "rule of development" in Darwin's day was that early life stages of different but related species were more similar in appearance than later life stages. Darwin argues that we should consider the environment experienced by an embryo to understand why this is true. Embryos develop inside an egg or in utero, which means they do not fend for themselves, nor are they fully exposed to the outside world. Therefore embryos will not be subject to natural selection in the same fashion as a free-living larvae or adults, so we should not expect them to show evidence of adaptation to a particular habitat. For this reason alone, we should expect that the embryonic life stage will be relatively stable as the descendants of some successful, diversifying lineage adapt to new environments. It is this stability that makes embryos useful for classification.

Things can be different if the organism is active and fends for itself in its early life stage. If this is the case, then it will be subject to natural selection at that stage, and its characters at that stage will evolve as it adapts to particular environments. In fact, it is possible for the early life stages of some species to be more different from one another than are the adults, but only if they are adapted to different lifestyles. We see an example of such diversity in the echinoderms, including starfish and sea urchins (fig. 29). Species in the same genus that are very similar as adults can look very different as larvae

Figure 29
Larvae and Adults of Two Species of Sea Urchin
Larval life stages can evolve independently of the adult life stage. Top left: *Heliocidaris tuberculata* larva; top right: *H. tuberculata* adult. Bottom left: *H. erythrogramma* larva; bottom right: *H. erythrogramma* adult. The adults of *H. tuberculata* produce free-swimming larvae that feed in the water column for a prolonged period before settling and metamorphosing. The adults of *H. erythrogramma* instead produce a smaller number of large, well-provisioned eggs that spend little time in the water column and do not feed as larvae but instead settle quickly and metamorphose.

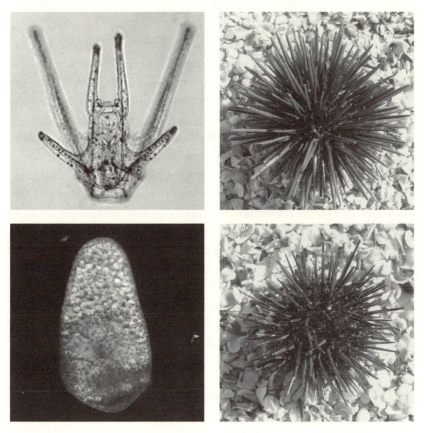

because of differences in the ecology of the larval life stage. Some species spawn many small eggs that hatch into free-living, actively feeding larvae. Others lay few, large eggs that hatch into larvae that spend only a short time in the water column as nonfeeding larvae, then settle and metamorphose into juveniles that look like miniature adults. The free-living, actively feeding larvae are structurally much more complex than the direct-developing larvae because they have to navigate, feed, and grow before they can metamorphose. Those species with direct development have lost all the features that were required for life as an independent larva. The adults of species that produce these two types of larvae are adapted to very similar environments, however, and are thus much more similar in appearance than are the larvae.

It is more often true that the morphology of larvae is much more stable than that of the adults. Darwin reminds us of his beloved barnacles, which Cuvier classified as mollusks, based on the morphology of the adults, but which were later revealed to be crustaceans because of the morphology of the larvae. This stability of larval versus adult characteristics reflects the very similar lifestyles of crustacean larvae and the very different lifestyles of the adults; since all larvae have a similar lifestyle, there is weak or no selection for them to evolve different morphologies. Their uniformity reflects uniform habits; it does not reflect some inherent constraint on their evolution.

A second "rule of development" was that organisms increase in complexity as they progress to later life stages. While Darwin does not say so here, this trend was interpreted by the advocates of recapitulation as the addition of new, more advanced life stages at the end of the developmental sequence of their less advanced ancestors. Darwin observes that a butterfly is more complex than a caterpillar, on the basis of its having many specialized structures. However, this is certainly not a rule. He notes that the typical first-stage larva of a barnacle has three pairs of legs, a rudimentary eye, and a functional mouth. They enter a second larval stage in which they actively seek a site where they will metamorphose into an adult. The second-stage larva has six pairs of legs, a pair of well-developed, compound eyes, and complex antennae, but it lacks a functional mouth because it does not feed. When they metamorphose into adults, the second-stage larvae adhere to a solid substrate and develop the protective shell. Their antennae disappear, their eyes degenerate into a single, simple structure, and they redevelop a functional mouth. Given the complexity of these changes, it is impossible to objectively describe one life stage as inherently more complex or by some measure "higher" than another. Furthermore, some of the nonfeeding second-stage larvae metamorphose into males, which have a structure far less complex than that of the larvae. The male "is a mere sack, which lives for a short time, and is destitute of mouth, stomach, or other organs of importance, excepting for reproduction" (*Origin*, p. 441). The males lead the life of a parasite, so their own life-support systems are degenerate in comparison to those of the females or hermaphrodites, which is why the males can be little more than sacks of sperm (see fig. 8, chap. 6). Other crustaceans metamorphose into adults that anchor onto and become parasites of other species; one example is the anchor worm parasites of fish. They too are highly simplified in comparison to their larvae, which reflects the adults' reliance on their host to provide for them. All these examples show that there is no necessary relationship between the anatomic complexity of early versus late life stages. Each stage evolves in accordance with its lifestyle.

The existence of larval and adult life stages, separated by a metamorphosis, may seem like a necessary part of development since it is so common, but it is not obligatory. Many organisms do not have a discrete larval stage and instead hatch or are born alive looking like miniature adults. Spiders and humans have much in common here because both begin life as miniatures of the adults. Most insects start life as wormlike larvae, but aphids, like spiders, lack this life stage and start with an adultlike morphology. The *Plethodon* salamanders (chaps. 9 and 10 above) lack the typical amphibian aquatic larval stage. Mothers lay eggs on land; juveniles that look like miniature adults hatch out. The frogs in the genus *Eleutherodactylus* have also deleted the aquatic larval (i.e., tadpole) phase. In the few species for which the development of the embryos has been studied, it was found that the embryos bypass the typical larval morphology as they develop directly into a froglet inside the egg. This diversity of early life stages indicates again that there is nothing sacrosanct about early development.

Darwin now lists the general facts of development that demand an explanation. They include

> the very general, but not universal difference in structure between the embryo and the adult;—of parts in the same individual embryo, which ultimately become very unlike and serve for diverse purposes, being at this early period of growth alike;—of embryos of different species within the same class, generally, but not universally, resem-

bling each other;—of the structure of the embryo not being closely related to its conditions of existence, except when the embryo becomes at any period of life active and has to provide for itself;—of the embryo apparently having sometimes a higher organization than the mature animal, into which it is developed. I believe that all these facts can be explained, as follows, on the view of descent with modification." (*Origin*, pp. 442–43)

In other words, all the diversity that we see in different life stages can be understood as a consequence of evolution by natural selection acting on individual life stages.

Darwin expands on these arguments with lessons learned from domestic animals. Breeders select on the appearance of or properties of the adults they are breeding. Darwin and others noticed that specialized breeds in which the adults had become very different from the adults of other breeds still produced young that were similar in appearance to the young of other breeds. The general rule seemed to be that, if selection acts on and causes some change in a particular life stage, we see the effects of that selection only in that same life stage in subsequent generations.

Darwin argues that two principles explain this observation. One is that early development takes place inside the mother's womb or inside an egg, so it is shielded from environmental influences that might act on juveniles or adults. The second is that experience shows that selection can act on individual life stages, either to modify them independently of others, to delete them entirely, or to add new life stages. It is possible that strong selection on one life stage could cause correlated changes in other life stages, but this is not necessarily what will happen.

Darwin tested these propositions. He compared the newborn of adults that were as different from one another as greyhounds from bulldogs, cart horses from racehorses, or the various breeds of domestic pigeons from the rock pigeon. In almost all cases, the differences between the young of the different breeds were much smaller than the differences seen in the adults. In the pigeons, the young displayed no differences at all in some traits, even though the adults were so different that they would likely be placed in different genera, were their origin as domestic breeds not known. The one exception was the short-faced tumbler. Here, the hatchings were almost as distinct from rock pigeon hatchlings as the adults were from rock pigeon adults even though this breed was the product of selection only on the appearance of adults. You may recall from chapter 2 above that this correlation between juvenile and adult morphology in the tumblers—having a short face at both life stages—could have fatal consequences, since the hatchling's beak was often so small that it could not break out of the egg. In a more natural setting this particular correlation would have been selected against.

Why do we see these trends? Breeders select adults or near adults for breeding. They do not consider the appearance of the hatchlings. As a consequence, the hatchlings are shielded from artificial selection, and any differences that we see in these early life stages will only be ones that are correlated with selection on adult morphology.

If we consider life in nature, the same rules should apply. Among birds, most nestlings are cared for by their parents, and so the most intense selection often falls on adults. For example, the large differences in bill dimensions that characterize the different species of the Galapagos finches (chap. 1 above) represent the feeding specializations of the adults, which feed on seeds of different sizes. The bill morphology of their dependent nestlings is irrelevant. As a consequence, the nestlings for all species within a genus of birds will generally be much more similar to one another than are the adults.

Imagine instead the embryonic development of mammals. As different orders adapted to different environments, the forelimb may have evolved into a bat's wing, a porpoise's paddle, or a human hand. In all cases, the embryo develops inside the mother, and the early life stages are under parental care. Among all mammals, the early embryos are much more similar in appearance than are the adults. It is only after birth that these animals begin to interact with their environment, and it is only then that they fully experience natural selection and fully display the specializations that characterize their species. Embryos will differ only to the extent that selection on the adults causes correlated changes in early development.

Why might selection favor direct development (i.e., the deletion of the larval life stage)? Here I will depart from Darwin's general arguments and present a specific one of my own, from frog- or salamander-centric perspective. Life as an aquatic larva can be tough. There are predators and parasites to contend with; sometimes the ponds can become terribly crowded and competitive; and sometimes they dry out before a larva can metamorphose. Direct development means being free of all these risks, but it comes with a price of producing fewer, better-provisioned eggs. If an amphibian evolves direct development, it also buys freedom from larval breeding sites and can invade habitats that do not have them, or enough of them to go around.

Some have speculated that direct development is the root of the success of *Plethodon* salamanders (see chaps. 9 and 10) and *Eleutherodactylus* frogs, the largest vertebrate genus (see above, this chap.), since both can be very abundant and have diversified into many species.

Why might natural selection favor the addition of a new life stage? It will do so if the new life stage fulfills some specialized function that increases the fitness of the species relative to an ancestor that lacked the added life stage. For example, the first larval life stage of barnacles is adapted to disperse, feed, and grow. The second, nonfeeding larval stage is specialized for finding a site to settle and metamorphose into an adult. Because adults glue themselves to their new home and no longer move, site choice is critical.

Darwin then returns to the role that embryonic traits play in classification. If common descent is the hidden bond behind the pursuit of a natural system of classification, then slowly evolving traits will be the most reliable in defining which organisms share a common ancestor. The reason that embryonic traits are so often of great value is that embryos are shielded from the effects of biotic interactions, which means that they are "in a less modified state; and in so far [as this state is retained,] it reveals the structure of its progenitor" (*Origin*, p. 449). If we consider animals like birds, mammals, and reptiles, which are wildly different in morphology as adults yet closely resemble one another as embryos, we can be confident that this resemblance is inherited from a common ancestor. However, the greater stability of early life stages is not a fixed rule, as illustrated above. Rather, the embryonic and larval life stages are instead more likely to be stable than the adult life stage because adults have a greater exposure to the biotic interactions that are the primary source of natural selection.

Agassiz's proposed law of nature was that "ancient and extinct forms of life resemble the embryos of their descendents, —our existing species" (*Origin*, p. 449). Darwin is intrigued by the possibility, but argues that this "law" will be true only if certain conditions are satisfied. There must not have been direct selection on early life stages resulting in the obliteration of the appearance of the common ancestor, nor should there have been strong selection on the adult causing correlated changes in morphology early in development, as Darwin found for the short-faced tumbler pigeon. The imperfection of the geological record will come into play here as well, since it may never permit us to see what the common ancestor of some of these ancient lineages looked like. If it was small and lacked "hard" tissues, then it will be unlikely ever to appear in the fossil record.

In conclusion, all the perceived rules of the embryologists are really just reflections of the relative importance of natural selection in shaping the different life stages.

Rudimentary, Atrophied, or Aborted Organs

Organs that appear to be degenerate and useless are common in nature. The existence of such organs is hard to explain under the hypothesis of special creation. Why create organs that serve little or no function? Darwin argues that his theory offers multiple explanations for such traits.

First, Darwin offers some examples of rudimentary traits. Many snakes have one large lung and a second that is reduced to a small, rudimentary lobe. Some also have rudiments of a pelvic girdle and hind limbs. Fetal baleen whales have the rudiments of teeth, even though the adults "have not a tooth in their heads" (*Origin*, p. 450). Fetal cows and deer have incisors in their upper jaws that never erupt; adults lack upper incisors. Beetles typically have functional wings that are protected by folding covers called "elytra." Many flightless beetles have only rudimentary, nonfunctional wings. In some, the elytra are fused and can no longer open, so the wings could not be used in flight even if they were present. Darwin is fascinated by the fact that male mammals have rudimentary teats and that, under some circumstances, they can even become functional. In flowering plants, a "perfect" flower has complete male and female reproductive organs and can produce both pollen and seeds. Some plants instead have flowers that function as either males only or females only. When this is the case, the flowers often contain rudiments of the organs that are typical of the opposite sex.

A second common pattern is that an organ serves two potential purposes in some organisms but, in related organisms, one of these functions dominates while the other becomes rudimentary. In fish, the swim bladder can function for either maintaining neutral buoyancy or for respiration. In some lineages it functions purely for buoyancy, while in others its only function is respiration.

Rudimentary organs often serve well as guides for classification because they, like embryonic traits, are stable. For example, the rudiments of the second and fourth digits in horses helped Richard Owen to recognize the affinity of horses to tapirs and rhinoceroses (chap. 18).

It is important to reflect on the significance of rudimentary organs. We always think of evolution as a process that progressively refines organs and organisms so that they are well adapted for particular functions, so we think of it as building complexity, not breaking it down. Natural selection can cause the evolution of an organ as exquisitely complex and refined as the eye, or an animal as specialized as a hummingbird. Yet we also clearly see all these examples of organisms retaining what appear to be the remnants of adaptations to some ancestral lifestyle led by some long-extinct ancestor.

There are two aspects of the development of rudimentary organs that give us clues to their origin. One is that they often appear early in development, but then are lost in the adult. This is true of the rudimentary teeth of whales and cows. The other is that they can be much larger and more prominent in the embryo but greatly reduced in the adult, as happens with the second and fourth toes in the limbs of horses.

Darwin proposes two general mechanisms for how a once-functional organ can be reduced to a rudiment or entirely lost from the adult life stage. The first is "use and disuse." An organ that is not used can actually become a liability; in an earlier chapter I recounted how Darwin found that the rudimentary eyes of his pet tuco-tuco, a subterranean rodent from South America, were infected because they had sustained an injury. For an animal that has no use for eyes, a retained eye can become a liability. A second mechanism is that loss is caused by the economy of nature and the cost of building and maintaining a structure. The resources that are used to build and maintain a structure are wasted if the structure serves no function, so there can be selection to favor the allocation of resources elsewhere. One example of such an economy is represented in the many species of insects that have both winged and wingless individuals in the same population. Individuals with functional wings also have large, metabolically active flight muscles. Those that lack wings have only rudimentary muscles. Their inability to fly is repaid by their being able to produce many more eggs than their winged counterparts, but they sacrifice the ability to disperse long distances and colonize new environments. Some insect populations or whole species have lost the ability to fly because they occupied a habitat where the advantage gained by producing more eggs outweighed the price paid for losing the ability to disperse, but they often retain rudiments of wings.

A more general view of Darwin's argument is that natural selection can cause any change that is attainable through a large number of small steps. For an organ that no longer serves some purpose, this may entail its loss through small steps or perhaps its conversion to some other function. If the organ serves multiple functions, then one function may be reduced as another replaces it.

The cause of the persistence of rudimentary organs may be the same as the cause of the tendency for different species to be more similar in early life stages than in the adult stage. If the incipient rudimentary organ is important only in adults, then selection against its development will be present only later in life. When animal breeders select for special attributes in adults, they produce breeds that are distinct from one another as adults but not necessarily as juveniles. When natural selection acts against the development of some trait in adults, it may still be present in some rudimentary fashion early in life, which is what we see in the teeth of baleen whales and ruminants, or in the first and fifth digits of horses. If all that remains of a trait is a rudiment expressed early in development, when it is not yet subject to natural selection, then there will be little or no selection in favor of its complete elimination. We imagine that this is the case for the rudimentary organs that are useful for classification, since the unifying property of all traits useful for classification is that they evolve slowly.

The hallmark of the process of evolution by natural selection is that it carries the imprint of history. It is this imprint that distinguishes it from special creation. A rudimentary organ is such an imprint because it is a fading echo of the life led by some distant ancestor of organisms alive today. The presence of such organs is a by-product of genealogy, or the fact that the organisms of today are modified versions of their ancestors. Darwin closes by saying: "On the view of descent with modification, we may conclude that the existence of organs in a rudimentary, imperfect, and useless condition, or quite aborted, far from presenting a strange difficulty, as they assuredly do on the doctrine of creation, might even have been anticipated, and can be accounted for by the laws of inheritance" (*Origin*, pp. 455–56).

Conclusion

Classification, morphology, and embryology all existed as large, well-developed disciplines before the *Origin*. These fields overlapped to some extent, and there was cause for suspecting a fundamental connection because the observations of morphologists and embryologists played such an important role in the development of a "natural" system of classification. What Darwin added with this chapter of the *Origin* was an explanation for why his contemporaries sensed (as we still do today) that these disciplines are connected. In doing so, he made evolution by natural selection the conceptual framework that unites them. When we recognize that our scheme of classifying organisms is actually a re-creation of their genealogical relationships, then we can also appreciate why those relationships are so well featured in the development and anatomy of every species. We can also appreciate why some traits are so effective in revealing relationships while others are not. The unifying feature of the former traits is that they evolve slowly, such that they can be recognized among related organisms even if they are adapted to very different environments. Some of the traits useful for classification are of high adaptive value, as is the case for lactation in mammals. Others, like the features of embryos early in development, are shielded from the selection experienced by free-living life stages. Yet others are the wrecks of structures that now serve little or no purpose in the adult life stages.

Darwin's theory also explains special properties of morphology and embryology. It tells us why we see what his predecessors called "unities of type" or "archetypes," which are the similarities in structure that unite large groups of organisms that are classified together. These are not Platonic ideals or visions of a creator, but are rather body plans inherited from some common ancestor, then modified as the descendants adapted to specialized lifestyles. The processes of cladogenesis and extinction, which shape the tree of life, also explain why all living organisms can be grouped into progressively fewer taxa at the top of the taxonomic hierarchy and why the phyla that represent the top of the hierarchy for animals can be distinguished by such different body plans. These patterns are simply a consequence of the fact that a few successful lineages multiplied, diversified, and inherited features of their common ancestor while most other lineages went extinct.

The special features of embryology, such as von Bear's observation that development proceeds by defining first the general, then the more specific properties of each organism, can also be understood as products of evolution by natural selection. To Darwin, all the trends we see are imprints of the evolutionary history of every organism. So, for example, the early stages of development vary less between species because they are far less likely to be subject to strong natural selection. This is so because the main agents of selection are the biotic interactions that an organism experiences throughout its life, but such interactions are weak or absent during the embryonic life stage.

At the end of the previous chapter I argued that plate tectonics became the unifying theory of the geological sciences because it gave us a unifying explanation for so many geological phenomena. Here, Darwin shows that evolution by natural selection does the same for the biological sciences. Here he binds together what were the largest branches of biology in his time. Subsequent developments have shown that his theory retains the same explanatory value in new scientific disciplines that he could not have anticipated.

Evolution Today

In the mid-1990s, Mark Springer began to address the origins of an order of mammals called the Insectivora, in collaboration with many colleagues, including Wilfried de Jong and Michael Stanhope. The insectivores were thought to be the living survivors of one of the first orders of placental mammals to appear in the fossil record. Their teeth are very much like those of the earliest placental mammals. We often rely on teeth for classifying mammals because they are so variable in structure among the orders, but also because they are the most durable tissue and are often all that remains in the fossil record. The living members of this order are particularly diverse, and some are quite peculiar. Springer was originally interested in the golden moles and tenrecs endemic to Africa, and wanted to know how they are related to other insectivores. What he found lay outside anyone's prior expectations.

If the insectivores were a natural grouping, then analyses of their DNA sequences should reveal that they were like branches off a single stem, meaning that they all shared a common ancestor. Springer found instead that the "insectivores" divided into two groups that were allied with very different stems (fig. 30). The golden moles and tenrecs were most closely related to an odd assortment of mammals that included the aardvark, sirens (manatees, dugongs), hyraxes, elephants, and elephant shrews. The other "insectivores" (moles, shrews, hedgehogs) were instead most closely allied with bats, carnivores, and hoofed mammals. What these two branches of "insectivores" really had in common was that they were slowly evolving members of different lineages that retained primitive skeletal features inherited from an ancestral placental mammal.

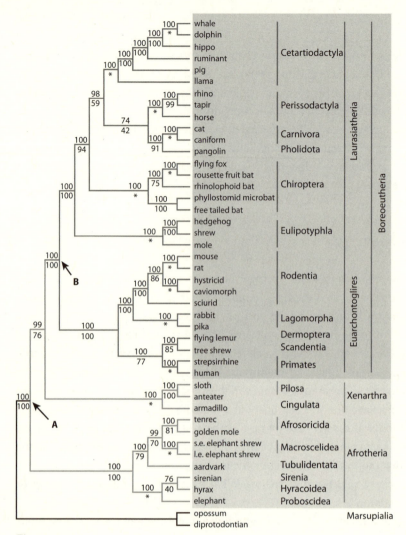

Figure 30
Mammal Family Tree
This tree diagram is based on DNA sequences from 19 nuclear and 3 mitochondrial genes and a total of 16,400 base pairs of DNA. The former "insectivores" are split into two distantly related lineages, one allied with the Afrotheria (tenrec and golden mole—at bottom of the diagram) and one with the Laurasiatheria (hedgehog, shrew, mole). The superordinal groupings (Afrotheria, Xenartha, Euarchontoglires, Laurasiatheria) are descended from a common ancestor that may have inhabited the Southern Hemisphere. Descendants were isolated on Africa (source of the Afrotheria), South America (Xenartha—the sloths, anteaters, and armadillos endemic to South America that so preoccupied Darwin), and in the Northern Hemisphere (Boreoeutheria, which split into the Laurasiatheria and Euarchontoglires). The numbers above and below each branch are the results of two different ways of evaluating the confidence that we can place on the conclusion that species downstream of the labeled branch really share a single common ancestor, with 100 being the highest possible degree of statistical certainty.

Springer noticed another oddity of the affinities between the golden moles, elephant shrews, and the others that they grouped with. All of them were either endemic to or had originated in Africa. A molecular-clock analysis of when they had shared a common ancestor dated to around 90–100 million years ago. Since Springer has degrees in geology and biology, he knew that this time frame carried special significance. It postdates the time when Africa broke away from the Gondwana supercontinent and drifted northward. Africa remained an island continent, as Australia is today, for tens of millions of years. The unexpected affinities that he found between the golden moles, elephant shrews, and other mammals thus represented a group of animals that were all descended from a common ancestor that had been stranded on this island continent and then diversified during this period of isolation. Springer and colleagues named this assemblage the "Afrotheria," which is a superorder since it contains animals so diverse that we now classify them into different orders.

This definition of a new mammalian superorder represented a synthesis of information from DNA sequences and what they revealed about the timing of diversification, from the paleontological record, and from the history of continental drift. It proved to be an opening salvo in what has become a redefinition of the classification of mammals above the level of order. In his thirteenth chapter of the *Origin*, Darwin predicted that sometimes the evidence of ancestry would be so obscured by extinction that it would be very difficult or impossible to reconstruct some complete family trees. Placental mammals have traditionally been one of these groups. It was easy to know which mammals were placental and which were not. And most placental mammals were easy to classify to the level of order. But there was little information in the anatomy of the living mammals to enable us to define the branching pattern that joined placental orders to an ancestral stem. Until recently, the fossil record revealed little about the transitions among orders. In fact, one of Simpson's motives when he wrote the modern synthesis classic *Tempo and Mode in Evolution* (1945; see chap. 19 above) was to explain why we see so little evidence for the origin of placental orders in the fossil record.

The definition of the superorder Afrotheria stimulated a new way of looking at the higher-order classification of mammals, since it showed that the combination of information from DNA sequences and historical biogeography could extract a sense of relatedness that had evaded prior investigators. As Darwin said, "when we have a distinct object in view, and do not look to some unknown plan of creation, we may hope to make sure but slow progress" (*Origin*, p. 434).

References

Browne, J. 1995. *Charles Darwin: Voyaging*. Princeton, NJ. Princeton University Press. (Pp. 475–46, Owen's urging Darwin to adopt the concept of the archetype in his barnacle studies.)

Gould, S. J. 1977. *Ontogeny and phylogeny*. Cambridge, MA: Harvard University Press. (Pp. 33–68, historical roots of embryology.)

Gould, S. J. 1989. *Wonderful life: The Burgess Shale and the nature of history*. New York: W. W. Norton & Co. (Pp. 97–107, details on invertebrate segmentation.)

Ibanez, R. D., and A. J. Crawford. 2004. A new species of *Eleutherodactylus* (Anura: Leptodactylidae) from the Darien Province, Panama. *Journal of Herpetology* 38:240–44. (Official description of one of the two new species collected in 2002.)

Mayr, E. 1982. *The growth of biological thought*. Cambridge, MA: Harvard University Press. (Pp. 171–80, details on Linnaeus.)

Murphy, W. J., E. Eizirik, S. J. O'Brien, O. Madsen, M. Scally, C. J. Douady, E. Teeling, O. A. Ryder, M. J. Stanhope, W. W. de Jong, and M. S. Springer. 2001. Resolution of the early placental mammal radiation using Bayesian phylogenetics. *Science* 294:2348–51.

Roff, D. A. 1986. The evolution of wing dimorphism in insects. *Evolution* 40:1009–20. (Trade-off between dispersal and egg production.)

Rudwick, M.J.S. 1976. *The meaning of fossils: Episodes in the history of paleontology*, 2nd ed. New York: Neale Watson Academic Publications. (Pp. 207–14, Owen's archetype.)

Springer, M. S., G. C. Cleven, O. Madsen, W. W. de Jong, V. G. Waddell, H. M. Amrine, and M. Stanhope. 1997. Endemic African mammals shake the family tree. *Nature* 388:61–64.

Stanhope, M. J., V. G. Waddell, O. Madsen, W. de Jong, S. B. Hedges, G. C. Cleven, D. Kao, and M. S. Springer. 1998. Molecular evidence for multiple origins of Insectivora and for a new order of endemic African insectivore mammals. *Proceedings of the National Academy of Sciences* 95:9967–72.

Villinski, Jeffrey T., Jennifer C. Villinski, Maria Byrne, and Rudolf A. Raff. 2002. Convergent maternal provisioning and life-history evolution in echinoderms. *Evolution* 56: 1764–75. (Direct vs. indirect development in urchins.)

Chapter 23

Recapitulation and Conclusion

As this whole volume is one long argument, it may be
convenient to the reader to have the leading facts and
inferences briefly recapitulated. (*Origin*, p. 459)

The last chapter of the *Origin* unfolds as if Darwin is a barrister arguing
his case before a jury. Darwin knew that his theory represented a radical departure from established science; here he defends it against his opponents' objections. He probably envisioned himself facing a jury box filled
with Cambridge dons, mentors, role models, and colleagues, including John
Herschel, William Whewell, Adam Sedgwick, Richard Owen, Louis Agassiz, Karl von Baer, Asa Gray, Charles Lyell, Joseph Hooker, Thomas Huxley,
Robert FitzRoy, and Georges Cuvier (in spirit), all leaning forward, staring
at him intently, most shaking their heads in disapproval.

The Opening Statements

Darwin opens by stating that his theory has solved the "mystery of
mysteries"—the origin of species. Then he goes far beyond, arguing that
one single mechanism, natural selection, accounts for phenomena as diverse as the appearance of design in nature, the structure of the taxonomic
hierarchy, comparative embryology, comparative anatomy, the structure of
the paleontological record, biogeography, and instinctive behaviors—and
these are just some of the highlights. Natural selection is as central to the
life sciences as gravity is to astronomy and physics.

A unifying theme of Darwin's defense, not stated here in the opening, will be to contrast what we should expect to see under his proposed theory versus what we should expect under the doctrine of special creation. It is easy to be anachronistic in interpreting this approach, since the same debate is going on today between different opponents. Darwin's opponents were the scientific establishment of his day. In taking on special creation, he was doing what scientists always do, which is to discriminate between alternative explanations. In Darwin's day, science and religion were confounded with one another. The science faculty of Cambridge and Oxford were ordained ministers in the Church of England, and "natural theology," as embodied in Archdeacon Paley's volume and later in the *Bridgewater Treatises*, was the explanation for the natural world. Because the prevailing belief was that each species was the product of an independent act of creation, special creation was the dogma Darwin was trying to disprove as he promoted his theory.

The arguments against his theory, on the other hand, will emphasize that the gradual, continuous process that is central to Darwin's theory is not what we see in nature. Everywhere, nature is riddled with discontinuities that defy a possible origin via a mechanism that capitalizes on small variations and gradual change. Species are separated by unbridgeable gaps, as we should expect if each is the product of an independent event of creation. Seemingly perfect organs, like eyes, appear without any trace of a transition. Where we see such design, there must be a designer. Species appear, then disappear, in the fossil record without any transition from one species to the next, as if this process of independent creation has occurred repeatedly throughout the history of life. Furthermore, Darwin would have us believe that all the design we see in nature is caused by small variations among individuals, and that these variations arise by chance and are random with respect to whether they are good or bad for the organism. How can order and design emerge from such chaos? How can these small differences between individuals, which Darwin claims to be the fuel for evolution, ever amount to making the difference between life and death, which is what Darwin's theory requires?

A modern reader may find such strong objections uncanny, as Darwin here turns his powerful intellect toward making the most forceful arguments he can muster against his own theory. His responses recapitulate the general argument he has made throughout the *Origin*, which is that, if species were independently created, then each species should be independent of all others; we should find no evidence of their genealogical interrelatedness. If, however, species are the products of evolution, there should be pervasive evidence of the genealogical relationships between them.

Darwin continues this quasi dialogue throughout the closing chapter. To a reader who is already convinced that evolution and speciation occur, the counterpoint of independent creation may be only of historical interest. For the significant segment of our society that is unconvinced by Darwin's theory, the point-counterpoint aspect of his final chapter will be of more immediate interest. For these readers, it is worth remembering that this "defense" of the theory was presented in 1859. Great progress has been made over the past 150 years. Those who support Darwin's theory can make a much stronger argument today. I have presented some elements of an updated argument in the "Evolution Today" segments above, but this book is about the *Origin* and not about the current state of evolutionary biology, so I do not present a comprehensive update of the evidence for evolution by natural selection.

The Case for the Prosecution

Darwin anticipated the most important arguments against his theory, so he confronted them throughout the *Origin*. These arguments represent the imagined case he builds against his own theory in his last chapter, and they remain elements of the opposition today, despite the quality of Darwin's rebuttals and all the progress made in support of the theory since 1859.

One key issue raised in opposition is the difficulty posed by complexity in nature. The features of organisms Darwin's theory seeks to explain as products of evolution include ones so complex, like the eye, that it seems they must be products of deliberate design. Darwin counters that every complex feature represents the assembly of what began as many small variations among individuals, each of which gave some advantage to the possessor. Darwin's defense rests on three propositions. First, there should be evidence of the steps required to attain a complex trait. Darwin found such evidence in the less complex versions we can sometimes see in living organisms, such as the range of photosensitive organs we see among species of mollusks (see fig. 16, chap. 14 above). Second, we should be able to find variation between individuals in such traits (chaps. 2 and 3). Third, all organisms produce more offspring than are required to sustain a stable population, causing them to be perpetually locked in a struggle for existence (chap. 4). The natural world is constantly burgeoning with an excess of most species, with their populations held in check by their interactions with other organisms. We only catch glimpses of this overproduction when some of the controls to abundance are removed, such as when the wild artichoke was transplanted from Europe to Patagonia, then exploded into being a dominant plant that claimed exclusive use of hundreds of square kilometers of land (chap. 4). A consequence of this struggle is that nature preserves only the fittest individuals—those having traits that enable them to prevail in the face of this struggle.

A second argument against the theory is that species are separated by "almost universal" reproductive barriers, which makes them distinct from varieties within species because varieties can freely interbreed; varieties are not separated by such barriers. Darwin counters that "species" are simply well-marked varieties, rather than being definable by some absolute boundary. Species cannot, in fact, be distinguished from one another by absolute reproductive barriers (chap. 11), as would be expected if each species were the product of some act of special creation. When two species are crossed, they sometimes produce viable offspring, and those offspring are sometimes fertile. When varieties are crossed, they usually produce viable offspring that can produce offspring of their own, but sometimes these crosses fail in the same way as crosses between species. The sterility we sometimes see when either species or varieties are crossed "is no more a special endowment than is the incapacity of two trees to be grafted together" (*Origin*, p. 460).

A third argument against Darwin's theory is that it predicts continuity throughout nature, but instead of seeing continuity, we see a natural world riddled with discontinuities. One obvious discontinuity is in the geographic distribution of many species. If, as Darwin argues, a species evolves in one location, disperses, and possibly diversifies into multiple species, then we should expect to see some continuity in their distribution. We instead sometimes see remarkably disjunct distributions, as if the organisms were independently created at each far-flung location.

Darwin offers two classes of explanation for how distributions came to be as they are today. One is that some organisms have surprising capacities for dispersal. Earlier in the *Origin* (chap. 20), he discussed many possibilities such as transport by wind, waves, or hitchhiking on rafts of flotsam, on icebergs, or, in the case of seeds, on the feet and feathers of birds, or in their crops and digestive tracts. Since the duration of any one species is usually millions of years, opportunities abound for emigration, even by means that seem improbable. We judge what is probable on very short timescales, like the duration of our lives, but evolution happens over vast intervals of time. The second class of explanations involves climate change. Darwin offers the ice ages as an example. Earlier in the *Origin* (chap. 20), he argued that the repeated advances and retreats of glaciers caused species ranges to shift to lower latitudes and altitudes as the climate cooled, then back again as it warmed. Climate warming could break up cold-adapted species' distributions into isolated pockets at higher elevations. Darwin imagined that such changes in climate, in concert with fluctuations in sea level and the emergence of new geographic barriers to dispersal, like mountain ranges, created discontinuities in how species were distributed.

But the opponents' focus on discontinuities highlights another problem: we also fail to see evidence of continuity between species. Darwin's theory creates the expectation that we should see a multitude of intermediate forms that span the gaps between species. We instead most often see species separated by what appear to be unbridgeable gaps (chaps. 10 and 14). The reason we do not see intermediates, Darwin counters, is that they are erased by extinction. If there are links to be found, they will have existed only in the past, during intervals of time between something that is alive today and a supplanted ancestor that is now extinct (chap. 10).

Even if Darwin's counter is true, his opponents still have grounds for objection: we should see evidence of such transitions between species preserved in the fossil record, but we do not. Why do whole groups of animals seem to appear at once in the fossil record with little or no evidence of their origin? Why do we see complex animals in the oldest-known strata of the Silurian, then rocks with no apparent life underneath them, when we should instead see evidence of these animals' origins and their gradual increase in complexity?

Darwin cites the substantial, multifaceted imperfections of the fossil record as the source of such apparent discontinuity (chaps. 17 and 18 above). The record contains only brief snapshots of time extracted from the long history of life. These snapshots represent only certain types of habitat and only certain types of organisms. They also represent only isolated "islands" of habitat, where sediments were accumulating, in the much larger "sea" of species distributions. We get to see only a tiny fraction of the real history of life on earth.

Consider, as well, how new species evolve. Each one is the spawn of a local population that evolves into a distinct variety, then a species. If a species originates in a small area and expands its range only after attaining special adaptations, then the site of origin is unlikely to be represented in the fossil record. What we are more likely to see instead is the sudden appearance of new species in areas beyond their points of origin, as if they were created in an instant. Furthermore, the time span represented by a single layer of sediment tends to be much shorter than the duration of a species, so a given stratum will rarely include a time when species transitions were occurring (chap. 17).

As for the lifeless rocks that lie beneath the complex life-forms preserved in the Silurian, Darwin offers a hypothesis, but this puzzling absence of pre-Silurian organisms has been resolved with subsequent research. The fossil record has now revealed a much longer history of life and has given us a much better understanding of the origins of complex, multicellular life (see chap. 19).

These arguments summarize the largest objections to Darwin's theory at the time of the *Origin*, and Darwin's responses to them. In his last response, he acknowledges the huge gaps in evidence that remain to be filled, or may never be filled: "it deserves especial notice that the more important objections relate to questions on which we are confessedly ignorant; nor do we know how ignorant we are. We do not know all the possible transitional gradations between the simplest and the most perfect organs; it cannot be pretended that we know all the varied means of Distribution during the long lapse of years, or that we know how imperfect the Geological Record is. Grave as these several difficulties are, in my judgment they do not overthrow the theory of descent with modification" (*Origin*, p. 466).

Darwin is strategically modest here. He knows that in the long run it is better to claim to little than too much. To win a theory's acceptance, one need not provide more evidence than exists, but only sufficient evidence—enough to expose the much graver difficulties of alternative explanations.

The Case for the Defense

The first phase of Darwin's case in defense of his theory repeats the argument he used at the start of the *Origin* (chap. 2): artificial selection (a process that many of his contemporaries knew and understood) provides a model for natural selection. In artificial selection, humans capitalize on small differences between individual animals or plants by breeding only those that have some desirable trait. Breeders then continue to accumulate change in the desired direction as variations continue to arise. The change attained by artificial selection is permanent, in a specified direction, and can be huge. The differences that we see between breeds of dogs or pigeons exceed those that distinguish naturally occurring species, perhaps even genera or families; yet domestic varieties continue to vary, so it always remains possible to select for new varieties.

These same processes can cause change in nature, but the difference is that breeders select on traits they can see, while nature scrutinizes everything that affects an individual's ability to survive and reproduce. The key to seeing this process in nature is to first realize that all organisms produce many more offspring than are required to replace themselves in the next generation and that there are subtle differences between individuals within a population. The ensuing struggle for existence favors those who are best suited for a given setting, as determined by these subtle variations. The most important elements of the struggle are interactions with other species, rather than the physical environment. The struggle is most severe between those who are most similar and hence compete for similar resources; the most able competitors are the closest relatives. Since the struggle includes any organism that contributes to regulating the population, be it a competitor, predator, or disease, it also involves organisms that are not at all closely related.

Darwin's contemporaries were skeptical that the small amounts of variation we see in nature were sufficient to support his proposed mechanism of evolution. Here he counters that we know that domestic varieties are produced by artificial selection on such small, individual variations. When we look to nature, we see evidence of very similar varieties within a species and can see that the differences between varieties sometimes grade continuously into subspecies and closely related species. There were constant disagreements among naturalists about whether a given population was a "mere variety" or a closely related species. Darwin sees the root of these disagreements in people's attempts to parse continuous variation into discrete categories. Such categories do not exist; species are just well-marked varieties.

Some of Darwin's contemporaries accepted that varieties could be products of "secondary" processes, by which they meant something other than the primary process: the special creation of distinct species. Darwin capitalizes on this concession by arguing that, if species are just well-marked varieties, then these same secondary processes can explain the origin of species as well. Darwin has already shown that the reproductive barriers between species are not absolute and are only more pronounced than the barriers that separate varieties (chap. 11). Anatomical differences between varieties and species are also differences of degree, rather than being absolute. If, by his argument, species are only well-marked varieties and no clear line of demarcation can be drawn between species and varieties, and if people agree that "secondary laws" can produce varieties, then there is no cause to invoke separate laws, such as special creation, to explain the origin of species.

Darwin sought evidence for a relationship between the formation of varieties and the formation of species by tabulating, for many genera, the number of species per genus, then the number of varieties per species (chap. 9). He found that the two were correlated; genera with many species tended to have species with many varieties. He reasoned that "where many large trees grow, we expect to find saplings" (*Origin*, p. 55). The success of the larger genera and the more variable species, he argued, was rooted in traits inherited from an ancestral species that was successful in producing many descendants. Here Darwin concludes: "These are strange relations on the view of each species having been independently created, but are intelligible if all species first existed as varieties" (*Origin*, p. 470).

Darwin next introduces his "principle of divergence" as the proposed mechanism that drives varieties apart and pushes them down the path toward the formation of new species. As a dominant species expands its population size and geographic range, the successful varieties will be the ones that are most different from all others because they will occupy new environments where the competition for resources is less severe. These varieties are candidates for the evolution of new species. Darwin predicts that successful lineages will continue to increase in abundance, produce new species, and diversify into new lifestyles. As successful lineages expand and diversify, the "weaker" ancestral lineages will go extinct. The longer-term consequence of his triumvirate—descent with modification, the principle of divergence, and extinction—is what we now recognize as the Linnaean taxonomic hierarchy: "This tendency in the large groups to go on increasing in size and diverging in character, together with the almost inevitable contingency of much extinction, explains the arrangement of all forms of life, in groups subordinate to groups, all within a few great classes, which we now see everywhere around us, and which has prevailed throughout all time. This grand fact of the grouping of all organic beings seems to me utterly inexplicable on the theory of creation" (*Origin*, p. 471).

This process of continual change by many small steps means that evolution will most often be recycling old parts for new functions. It only rarely results in the origin of something truly novel, like the vertebrate eye, and even such novelty is a long-term product of gradual change. For this reason, nature has been described as "prodigal in variety, though niggard in innovation" (*Origin*, p. 471). There is no reason for this to be true if each species was independently created, since creation should have no restraints on achieving novelty, nor should it yield such evidence of common ancestry.

A different argument for evolution is to consider the mismatches between organisms and their environment that we often see, which suggest a recent transition to a new lifestyle (chap. 14). The upland goose almost never sees water, but it retains the same webbed feet that we see in its aquatic relatives. The water ouzel looks like a thrush but feeds underwater, using its wings to steer. Darwin saw a woodpecker that lived in a treeless region of Patagonia and preyed on insects on the ground, yet retained the same appearance and anatomy as other woodpeckers. All these examples point to a change in lifestyle that was sufficiently recent for these animals to retain features of an ancestor with a very different lifestyle. All are products of the struggle for existence and the constant probing of the environment for vacancies. Such mismatches are predicted by Darwin's theory but inexplicable if each animal is the product of an independent act of creation.

Darwin argues that the intensity of the struggle for existence varies with locality. The intensity should be greater on large landmasses, such as Eurasia, than on smaller landmasses, such as Australia or oceanic islands, because larger areas are host to more species, and to species with larger populations. Having more species intensifies the struggle for existence. Larger populations produce more variations for natural selection to act upon. Organisms native to Eurasia should thus be more competitive and tend to dominate when introduced to smaller landmasses. This is the pattern we see (chap. 4 above). If species were all created independently, then they should all be best suited to wherever they are found.

Darwin's proposed mechanism of speciation is evident in other properties shared by varieties and species. For example, many species of invertebrates have some populations that live aboveground and others that live inside caves. The cave populations often have reduced eyes and enhanced antennae similar to the eyes and antennae that characterize species found exclusively in caves (chap. 6). There would be no reason for such similarities between intraspecies variation and interspecies variation if each species was the product of an act of creation but each variety was the product of some secondary process. Under Darwin's theory, the differences that enable us to distinguish between species are the same as those that we use to distinguish between varieties. The only difference is in degree; the differences between species tend to be larger than those between varieties. These trends argue that a single process can explain the origin of both varieties and species.

"If we accept that the geological record is imperfect in an extreme degree" (*Origin*, p. 475), we can find strong support there for Darwin's theory. His theory predicts that there are no fixed rules in how long a given species will persist, how quickly it will evolve, or how quickly it will diversify into new species. Local populations of a species may evolve into new varieties. Most go extinct, but some persist and diversify. The rate of diversification will vary between the surviving lineages, with some giving rise to few descendants that may all be classifiable into one genus while others diversify into many lineages that may be recognized as distinct genera, families, or orders. As they diversify, they will extend their geographic range. Their success in the struggle for existence will be gained at the expense of other species, which will be driven to extinction. What we see in the fossil record is consistent with this broad-brush scenario for the long-term consequences of evolution; lineages vary in their rate of change in morphology and rate of diversification into new species. Some lineages show little change over vast intervals of time. Some expand and diversify while others decline and disappear. There are no fixed rules of either diversification or replacement.

We can see clear evidence in the similarities between fossil and living species that the fossil record represents the history of life. Recent fossils are sometimes the same as living species or may be classifiable as extinct species from living genera. More ancient fossils are more different from species alive today and are sometimes difficult to classify because they share traits with what we now recognize as two or more distinct groups of organisms. Archaeopteryx shares traits with birds and dinosaurs. We also see affinities between the living and fossil species found in different parts of the world. Living and fossil mammals in the order Edentata (armadillos, anteaters) are largely confined to South America, while some living and fossil orders of marsupials are largely confined to Australia. Finally we find that once a species disappears from the fossil record, it never reappears. These are all facts we would expect if fossils and living species are part of the same family tree. If species are instead products of independent acts of special creation, then we should expect no such affinities between the past and the present. Furthermore, a species that was created once could be created again.

Darwin's theory of descent with modification can also explain the geographic distributions of living species (chaps. 20 and 21). The factors that determine where we find species today trace to a single ancestral species that originated in one location, then migrated outward as its population expanded and diversified into new varieties, then species, then higher levels in the taxonomic hierarchy. Where its descendants can be found today is determined by their abilities to migrate, by the barriers that confine their dispersal, and by climate change. Darwin observes that "on the same continent, under the most diverse conditions, under heat and cold, on mountain and lowland, on deserts and marshes, most of the inhabitants within each great class are plainly related; for they will generally be descendants of the same progenitors and early colonists" (*Origin*, p. 477).

These affinities over broad geographical areas, regardless of the nature of the habitat, reflect descent from some common ancestor that inhabited that continent. If we include patterns of climate change into this "principle of former migration," we can account for many of the discontinuities we see in plant and animal distributions. Darwin concentrated on the effects of glaciers, since they represented a recent phenomenon for which there was overwhelming physical evidence and which left a strong imprint on species distributions. The cyclical advance and retreat of glaciers represents one explanation for disjunct species distributions, even related species found in the Northern and Southern hemispheres on land masses as distant from one another as New Zealand and Europe. Such species are descendants of ones that had ranges straddling the equator during a period of glaciation, then shifted both north and south as the climate warmed.

Migration with subsequent adaptation explains the peculiar nature of the flora and fauna of oceanic islands. The difficulty of reaching remote places creates a filter for the type of organism that is likely to colonize. Terrestrial mammals and amphibians cannot reach remote oceanic islands, but reptiles, birds, bats, and certain types of plants and invertebrates can. Those organisms that colonize islands will, if they are well isolated from the mainland, often evolve into new, locally adapted species that may be quite different from their closest relatives on the mainland. Darwin's theory also explains the relationships we see between organisms found in neighboring habitats, such as South America and the Galapagos Islands or Africa and the Cape Verde Islands. The closest landmass will be the most likely source of colonists of a new island.

Darwin's theory of descent with modification, in combination with dispersibility and climate change, explains the affinities between species within an area in spite of differences in climate, and it explains the differences in species composition between areas, in spite of similarities in climate. It explains the affinities between the living species in a region and those represented in the fossil record. It also explains why species found on islands are most closely allied with those on the nearest landmasses. At the same time, all these facts "are utterly inexplicable on the theory of independent acts of creation" (*Origin*, p. 478). If species were products of special creation, we would expect instead that a species well adapted to a given habitat would occupy that kind of habitat wherever it is found throughout the world. There would be no reason for us to see continuity between species found on islands and the neighboring continents, nor the peculiarities we associate with the plant and animal communities found on islands. Nor should there be a relationship between the fossils and the living species found in any one part of the earth.

The hierarchical classification of living organisms and the way fossil organisms sometimes fall between living species is also explicable by Darwin's theory, given the contingencies of the principle of divergence and extinction (chaps. 10, 18, and 22 above). These same contingencies also explain why it is sometimes so hard to define how some groups are related, such as how the different orders of placental mammals are related to one another (chap. 22). A long period of separation, continued divergence, extinction, and the imperfection of the fossil record can combine forces to erase all evidence of some branches on the tree of life.

Darwin's theory helps us to understand why some characters are of great value in classifying organisms while others are not. Rudimentary characters, characters that have little adaptive significance, and embryonic characters all tend to be useful in classification, whereas ones that are closely allied with local adaptation rarely are. The characters useful in classification all share the property of evolving slowly, which is important because they carry more persistent information about the "community of descent," or the genealogical relationships among organisms.

His theory accounts for the "unity of type," or the structural similarities that unite different organisms. It explains why mammals as different as a giraffe, an elephant, and a mouse all have seven cervical vertebrae, and why all quadrupeds have the same arrangements of bones in their limbs, regardless of how their limbs are used. The "unity of type" represents traits they all inherited from a common ancestor that persisted through later modification by natural selection.

His theory explains the similarities we see between different organs in the same animal: "The similarity of pattern in the wing and leg of a bat, though used for such different purpose,—in the jaws and legs of a crab,—in the petals, stamens, and pistils of a flower, is likewise intelligible on the view of the gradual modification of parts and organs, which were alike in the early progenitor of each class" (*Origin*, p. 479).

Darwin's theory explains why organisms that are quite different as adults can be quite similar as embryos. Organisms in later stages of life are exposed to the biotic interactions that are the chief source of natural selection, so they will be subject to strong selection during those life stages and will evolve as a consequence. Embryos develop in an egg or the womb, where they are exposed to few biotic interactions, so they experience little or none of the selection that is imposed on free-living life stages. As a consequence, embryonic traits will evolve slowly and will retain the evidence of common ancestry among distantly related organisms. Darwin argues that when we see these similarities as traits inherited from a common ancestor, they cease to be a mystery: "We may cease marveling at the embryo of an air-breathing mammal or bird having branchial slits and arteries running in loops, like those in a fish which has to breathe the air dissolved in water, by the aid of well-developed branchiae" (*Origin* p. 479).

Rudimentary organs are also easily explained by his theory. These are organs that served some purpose in an ancestor but lost their function as the lifestyles of descendant species changed. As the ancestors of whales became aquatic, their hind limbs became unnecessary. As moles adapted to living underground, their eyes became unnecessary. Such organs may deteriorate because of lack of use or because of more efficient resource allocation, or they may be actively selected against because the new lifestyle of their possessor makes them a liability. In all cases, the traits are ones that were expressed in the later life stage of the ancestor, so they will disappear more readily in the late life stages, but their rudiments are often still seen in embryos. "On the view of each organic being and each separate organ having been specially created, how utterly inexplicable it is that parts, like the teeth in the embryonic calf or like the shriveled wings under the soldered wing-covers of some beetles, should thus so frequently bear the plan stamp of inutility! Nature may be said to have taken pains to reveal, by rudimentary organs and by homologous structures, her scheme of modification, which it seems that we wilfully will not understand" (*Origin*, p. 480).

The Summation

Darwin begins the concluding section of his defense as follows:

> I have now recapitulated the chief facts and considerations which have thoroughly convinced me that species have changed, and are still slowly changing by the preservation and accumulation of successive slight favourable variations. Why, it may be asked, have all the most eminent living naturalists and geologists rejected this view of the mutability of species? It cannot be asserted that organic beings in the state of nature are subject to no variation; it cannot be proved that the amount of variation in the course of long ages is a limited quantity; no clear distinction has been, or can be, drawn between species and well-marked varieties. It cannot be maintained that species when intercrossed are invariably sterile, and varieties invariably fertile; or that sterility is a special endowment and sign of creation. The belief that species were immutable productions was almost unavoidable as long as the history of the world was thought to be of short duration; and now that we have acquired some idea of the lapse of time, we are too apt to assume, without proof, that the geological record is so perfect that it would have afforded us plain evidence of the mutation of species if they had undergone mutation. (*Origin*, pp. 480–81)

The main problem Darwin's contemporaries had with his theory of evolution was that its actual workings in nature seemed imperceptible. If species can be modified so much by variations so small—requiring vast amounts of cumulative change on their way to becoming entirely different species—then we should see evidence that this is happening, they objected. But instead we see large gaps of discontinuity and little evidence of gradual change. Here Darwin draws an analogy between his theory and Lyell's uniformitarian theory for geological processes. Lyell argued that the ordinary processes we see going on today, such as winds, rain, and tides, have shaped the major geological features of the planet, even though they seem to have so little consequence on a day-to-day basis. The key to the disconnect between the processes we see and the results they can achieve is the immense age of the earth, and our inability to grasp the magnitude of a quantity of time as long as 100 million years.

Darwin's opponents confronted him with the alternative that all he sees can be accounted for by a "plan of creation" or "unity of design," but these are just phrases that have no explanatory value. They simply restate the facts, Darwin says. He acknowledges that there are loose ends associated with his theory, so "any one whose disposition leads him to attach more weight to unexplained difficulties than to the explanation of a certain number of facts, will certainly reject my theory" (*Origin*, p. 482). He looks instead to the new generation of naturalists, who may have already wondered about the immutability of species and hence may be more receptive to his theory. He asks that they weigh impartially the evidence he has presented and, if they find his arguments convincing, speak out in support of them. Only a chorus of support for his theory will be able to remove the prejudices associated with the acceptance of special creation.

Darwin sees change afoot in the form of disagreements over how to classify species. Systematists are actively revising the classification of some groups, declaring what have been seen as different species to be varieties instead, or vice versa. If they accept that varieties are products of secondary processes while species are products of independent acts of creation, then why is it not possible to clearly distinguish between the two? "They admit variation as a *vera causa* in one case [varieties], they arbitrarily reject it in another [species], without assigning any distinction to the two cases" (*Origin*, p. 482). Darwin predicts that history will not remember them kindly, since they are highly critical of his proposal but not at all critical of their acceptance of special creation. If we accept special creation, then how can we explain how life began, either in terms of chemistry or form? Was it created as a seed or egg or adult? And were the first created individuals in each mammal species given "false marks of nourishment from the mother's womb"—a belly button? If we are to accept special creation, then we must put the same critical spotlight on how it happened.

Some may wonder, Darwin supposes, how far he is willing to push his theory in the direction of common ancestry. How many original progenitors might account for all the life on earth today? When he applies the same rules he used conceptually in his "tree of life" figure—looking at shared features of organisms that show their common ancestry, and tracing these backward in time—he can foresee, as the result, all living animals having been descended from at "most four or five progenitors," which must correspond to Cuvier's embranchments of life, and "plants from an equal or lesser number" (*Origin* p. 484). But the exercise need not stop there. Darwin can see substantial similarities among all living things in terms of their chemical composition, cellular structure, and other aspects of their biology. "Therefore I should infer from analogy that probably all the organic beings which have ever lived on this earth have descended from some one primordial form, into which life was first breathed" (*Origin*, p. 484).

The Future

Looking ahead, Darwin speculates, "we can dimly foresee that there will be a considerable revolution in natural history" (*Origin*, p. 484) when his theory is accepted. One positive change, he thinks, will be that systematists will not need to worry about whether some life-form is a variety or a species. It will instead be possible for them to treat species as they once treated genera. They view genera as combinations of species similar enough to be grouped together, although in Darwin's view such species are similar because they share a common ancestor. Likewise, a species will become an aggregate of varieties that are distinct enough to be recognized as different from other such aggregates and that also share a common ancestor.

This prediction has not been upheld, perhaps for two reasons. One is that we now have a more refined definition for what a species is, based on a more biological rationale for distinguishing species (chap. 8). The second is that speciation is a continuous process, so the same gray zone between varieties and species that Darwin identified still exists to fuel arguments about where to draw the line.

Darwin also argues that all areas of biology are now united under a single explanatory framework, making them more meaningful and interesting subjects for study:

> The terms used by naturalists of affinity, relationship, community of type, paternity, morphology, adaptive characters, rudimentary and aborted organs, &c., will cease to be metaphorical, and will have a plain signification. When we no longer look at an organic being as a savage looks at a ship, as at something wholly beyond his compre-

hension; when we regard every production of nature as one which has had a history; when we contemplate every complex structure and instinct as the summing up of many contrivances, each useful to the possessor, nearly in the same way as when we look at any great mechanical invention as the summing up of labour, the experience, the reason, and even the blunders of workmen; when we thus view each organic being, how far more interesting, I speak from experience, will the study of natural history become! (*Origin*, pp. 485–86)

His theory also opens up new areas of inquiry "on the causes and laws of variation, on the correlation of growth, on the effects of use and disuse, on the direct action of external conditions, and so forth" (*Origin*, p. 486). Among the many beckoning areas of inquiry not mentioned, but implied throughout the book, are the laws of inheritance. The production of new varieties under artificial selection is also now of interest, since we can envision new breeds as reflecting the same variation seen in nature, with our selection seen as a surrogate for the natural process. In the yet more distant future, "psychology will be based on a new foundation, that of the necessary acquirement of each mental power and capacity by gradation. Light will be thrown on the origin of man and his history" (*Origin*, p. 488).

The classification of plants and animals can now be seen as the reconstruction of genealogies. Even though we do not have a family tree in hand, we can see its reconstruction as the goal of classification. We now have a rationale for choosing characters that are useful for classification—those that are less subject to the immediate action of natural selection and hence evolve less rapidly and retain more information about common ancestry. Traits with little apparent significance, as well as embryonic traits and rudimentary traits, all have this stability in common. Aberrant species such as the lungfish or duck-billed platypus, or living fossils such as *Lingula* brachiopods, will help us to envision what ancient forms of life were like. "Embryology will reveal to us the structure, in some degree obscured, of the prototypes of each great class" (*Origin* p. 486)

Understanding the distribution of the species we see today is possible when we realize that each genus and each species arose from what was a single population of some variety that itself had evolved into a distinct species, then emigrated to new localities as it diversified. We can integrate these distributions with what we learn about their means of dispersal, about changes in the main geological features of the planet, and about climate change to reconstruct a more complete picture of the history of life as it diversified and dispersed to new locations.

"The noble science of Geology," Darwin continues, "loses glory from the extreme imperfection of the record" (*Origin*, p. 487), which is far from being a complete history of life. "The crust of the earth with its embedded remains must not be looked at as a well-filled museum, but as a poor collection made at hazard and at rare intervals" (*Origin*, p. 487). Nevertheless, we can now make inferences about the extent of what is missing, because his theory suggests that "the amount of organic change in the fossils of consecutive formations probably serves as a fair measure of the lapse of actual time" (*Origin*, p. 488). Such inferences about gaps in the record must be made with caution, since at any one time we can expect that only a few species will be experiencing selection and will be evolving. This means that during the interval of time that separates two formations, some species may have changed little or not at all while others may have changed a great deal. Darwin also speculates that the complete history of life must be much longer than is known to him, since the oldest-known fossils (of his day) are already complex organisms. During the earliest stage of the history of life, "when the forms of life were probably fewer and simpler, the rate of change was probably slower; and at the first dawn of life, when very few forms of the simplest structure existed, the rate of change may have been slow in an extreme degree" (*Origin*, p. 488).

Now that Darwin sees all life as part of the same grand family tree, rather than as the products of many individual acts of creation, he finds that view ennobling. Under his theory, we can understand how the present is the product of past events, but we can also look toward the future: "Judging from the past, we may safely infer that not one living species will transmit its unaltered likeness to a distant futurity. And of the species now living very few will transmit progeny of any kind to a far distant futurity" (*Origin*, p. 489).

The way that species are grouped together into genera and genera into families, all so distinct from one another, tells us that over time most of the species that have ever lived have gone extinct, leaving no descendants. It is only this complete loss of most species that can convert the continuous process of evolution into the discontinuities between living organisms that we see today. The fact that large genera contain dominant species with many varieties tells us that it is the widespread and abundant species of today that are most likely to be the progenitors of the species in the future. The fact that all species alive today can be seen as descendants of those that lived before the oldest-known fossils of Darwin's day tells us that there has never been a break in the history of life, "and that no cataclysm has desolated the whole world" (*Origin*, p. 489). Likewise, it says that we can look toward a future in which life continues to evolve.

I end with Darwin's final paragraph, since I feel that it is self-explanatory and I cannot hope to improve on it. Darwin concludes by imagining how differently we may see the world of the living once we appreciate the role that evolution by natural selection has played in shaping it:

> It is interesting to contemplate an entangled bank, clothed with many plants of many kinds, with birds singing on the bushes, with various insects flitting about, and with worms crawling through the damp earth, and to reflect that these elaborately constructed forms, so different from each other, and dependent on each other in so complex a manner, have all been produced by laws acting around us. These laws, taken in the largest sense, being Growth with Reproduction; inheritance which is almost implied by reproduction; Variability from the indirect and direct action of the external conditions of life, and from use and disuse; a Ratio of Increase so high as to lead to a Struggle for Life, and as a consequence to Natural Selection, entailing Divergence of Character and the Extinction of less-improved forms. Thus, from the war of nature, from famine and death, the most exalted object which we are capable of conceiving, namely, the production of the higher animals, directly follows. There is grandeur in this view of life, with its several powers, having been originally breathed into a few forms or into one; and that, whilst this planet has gone cycling on according to the fixed law of gravity, from so simple a beginning endless forms most beautiful and most wonderful have been, and are being, evolved. (*Origin*, pp. 489–90)

Epilogue

And what was the opinion of the jury? As Georges Cuvier was dead, we cannot know how he would have received Darwin's defense of his theory; I had to include his name in my imagined jury because he played such a prominent role in the run up to the *Origin*. He was an active opponent of evolutionary theory during his life, but that was the evolution of Jean-Baptiste Lamarck.

Charles Lyell, Joseph Hooker, Asa Gray, and Thomas Huxley were all convinced and became advocates of Darwin's theory, although they were not necessarily complete converts. Huxley, who proclaimed himself to be Darwin's bulldog and was most active in defending the theory, was nevertheless skeptical of natural selection. Gray found his own way of reconciling evolution with his religious beliefs. Lyell could never bring himself to apply evolution to the origin of man. All the other members of the jury actively opposed Darwin. Recall that John Herschel and William Whewell together shaped Darwin's philosophy of science. Herschel described natural selection as "the law of higgledy-piggledy." William Whewell is rumored to have banned the *Origin* from the library of Trinity College at Cambridge. Adam Sedgewick, the Cambridge geologist who had been Darwin's field geology tutor before Darwin's departure for the voyage of the *Beagle*, wrote to him: "You cannot make a good rope out of a string of air bubbles." I have already mentioned Richard Owen's contempt for the *Origin*. Louis Agassiz, by then a professor at Harvard, was an active and outspoken opponent of Darwin's theory. (Asa Gray, also a professor at Harvard, was the North American counterpart of Thomas Huxley in his support of the *Origin* and opposition to Agassiz.) Karl von Baer, the embryologist who did most to shape Darwin's evolutionary interpretation of embryology, opposed Darwin's theory in some of his last essays. Robert FitzRoy, the captain of the *Beagle*, wrote to Darwin that "I, at least, *cannot* find anything 'ennobling' in the thought of being a descendent of even the *most* ancient Ape." Darwin was correct in predicting that the future of his theory lay in the hands of a new generation.

References

Browne, J. 2002. *Charles Darwin: The power of place*. Princeton, NJ: Princeton University Press. (Pp. 82–125, source of material on the reception to the publication of the *Origin*.)

Chapter 24

Evolution Today: The Witness Has Been Found, Again and Again

In the imaginary murder investigation presented in the introduction, we argued that, with the right evidence, one can know who committed the crime, and there is no need for an eyewitness. Darwin's argument in the *Origin* fulfills this logic. He did not have direct proof that evolution by natural selection happens nor that it causes speciation. He assumed that these processes work on a timescale so much longer than our lives that they cannot be directly observed; we instead have to make inferences from the evidence we can see, in the same way we can infer that footprints in the sand tell us that some animal passed this way. His logic was inspired by Charles Lyell, who argued that while we cannot witness the origin of mountain ranges, we can infer their origin from observable processes. Darwin witnessed how an earthquake elevated the land in the harbor at Concepcion and saw fossils of marine origin high in the Andes. He inferred that many such earthquakes over millions of years can build mountain ranges.

Darwin built his argument for evolution on processes that can be seen and measured: the potential geometric increase in population size; the ensuing struggle for existence; the existence of individual variations faithfully transmitted to offspring; and the impact of these variations on who survives to reproduce. He saw the combined action of these processes at work in artificial selection. He then considered their long-term consequences.

If Darwin's theory is true and if it is the general law that shaped life on earth, then we should find evidence for it throughout the history of life and in all features of living organisms. This is a tall order. Darwin fulfilled it to the extent possible given the state of knowledge in the mid–nineteenth century. He argued that observations from diverse disciplines—embryology, comparative anatomy, classification, the fossil record, the geographical distribution of plants and animals, and more—were all consistent with, and could be explained by, his theory of evolution by natural selection. It is this reconciliation of such diverse disciplines that constitutes the evidence in our investigation.

The biologists inspired by Darwin's long argument never saw it as dogma that should be accepted without question. It was instead seen as a challenge. A scientist can gain fame more quickly by successfully disproving a cherished hypothesis than by supporting it. Many scientists have set their sights on Darwin's theory. The *Origin* contains a complex network of hypotheses that were either explicitly stated by Darwin or implicit in the structure of the theory. Those who followed Darwin have pursued virtually all these hypotheses, and many of them have grown into independent disciplines. It is these modern research programs, which rigorously investigate the questions left open by Darwin's far-ranging hypotheses, that have led to the discovery of more and more evidence, the eye witnesses, that confirm his theory.

An interesting feature of most of the research spawned by the *Origin* is that it did not begin to develop in earnest until after the "modern synthesis" period, which began in the 1930s. The modern synthesis not only reconciled the discovery of Mendelian inheritance with Darwin's theory but also heralded the modernization of many aspects of the theory, such as the formal definition of "species" and the study of the process of speciation. This hiatus of more than seventy years between the publication of the *Origin* and the advent of serious research programs that developed the implications of Darwin's theory is a measure of the scrutiny Darwin's ideas received before they were widely accepted.

A missing eyewitness to Darwin's theory when the *Origin* was published was natural selection. Darwin thought that this process was too slow to be directly observed. He was wrong. Chapter 7 presents one example of the modern study of natural selection. There, I showed how modern theory refined Darwin's concept of fitness and predicted how organisms should evolve in response to changes in mortality rates. Now we can claim to have developed a network of quantitative theory that predicts how organisms will evolve. We have performed many critical experiments that test the predictions of these theories in laboratory and natural settings. We can even quantify how fast evolution by natural selection can occur in a natural setting. Perhaps Darwin's biggest mistake was to underestimate the speed and potency of natural selection.

A second missing eyewitness was speciation. We now have developed a formal definition of species. We have defined a variety of mechanisms that can cause speciation, developed a diverse body of theory that explores the conditions that facilitate speciation, have documented speciation in action (chap. 12) and have also documented the complex relationship between natural selection, the formation of varieties, and the continuity between varieties and species.

Two additional missing witnesses were proofs of the antiquity of the earth and the antiquity of life on earth. Darwin's theory demanded that the earth be immensely old and that the record of life on earth be far longer than the record that was known to him. We can now directly estimate the age of the earth (chap. 17) and have documented a history of life that is seven to eight times longer than the history known in Darwin's time (chap. 19). In Darwin's day, the record was just a chronological ordering from old to young rock strata based on the correlation of fossils found in different parts of the world. Their relative ages were inferred. We can now often associate strata with an estimate of real time using radiometric dating. The dating confirms that the record really is chronological, plus we can approximately date the age of fossils and combine this information with family trees constructed from DNA sequences to estimate the time periods when lineages diverged from one another.

Yet more missing eyewitnesses were the fossils that document the common ancestry of distinct groups of living organisms. The fossil record known to Darwin had revealed some evidence for such transitions among taxa, such as *Archaeopteryx*. Now we have hundreds of fossils that define the origin of different modern taxa from common ancestors. Some of the best-documented transitions have come from recent finds, most after 1990, and include the transition from fish to amphibians, reptiles to mammals, dinosaurs to birds, and terrestrial mammals to whales.

Other missing eyewitnesses were the mechanism of inheritance, the source of the individual variants that are the fuel for evolution, the discovery of how this variation is maintained, and the quantification of variation present in natural populations. These witnesses are represented by the discovery and full characterization of Mendelian inheritance, the invention of population and quantitative genetics, and the discovery of DNA as the molecular basis of inheritance.

The discovery of DNA provided an unexpected tool for documenting evolution and the relationships between living organisms. In earlier chapters, I showed how the differences in DNA sequences between species can be used to construct "family trees" and to provide information about when different species shared a common ancestor or how quickly complex organs evolved. The utility of changes in DNA sequences represents a powerful and unanticipated test of Darwin's principles of descent and the gradual divergence of character. He coined these principles to describe how anatomy would change over time and to explain why some anatomical features are more useful than others in defining how organisms are related to one another. He could not have imagined the advent of our understanding of the molecular basis of inheritance and evolution, yet his criteria apply equally well to DNA sequences.

The virtue of DNA is that it contains so many more characters to study than does anatomy. Organisms have thousands of individual genes, and each gene contains hundreds to thousands of base pairs of DNA. Genomes can also contain a huge diversity of DNA that does not code for gene products. DNA from any of these segments can be used as characters for making inferences about evolution in the same way that we use anatomy.

The modern study of evolutionary biology has presented us with many more eyewitnesses, but I will describe only one more group here: evidence for human evolution. Charles Lyell, who began his interactions with Darwin as a mentor and nonbeliever in evolution but ended as an advocate of Darwin's theory, once commented that he was not ready to "go the whole orang." By this, he meant that he was not ready to deal with humans as members of the animal kingdom. This same hesitation prevails today. I will conclude this book by showing how we can apply Darwin's principles and inductions to our own evolutionary history and study humans as members of the animal kingdom. As was the whole *Origin*, what follows is only an abstract, since research on the evolution of humans has become a very large endeavor.

Darwin studiously avoids the topic of human origins in the *Origin* and simply concludes that, as one consequence of his theory, "Light will be thrown on the origin of man and his history" (*Origin*, p. 488). One reason to avoid the topic was that it would add too much fuel to what was already destined to be an incendiary book. Another reason was that very little was known about human origins at the time. The fact that humans were so conspicuously absent from the fossil record, as it was then known, was seen as

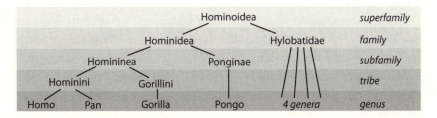

Figure 31
Tree of the Superfamily Hominoidea

evidence that we were a recent, ultimate act of creation. Thomas Huxley (1863) proposed, on the basis of comparative anatomy, that the closest relatives to humans were the great apes. He also inferred that humans originated in Africa because our closest relatives, the chimps and gorillas, were found there. These proposals were both specific applications of Darwin's theory. The recognition of the affinity between humans and chimps was not new. Linnaeus classified humans and chimps as being in the same genus. Huxley's innovation was to apply a new interpretation to an old anatomical tradition. Now we can integrate Huxley's "unity of type," or inference of relatedness from anatomy, with many other kinds of data.

Who are we? As with all organisms, our classification is a synthesis of information from DNA sequences and the anatomy of living and fossil species. Different "family trees," or taxonomic divisions, have been proposed for classifying the apes. The one I present here (fig. 31) obeys modern rules of classification, the most important being that organisms classified together must always be descended from one common ancestor. Humans are most closely related to chimps. Humans and chimps (there are at least two species of chimps) are placed together in the tribe Hominini. Humans, chimps, and gorillas are grouped together in the subfamily Homininae. Humans, chimps, gorillas, and orangutans are grouped together in the family Hominidae. Finally, all these "great apes" are classified together with the family Hylobatidae (gibbons, or lesser apes) in the superfamily Hominoidea, which is in turn in the primate order. I will use the term "hominin" to refer to all those species that fall between humans and their common ancestor with chimps.

The key to seeing this as a "natural" classification is that each of these taxa contains a cluster of species that can be traced to a single common ancestor. A common alternative to this classification is to group the chimps, gorillas, and orangutans in one family (Pongidae) and humans in a different family (Hominidae). However, data derived from DNA sequences tell us that humans, chimps, gorillas, and orangutans were all derived from a single common ancestor. Grouping chimps, gorillas, and orangutans together in exclusion of humans creates a classification that is not "natural" because it excludes one member, humans, that shares the same common ancestor as the other three genera.

One "eyewitness" to the origin of humans is the information contained in our genes. Humans and chimps were among the first organisms to be compared in the nascent discipline of "molecular evolution." Sarich and Wilson (1967) compared the structure of albumin, a blood serum protein, from humans and the other great apes. Albumin from chimps, gorillas, and humans was remarkably similar, and the degree of similarity led them to conclude that the common ancestor of these species lived just 5 million years ago. This was a much more recent date than that suggested by physical anthropologists. Researchers have since compared many individual genes of humans, chimps, gorillas, and more distantly related organisms and arrived at a similar conclusion, which is that humans and chimps are much more similar to each other than either of them is to any other organism. Gorillas are more distantly related to chimps and humans. The most recent analyses place the divergence of the human and chimp lineages at approximately 6 million years ago.

We now have complete draft genomes for humans and one species of chimp and are in the process of characterizing the differences between them. This can be done by comparing proteins, since genetic codes are blueprints for the construction of proteins. It turns out that 29% of the proteins coded for by chimp genes have identical amino acid sequences to those coded for by human genes. The remaining proteins differ by an average of only two amino acids. Many of the established genetic differences between humans and chimps have a clear functional significance. Humans and chimps differ in genes that contribute to disease resistance, in one gene known to be associated with the development of speech, and in others associated with hearing, which is necessary to decode speech.

Another eyewitness that documents the origin of humans and their affinity with the great apes is the fossil record. Darwin argued that if the geological record contains the history of evolution, then recent fossils will look more like living organisms than more ancient ones will (principle of descent with gradual modification). The record should also reveal evidence of relationships between living groups in the form of fossils that share traits with both, but because of the extreme imperfection of the record, it will not contain the complete history of these relationships. Rather than seeing a complete sequence of ancestors and descendants, we will instead tend to find only some representatives of the former diversity of species, which will tend to be members of extinct lineages that are related only indirectly to species that are alive today. Finally, Darwin observed that there was often an affinity between the living species found in some region and the fossils found in the same region. All these trends are predicted by his principle of descent with modification.

Paleoanthropologists have discovered a series of early hominin fossils that can be used to test Darwin's inductions, but also to test the alignment between the estimated chronology of human evolution based on radiometric dating and that predicted from comparisons of DNA sequences. These fossils display a mix of characters that are intermediate between humans and the living great apes, which is what we would expect of species that are among the lineages that include common ancestors of humans and chimps.

Many details of these fossils define them as part of the hominin lineage. I will describe just two of them. Teeth often provide the best evidence for how extinct and living mammals are related because they are the most durable part of our bodies, and therefore are the most likely to be preserved as fossils. They also have a complex anatomy that has proven useful for classification. Individual teeth from adults often have wear patterns that tell us how they fit together and were used. Many of the key characters that define the oldest fossils as being part of the hominin lineage come from teeth that are in some way intermediate between those of living humans and living great apes. For example, great apes and humans differ in their canine teeth. The living great apes and the older fossil apes have long, pointed canine teeth and a gap, or diastema, in the opposite jaw. When the jaw is closed, the canine fits into the gap like a knife in a sheath. The wear on the canine is on the side, where it slides against the tooth in the opposite jaw when the mouth is closed. In humans, the canine is "incisiform," which means it is shaped like our middle wedge-shaped incisors. The canines and incisors all form a straight row, and there is no diastema. Our canine is worn on the tip, where it grinds against the tooth on the lower jaw. The oldest fossils in our lineage all have reduced canines and reduced diastemas on the lower jaw relative to chimps; most of them have canines that are worn on the tip rather than or in addition to being worn on the side. In living primates, the males have larger canines than females. They are used in threat displays and are thought to be a product of sexual selection. The reduction in the canine teeth of the human lineage may have accompanied changes in behavior and sex roles.

The second character associated with recognizing a fossil as a hominin is an erect, bipedal gait. Gorillas and chimps can walk on their hind legs for short distances, but humans are distinct in having a fully upright posture, with our legs aligned directly beneath us, and the ability for sustained, balanced locomotion on two legs. This "character" is actually a complex of characters involving changes to the structure of the foot, the joints and bones of the leg, the pelvis and lower back, and the way the skull is aligned on the vertebral column. Changes that would favor bipedality in any of these characters may constitute evidence that an animal was habitually bipedal, although the quality of such evidence varies among characters. The knee joint, for example, is much more informative than is the way the skull aligns on the vertebral column. Because fossils are only fragments of the whole animal, they do not always present the evidence that we need to determine if the animal had bipedal gait.

The oldest-known fossils now classified by at least some paleoanthropologists as being in the tribe Hominini include *Sahelanthropus tchadensis* (6–7 myr ago) from Chad, *Orrorin tugenensis* (6 myr ago) from Kenya, *Ardipithecus kadabba* (5.2–5.7 myr ago), and *Ardipithecus ramidus* (4.5 myr ago), both from Ethiopia. All these fossils share changes in dentition and other characters that align them with the hominins, but in other ways have traits that could also align them with chimps. For example, *Sahelanthropus* had the prominent brow ridge of an ape, and was also apelike in the size and shape of its braincase, but had a reduced canine with wear on the tip and other tooth features that are typical of our lineage. Its face was shorter than is typical of apes but characteristic of hominins. Apes have a jaw that projects far in front of their eyes, but the trend is for the projection of the jaw to be reduced in the hominins. In fact, *Sahelanthropus* had a shorter jaw projection than seen in later species of hominins. There is not sufficient evidence for us to know if *Sahelanthropus* had a bipedal gait. *Ardipithecus kadabba* had reduced canines, but less so than *Sahelanthropus*. It appears to have had a bipedal gait. The ecological setting of the *Ardipithecus* fossils suggests that this genus lived in a forest or a forest-grassland transition zone. It had been thought that bipedality evolved as the early hominins adapted to a grassland habitat. East Africa was becoming drier then, and forests were being replaced by savannas. Many of the mammals that adapted to this new habitat did so by evolving longer limbs that were better adapted for efficient long-distance travel. The evolution of bipedality in the hominin lineage is traditionally interpreted as being part of this same trend—as a species, we are remarkably good at running long distances at moderate speeds. However, the ecological setting of the *Ardipithecus kadabba* fossils suggests that the evolution of bipedality may have begun before the hominins shifted to this new environment.

There is ongoing debate about whether all these fossils really are hominins, let alone direct ancestors of humans. Darwin predicted such uncertainty in more general terms when he used his "tree of life" figure to illustrate how his theory would play out as lineages diversified or went extinct. The distant common ancestors of humans and the other great apes can be expected to share only general traits with the living species they are related to, and should not be expected to be anatomically intermediate between their living descendants (i.e., with only a "medium" brow ridge, only a quasi-upright stance, etc.). It is unlikely that any of these hominins are the real "missing link" between humans and the other great apes. It is more likely that they are instead representatives of the biological diversity of great apes that could be found between 4 and 7 million years ago in Africa. We must assume that most of this diversity has been lost without being represented in the fossil record and that the true common ancestor is among those not seen.

We have a larger and more diverse collection of hominin fossils that date between 4.2 myr ago and the present. A common interpretation of these fossils is that they fall into four genera (*Kenyanthropus, Australopithecus, Paranthropus, Homo*) with two to five species described so far for each of them. Their geographic range extended over the full north-to-south axis of East Africa. They have turned up, albeit in low numbers, in many of the exposed rock formations of appropriate age. One species was recently found in Chad, much farther to the west than all other members of this lineage, which suggests that the hominins may have had a much larger geographic range. There were often species from multiple lineages alive at the same time that possibly co-occurred with one another. For example, between 2 and 3 myr ago species from three genera could all be found in East Africa, each with a distinct ecological niche.

There is again argument about which of these more recent lineages is the true ancestor to the genus *Homo* and which species of *Homo* is the true ancestor to *Homo sapiens*. It was once thought that the genus *Homo* was represented by a single lineage—*H. habilis* to *H. erectus* to *H. sapiens*—with each species simply replacing its ancestor. It is now clear that the genus was a branching lineage that contained more than three species and probably had multiple living species present at the same time. We have just discovered another species of hominin, currently classified as *Homo floresiensis*, from the island of Flores in the Malay Archipelago, that was still alive as recently as 18,000 years ago.

A literal interpretation of the fossil record is that the Hominini are a recently evolved branch of great apes that reached a peak of diversity 2–3 myr ago, then declined to a single, surviving species—*Homo sapiens*. We defy the usual trend of a lineage that is declining in that this sole surviving species is "dominant"; we are abundant, widespread, and diverse. The most likely source of inaccuracy in this interpretation is that we have underestimated the true diversity of the hominins at the time of their origin, between 5 and 7 myr ago. We have only recently begun to explore rock formations that were deposited during that window of time.

We can see in these fossils the development of all the traits that characterize modern humans and distinguish them from chimps. These traits include the evolution of all the characters that contribute to a bipedal gait. They also include progressive change in the structure and arrangement of teeth and other features of our skulls, a hand that is capable of the precision grip that characterizes modern humans, increased height, and increased cranial capacity. The fossil record implies that these traits evolved at different times and rates. Bipedality and tooth evolution began to emerge more than 5 myr ago. Toolmaking ability is first associated with *Homo habilis* (handy man), which is the earliest-known member of our genus, around 2.5 million years ago. Tools may also be associated at the same time with *Australopithecus garhi*. The dramatic increase in brain size came later, in association with *Homo erectus* and more recent fossils.

The aggregate of these results addresses Darwin's inductions about the fossil record. The fossil record has revealed species that show evidence of shared ancestry to the living apes and humans. The oldest fossils share traits that straddle those now seen in humans and chimps. The more recent fossils are progressively more similar to living humans. The record yields estimated ages of fossils that correspond well with the time when molecular genetics predicts humans and chimps shared a common ancestor. It also addresses Darwin's induction about geographic distribution. The biogeography of the hominins is strikingly coherent and supports Huxley's inference that our origin was in Africa because that is where our closest living relatives are found. All the hominin fossils that date between 7 myr and 2 myr ago are from Africa, almost all from East Africa. The first hominin found outside Africa was *Homo erectus*, which appears to have migrated out of Africa during a period when the climate was wetter than today, so a walk across the now-dry deserts of the Middle East was more feasible. *H. erectus* was also the first fossil hominin to be found (by Eugene Dubois in Java during the 1890s). Specimens that date between 1.8 and 1.0 myr ago have now been found in other parts of Europe and Asia.

If we look only at species alive today to make inferences about our past, then we also see a strong imprint of our evolutionary history. The richest imprint is found in DNA sequences, which confirm what had already been concluded by nineteenth-century anatomists—chimps are our closest living relatives, then gorillas, then orangutans. Some features of our anatomy make no sense unless we view them as inherited from an ancestor that was adapted to a different environment—they are like the webbed feet of the upland goose. One such complex of features is the shape of our torso and the structure of our arms including the shoulder, wrist, and elbow joints. Our torso is short from front to back and wide from shoulder to shoulder. Our shoulder blades are aligned along our back, and the shoulder joints point out to the side. All the joints in our arms allow a wide range of movement—picture an Olympic gymnast doing a routine on the rings and the range of motion we see in his arms. Contrast the structure of our torso and forelimbs with those of a mammal that has quadrupedal as opposed to bipedal locomotion. Dogs and cats, for example, have a torso that is narrower from side to side than it is from front to back. Their shoulder blades are aligned along the sides of the body. Their forelimbs can swing fore and aft along with the hind limbs during locomotion, but have a very limited range of motion.

It is tempting to think of our torso as having evolved in association with our bipedal gait. Comparative anatomy tells a different story. All species in the superfamily Hominoidea share this structure of the upper body. It originally evolved as an adaptation to arboreality and brachiation, or using arms to swing from branch to branch in a tree. It is this lifestyle that shaped our torso, but also established distinct functions for the fore- and hind limbs and thus set the stage for the later evolution of our bipedal gait. (If you ever have occasion to see a gibbon in action at a zoo or in a nature film, note the shape of its torso and the range of motion in its forelimbs as it travels from branch to branch. The anatomy of its upper body is like ours.

Its locomotion will be reminiscent of an Olympic gymnast, but the gibbon will amaze you with its speed and grace.) As great apes adopted more terrestrial lifestyles, they retained the tree dwellers' torso and the differences in function between the fore- and hind limbs. Our upper body is thus an example of evolution as tinkering, or the progressive modification of old parts for new functions.

Darwin also inferred that evolution by natural selection often results in the early stages of development showing greater similarity between related species than the adult stages do, because early development takes place in a sheltered environment. It is mainly during the free-living stages of life, when they must fend for themselves, that organisms will experience strong selection. It is therefore in adults that we most often see the strongest divergence between species. Humans and chimps show such patterns. They are far more similar to one another early in development than they are as adults. Since we usually think of our big brain as our most distinctive feature, I will use relative brain size and skull shape to illustrate the application of Darwin's induction to human origins. The skull shape and the relative brain size of infant chimps and humans are far more similar than are those of adult chimps and humans (fig. 32). The underlying mechanism for the divergence in skull shape as chimps and humans mature is in the relative rates of growth in different parts of the skull. It is convenient to think of the skull as consisting of two regions—the "facial" region, which is anterior to a line from the back of the eye to the hinge of the jaw, and the "cranial" region, which is posterior to this line. The facial region includes the eyes, nose, and jaw, while the cranial region includes the braincase. In chimps, the facial region grows more rapidly than the cranial region, so adults have a projecting snout and proportionally smaller braincase relative to the juveniles. In humans, the two rates of growth are approximately equal throughout development, so the skull proportions remain similar from infancy to adulthood. Because the growth of the cranial region keeps pace with that of the facial region, adult humans also have larger brains than adult chimps.

Many eyewitnesses to human origins can thus be found in the fossil record, and what the fossils reveal about our ancestors and their biogeography correlates well with the testimony of other eyewitnesses—such as those we find in our anatomy, our development, and our genetic codes.

Not all of Darwin's proposals have stood up to close scrutiny. His ideas about inheritance, for example, proved to be wrong (chap. 2); however the true mechanism of inheritance strengthened rather than weakened

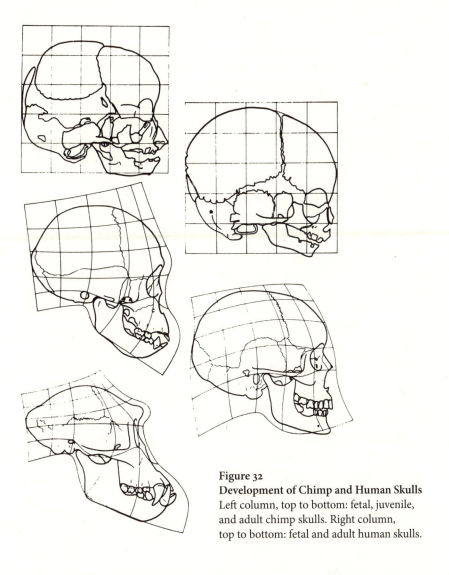

Figure 32
Development of Chimp and Human Skulls
Left column, top to bottom: fetal, juvenile,
and adult chimp skulls. Right column,
top to bottom: fetal and adult human skulls.

his argument. He thought that inheritance was a process of blending the contributions from each parent. Such blending would always remove the heritable variation that is required for evolution to work, so he had to come up with a mechanism for replenishing variation, which turned out to be wrong. The mechanism of Mendelian inheritance generates variation via mutations and retains this variation, so Darwin need never have worried. Extinction (chap. 19) has proven to have multiple causes, so Darwin's proposed cause of extinction is only a subset of those that are recognized today. Natural selection is no longer the sole cause of evolution; we now recognize genetic drift, mutation, and migration as components of evolution. Speciation has also been found to have multiple causes, rather than being purely a product of natural selection. The discovery of reticulate evolution, as when two different types of single-cell organisms somehow fused to create a new type of organism (chap. 19), represents a significant deviation from Darwin's principle of descent with modification. A second unexpected twist was the discovery that continents drift (chap. 19). In this case, some of the anomalies of biogeography turned out to have a much easier explanation than imagined by Darwin, and his theory was strengthened in the end. Although not perfect at its inception, Darwin's theory of evolution has proven to be a sufficiently good description of reality, which has allowed it to be reconciled with all these new discoveries and to be retained.

The eyewitnesses to Darwin's theory are now diverse and abundant. The integrated information that defines the origin of humans and so many other organisms represents how the integration of data from DNA sequences, anatomy, development, the fossil record, and biogeography can be combined to inform us about the history of life. It is now within our reach to do what Darwin thought impossible given the evidence of the fossil record alone, which is to define the complete tree of life. The eyewitnesses include our ability to apply absolute dates to a fossil record that is seven to eight times older than the one known to Darwin, and now includes far more evidence of the origins of living taxa than was known in his time. We also have proof and quantification of natural selection and speciation as processes, as well as the use of genetics to study the mechanisms that cause adaptations and speciation. The host of eyewitnesses continues to increase in the form of the many well-developed subdisciplines of evolutionary biology that were spawned by the *Origin*.

Given the explanatory value of Darwin's theory of evolution and all the evidence that we have in its favor, why should we still argue about whether or not it is true? When Galileo found empirical support for Copernicus's theory that the earth was not the center of the universe, around which all else revolves, but rather that the earth is just one of the planets that revolves around the sun, he was opposed by the Catholic Church because his results defied the prevailing interpretation of the Bible. It seems that the real motivation of the church was that such a challenge, if it proved to be true, would cause its followers to lose faith. Faith was not so fragile. Darwin's theory presents similar challenges. Faith will survive them as well.

References

Brunet, M., et al. 2002. A new hominid from the upper Miocene of Chad, Central Africa. *Nature* 418:145–51.

Campbell, B. G., J. D. Loy, and K. Cruz-Uribe. 2006. *Humankind emerging*, 9th ed. Boston: Pearson.

Gould, S. J. 1977. *Ontogeny and phylogeny*. Cambridge, MA: Harvard University Press.

Green, R. E., J. Krause, S. E. Ptak, A. W. Briggs, M. T. Ronan, J. F. Simons, L. Du, M. Egholm, J. M. Rothberg, M. Paunovic, and S. Paabo. 2006. Analysis of one million base pairs of Neanderthal DNA. *Nature* 444:330–36.

Haile-Selassey, Y. 2001. Late Miocene hominids from the Middle Awash, Ethiopia. *Nature* 412:178–81.

Huxley, T. 1863. *Evidence as to man's place in nature*. London: Williams and Norgate.

Kumar, S., A. Filipski, V. Swarna, A. Walker, and S. Blair Hedges. 2005. Placing confidence limits on the molecular age of the human-chimpanzee divergence. *Proceedings of the National Academy of Sciences* 102:18842–47.

Mikkelsen, T. S., et al. [the Chimp Sequencing Consortium]. 2005. Initial sequence of the chimpanzee genome and comparison with the human genome. *Nature* 437:69–84.

Sarich, U. M., and A. C. Wilson. 1967. Immunological time scale for hominid evolution. *Science* 158:1200–1202.

Senut, B., N. Pickford, D. Gommery, P. Mein, K. Cheboi, and Y. Coppens. 2001. First hominid from the Miocene (Lukeino Formation, Kenya). *Earth and Planetary Sciences* 332:137–44.

Illustration Credits

Frontispiece © Mark Chappell.

Figure 1. Charles Darwin, *Variation of Animals and Plants under Domestication*, fig. 24.

Figure 2. Rock pigeon courtesy of Andreas Trepte, www.photo-natur.de. Three domestic pigeons courtesy of National Pigeon Association GB.

Figure 3. Early twentieth-century dachshund © The Natural History Museum, London. Early twenty-first-century dachshund © Ellen Levy Finch.

Figure 4. Photograph by Dr. W. Landauer, University of Connecticut.

Figure 5. Redrawn from Steven Salmony, "Is the Human Population Bomb Exploding *Now*?" guest editorial, 22 March 2005, Fragilecologies, http://www.fragilecologies.com/mar22_05.html.

Figure 6. Sailfin molly © Noel M. Burkhead. Pair of swordtails in hand © Christiane Meyer, Belize Foundation for Research and Environmental Education (BFREE).

Figure 7. Photograph by L. T. Wasserthal, 1992, published in L. T. Wasserthal, "The pollinators of the Malagasy star orchids *Angraecum sesquipedale*, *A. sororium* and *A. compactum* and the evolution of extremely long spurs by pollinator shift," *Botanica Acta* 110 (1997): 343–430.

Figure 8. Charles Darwin, *A Monograph on the sub-class Cirripedia: The Lepadidae*, plate IV, fig. 8 and 8a; plate V, fig. 1.

Figure 9. Redrawn from Scott Freeman and Jon C. Herron, *Evolutionary Analysis*, 3rd ed. (Old Tappan, NJ: Pearson Prentice Hall, 2004), fig. 6.15.

Figure 10. © Boris Sket.

Figure 11. Guppies © Paul Bentzen. Waterfall © Andrew Hendry.

Figure 12. Charles Darwin, *On the Origin of Species by Means of Natural Selection, or the Preservation of Favoured Races in the Struggle for Life*, chap. 4.

Figure 13. © Arnaud Jamin.

Figure 14. From C. A. Palmer et al., "Lineage-specific differences in evolutionary mode in a salamander courtship pheromone," *Molecular Biology and Evolution* 22 (2005): 2243–56.

Figure 15. Photographs by Jun Kitano. From C. L. Peichel, "Fishing for the secrets of vertebrate evolution in threespine sticklebacks," *Development Dynamics* 234 (2005): 815–23.

Figure 16. From Francisco J. Ayala, "Darwin's greatest discovery: Design without designer," *Proceedings of the National Academy of Sciences* 104 (2007): 8572. Reprinted with permission from *Encyclopaedia Britannica* © 2005 by Encyclopaedia Britannica, Inc.

Figure 17. From C. A. Boisvert et al., "The pectoral fin of Panderichthys and the origin of digits," *Nature* 456 (2008): 636–38.

Figure 18. From D. Reznick et al., "Independent origins and rapid evolution of the placenta in the fish genus *Poeciliopsis*," *Science* 298 (2002): 1018–20.

Figure 19. From William Morton Wheeler, *Ants: Their Structure, Development, and Behavior* (New York: Columbia University Press, 1910).

Figure 20. Photograph of unconformity in Zanskar Valley, India, by Nigel Hughes.

Figure 21. Redrawn from Martin Rudwick, *The Meaning of Fossils: Episodes in the History of Paleontology*, 2nd ed. (New York: Neale Watson Academic Publications, 1976), fig. 4.11.

Figure 22. Redrawn from Martin Rudwick, *The Meaning of Fossils: Episodes in the History of Paleontology*, 2nd ed. (New York: Neale Watson Academic Publications, 1976), fig. 4.5a.

Figure 23. Courtesy of Sören Jensen.

Figure 24. Courtesy of Adrian Pingstone.

Figure 26. From E. C. Schmidt-Ehrenberg, "Die Embryogenesie des Extremitaten-skelettes der saugetiere," *Revue Suisse Zoologique* 49 (1942): 33–132.

Figure 27. From Blaire Van Valkenburgh, "Déjà vu: The evolution of feeding morphologies in the Carnivora," *Integrative and Comparative Biology* 47, 1 (2007): 147–63.

Figure 28. From Edward S. Russell, *Form and Function: A Contribution to the History of Animal Morphology* (London: J. Murray, 1916).

Figure 29. © Elizabeth and Rudolf Raff.

Figure 30. From W. J. Murphy et al., "Resolution of the early placental mammal radiation using Bayesian phylogenetics," *Science* 294 (2001): 2348–51.

Figure 31. Hominoid taxonomy redrawn from Wikipedia, the free encyclopedia, http://en.wikipedia.org/wiki/File:Hominoid_taxonomy_7.svg

Figure 32. From D. Starck and B. Kummer, "Zur Ontogenese des Schimpansen-schadels," *Anthropologischer Anzeiger* 25 (1962): 204–15, http://www.schweizerbart.de

Index

Page numbers in **bold** indicate illustrations

abundance: as factor in evolution, 68–69, 91, 230–31, 288; limits to population growth, 291, 384; overproduction of offspring, 124–25, 180, 208, 223, 384; rarity linked to extinction, 170, 290; variation and, 157, 170–71, 298

adaptation, 36; accelerated maturation as, 127–30; biotic interaction and, 67, 72; blindness as, 105–7; complex structures or organs and, 227, 245–49; convergent evolution and, 356–58, **357**; counteradaptation, 67; degeneration of structures (wing, eye, etc.) as, 105–7, 389–90; divergence and, 142; integrated adaptations, 67–68, 79; life history evolution and, 122, 124–33; local environment and, 91–92, 104–5, 155, 214–15; of mosquitoes to the London Underground, 209–12; natural selection and, 91–92, 104–5, 142, 155, 214–15, 389; as non-random response, 17; physical environment and, 91–92, 104–5, 155, 209–15, 389; preadaptations, 211–12, 240–41, 251, 373; rate of, 92; resource allocation and, 124–25; as "tinkering" with pre-existing features, 95, 224, 389; value of "trivial," 241–43; variation and, 64

adaptive radiation, 31–32, 36, 185–86, 285

affinity of species: biogeography and, 317–18, 325–26, 328, 341–42; classification systems and, 111–12, 347–48, 353, 358, 373, 377–79, 392; developmental biology and, 346; as evidence of common ancestry, 360–63, 396–97; between fossil and living forms, 295–97, 390–92; genetic evidence and, 377–79; humans and apes as related species, 12–13, 84, 189, 404–12, **413**; plate tectonics as explanation for, 344–45; as problematic, 341–42

Afrotheria superorder, 377–79

Agassiz, Louis, 297, 326–27, 363, 365, 372, 381, 400

age of the earth, 276–79, 343, 403

allopatric (geographic) speciation, 148, 162, 167–69, 201–3

altruism, instinctive, 251–52

ammonites, 291, 292, 310

amphibians: dispersal barriers and, 331, 337–38, 391; life stages of, 160, 369, 371–72. *See also* frogs; salamanders

amphipods, 106–7

anagenesis, **174–75**, 178–79, 224, 267

analogous variations, 113–18

ancon sheep, **49**, 49–51, 153

Anglican church, 4

Animal Kingdom (Cuvier), 271

Anser sp. (geese), 195

anteaters, 298, 300, 318, 390

antennae, 106, **107**, 389

ants: castes and specialization of, 255–56; coevolution of slave-making and slave ants, 253–55; instinct and behaviors as acquired through evolution, 250–60; interactions with aphids, 251–52; profound morphological differences between castes, 257–60, **258**; reproduction among, 256–57

apes, 12–13, 84, 189, 405–10, 412, **413**

aphids, 251–52

apple maggot flies *(Rhagoletis pomonella)*, 214

Archaeopteryx, 390, 403